Diseases of Horticultural Crops and their Management

THE AUTHORS

Dr. G. Arjunan, *Ph.D.,* Former Professor and Head, Department of Crop Protection, Anbil Dharmalingam Agricultural College and Research Institute, Tamil Nadu Agricultural University, Tiruchirapalli. He worked for over 38 in years in rice, pulses, sugarcane and cassava disease resistant programs. Associated scientist in the development of virus resistant varieties in pulses. He published 26 books, 95 research papers and 350 extension articles. Recipient of Fellow of Phytopathological Society of India and Fellow of Plant Protection Association of India.

Dr. D. Dinakaran, *Ph.D.,* Professor and Head, Agricultural Research Station, Tamil Nadu Agricultural University, Virinjipuram. His area of specialization is biological control of crop diseases and integrated pest management. He published 4 books, 10 booklets, 45 research papers, 65 extension articles and authored 6 chapters in books. Life member in Society for Biological Advancement, Indian Society of Plant Pathologists and Madras Agricultural Students Union.

Dr. S. Parthasarathy, *Ph.D.,* Assistant Professor (Plant Pathology), Amrita School of Agricultural Sciences, Amrita Vishwa Vidyapeetham, Coimbatore. He completed his Ph.D. from Tamil Nadu Agricultural University, Coimbatore. He is a recipient of BIRAC UIC Innovation Fellow during 2015 and also fellowships from MHRD, UGC and IDRC. He received several medals and awards in International and National conferences. He has published over 50 research articles in refereed journals, authored 10 books, 30 book chapters and 42 extension articles.

Diseases of Horticultural Crops and their Management

Dr. G. Arjunan

Former Professor and Head (Crop Protection),
Anbil Dharmalingam Agricultural College and Research Institute,
Tamil Nadu Agricultural University, Tiruchirapalli , Tamil Nadu

Dr. D. Dinakaran

Professor and Head,
Agricultural Research Station,
Tamil Nadu Agricultural University, Virinjipuram , Tamil Nadu

Dr. S. Parthasarathy

Assistant Professor (Plant Pathology),
Amrita School of Agricultural Sciences,
Amrita Vishwa Vidyapeetham, Coimbatore, Tamil Nadu

2022

Daya Publishing House®
A Division of
Astral International Pvt. Ltd.
New Delhi – 110 002

ISBN 978-93-5461-341-8

Published by : **Daya Publishing House®**
A Division of
Astral International Pvt. Ltd.
– ISO 9001:2015 Certified Company –
4723/23, Ansari Road, Darya Ganj
New Delhi-110 002
Ph. 011-43549197, 23278134
e-mail: info@astralint.com
Website: www.astralint.com

Preface

Management of plant diseases are greatest importance for successful and profitable cultivation of horticultural crops. Under these environments, important and relevant information were collected and compiled in a textbook form titled "Diseases of Horticultural Crops and their Management". This book is mainly intended for the Plant Pathology courses of graduate students in the field of Agriculture, Horticulture, Botany, Forestry and Mycology. It is clear that young students are suffering from cultural shocks to shift from their environment. Semester system of education of B.Sc./M.Sc. (Ag.), B.Sc./M.Sc. (Horti.), B.Sc./M. Sc. (Forestry), B.Sc./M.Sc. (Botany), and B. Tech. (Horti.), students are quite dynamic for which the students are to be helped for changeover. We can identify their difficulties for comprehension of language, non-availability of textbooks for their semester system. There is a need to use simple language. The present book titled "Diseases of Horticultural Crops and their Management" suite to the need of students. This book is written in simple understandable language dealing with various subject matters of Plant Diseases. This book has been prepared with a specific purpose of importing complete comprehensive information about major diseases of horticultural crops in India and we hope that the students and readers will find this with much utility.

We thank all the authors/publishers from which references were collected on various aspects of plant pathological aspects. We are sure that this book will serve as valuable text cum reader friendly textbook to the graduate and post-graduate students of agricultural universities.

Dr. G. Arjunan

Dr. D. Dinakaran

Dr. S. Parthasarathy

Contents

Preface v

1. Fruit Crops **1**

1. Apple (*Malus × domestica* Borkh.)
2. Banana (*Musa* sp.)
3. Ber (Jujube) (*Ziziphus mauritiana* Lam.)
4. Citrus (*Citrus* sp.)
5. Custard Apple/Sugar Apple (*Annona squamosa* L.)
6. Grapevine (*Vitis vinifera* L.)
7. Guava (*Psidium guajava* L.)
8. Indian Gooseberry/Aonla (*Phyllanthus emblica* L.)
9. Jackfruit (*Artocarpus heterophyllus* L.)
10. Mango (*Mangifera indica* L.)
11. Papaya (*Carica papaya* L.)
12. Peach (*Prunus persica* L.)
13. Pear (*Pyrus communis* L.)
14. Pineapple (*Ananas comosus* L.)
15. Plum (*Prunus domestica* L.)
16. Pomegranate (*Punica granatum* L.)
17. Sapota (*Manilkara zapota* L.)

2. Vegetable Crops **71**

1. Bean (*Phaseolus vulgaris* L.)
2. Beetroot (*Beta vulgaris* sub sp. *vulgaris* L.)
3. Bhendi [*Abelmoschus esculentus* (L.) Moench]
4. Brinjal (*Solanum melongena* L.)
5. Cabbage and Cauliflower (*Brassica oleracea* L.)
6. Carrot [*Daucus carota* sub sp. *sativus* (Hoffm.)]
7. Cassava/Tapioca (*Manihot esculenta* Crantz)
8. Celery (*Apium graveolens* L.)
9. Curcurbits (*Cucurbita* L.)
10. Drumstick/Moringa (*Moringa oleifera* Lam.)
11. Pea (*Pisum sativum* L.)
12. Potato (*Solanum tuberosum* L.)
13. Radish [*Raphanus raphanistrum* sub sp. *sativus* (L.) Domin]
14. Sweet Potato [*Ipomoea batatas* (L.) Lam.]
15. Taro/Dasheen/Cocoyam [*Colocasia esculenta* (L.) Schott.]
16. Tomato (*Solanum lycopersicum* L.)
17. Yam (*Dioscorea* spp.)

3. Flower Crops **155**

1. Anthurium (*Anthurium andraeanum* Linden ex André)
2. Aster (*Aster amellus* L.)
3. Carnation (*Dianthus caryophyllus* L.)
4. China Aster [*Callistephus chinensis* (L.) Nees.]
5. Chrysanthemum (*Chrysanthemum indicum* L.)
6. Crossandra [*Crossandra infundibuliformis* (L.) Nees]
7. Dahlia (*Dahlia pinnata* Cav.)
8. Gerbera (*Gerbera aurantiaca* Sch. Bip.)
9. Gladiolus (*Gladiolus communis* L.)
10. Jasmine (*Jasminum* sp. L.)
11. Lilium (*Lilium candidum* L.)
12. Marigold (*Tagetes erecta* L.)
13. Oleander (*Nerium oleander* L.)
14. Orchids
15. Pyrethrum [*Chrysanthemum cinerariifolium* (Trevir.) Vis.]
16. Rose (*Rosa* sp.)

17. Tuberose (*Polianthes tuberosa* L.]
18. Zinnia (*Zinnia elegans* Jacq.)

4. Spices and Condiments **247**

1. Betelvine (*Piper betle* L.)
2. Chilli (*Capsicum annuum* L.)
3. Cinnamon/Darchini (*Cinnamomum zeylanicum* Blume)
4. Clove (*Syzygium aromaticum* L.)
5. Coriander (*Coriandrum sativum* L.)
6. Cumin (*Cuminum cyminum* L.)
7. Curry Leaf [*Murraya koenigii* (L.) Spreng.]
8. Fennel (*Foeniculum vulgare* Mill.)
9. Fenugreek (*Trigonella foenum-graecum* L.)
5. *Cercospora* Leaf Spot
10. Garlic (*Allium sativum* L.)
11. Ginger (*Zingiber officinale* Roscoe)
12. Nut Meg (*Myristica fragrans* Houtt)
13. Onion (*Allium cepa* L.)
14. Tamarind (*Tamarindus indicus* L.)
15. Turmeric (*Curcuma longa* L.)
16. Vanilla (*Vanilla planifolia* Andrews)

5. Plantations Crops **299**

1. Arecanut (*Areca catechu* L.)
2. Cashew (*Anacardium occidentale* L.)
3. Cocoa (*Theobroma cacao* L.)
4. Coconut (*Cocos nucifera* L.)
5. Coffee (*Coffea arabica* L.)
6. Large/Black Cardamom (*Amomum subulatum* Roxb.)
7. Pepper (*Piper nigrum* L.)
8. Rubber (*Hevea brasiliensis* Mull. Arg.)
9. Small Cardamom (*Elettaria cardamomum* Maton)
10. Tea [*Camellia sinensis* (L.) Kuntze.]

6. Aromatic Plants **359**

1. Camphor/Camphor Laurel (*Cinnamomum camphora* L.)
2. Citronella (*Cymbopogon winterianus* Jowitt.)

3. Davana (*Artemisia pallens* Walker)
4. East Indian Rosebay (*Tabernaemontana coronaria* R. Br.)
5. Garden Geranium (*Pelargonium × hortorum* L.H. Bailey)
6. Ginseng (Asian Ginseng - *Panax ginseng* Baill.)
   (American Ginseng - *Panax quinquefolium* L.)
7. Gymnema (*Gymnema sylvestris* R.Br.)
8. Haritaki/Chebulic Myrobalan (*Terminalia chebula* Retz.)
9. Henbane (*Hyoscyamus niger* L. and *H. muticus)*
10. Indian Ipecac/Nanjaruppan [*Tylophora indica* (Burm.f.) Merr.]
11. Khus Khus/Vetiver [*Chrysopogon zizanioides* (L.) Roberty]
12. Lemongrass [*Cymbopogon flexuosus* (Nees ex.Steud.)]
13. Liquorice (*Glycyrrhiza glabra* L.)
14. Malabar Nut/Vasaka (*Justicia adhatoda* L.)
15. Mayapple (*Podophyllum peltatum* L.)
16. Mint (*Mentha arvensis* L.)
17. Opium Poppy (*Papaver somniferum* L.)
18. Palmarosa [*Cymbopogon martini* (Roxb.) Wats.]
19. Patchouli (*Pogostemon cablin* Benth.)
20. Physicnut (*Jatropha curcas* L.)
21. Rose Scented Geranium [*Pelargonium capitatum* (L.) Ait., *Pelargonium graveolens* L'Herit.]
22. Rosemary (*Rosmarinus officinalis* L.]
23. Safed Musli (*Chlorophytum borivilianum* Santapau and R.R.Fern.)
24. Saffron (*Crocus sativus* L.)

7. Medicinal Plants 409

1. Aloe (*Aloe barbadensis* Mill.)
2. Ashwagandha/Indian Ginseng [*Withania somnifera* (L.) Dunal]
3. Asparagus (*Asparagus racemosus* L.)
4. Bael or Wood Apple (*Aegle marmelos* Correa)
5. Belladonna/Deadly Nightshade (*Atropa belladonna* L.)
6. Black Nightshade/Makoi (*Solanum nigrum* L.)
7. Cinchona (*Cinchona officinalis* L.)
8. Costus (*Costus speciosus* Koen ex. Retz)
9. Dioscorea (*Dioscorea alata* L.)
10. Fox-Glove [*Digitalis purpurea* L. (*Digitalis lanata, D. grandiflora* and *D. thapsi*)]

11. Glory Lily (*Gloriosa superba* L.)
12. Indian Coleus (*Coleus forskohlii* L.)
13. Indian Pennywort (*Centella asiatica* L.)
14. Isabgol/Ispaghul (*Plantago ovate* Forssk.)
15. Kalmegh [*Andrographis paniculata* (Burm.f.) Wall. ex Nees]
16. Neem (*Azadirachta indica* A. Juss.)
17. Periwinkle [*Catharanthus roseus* (L.) G.Don]
18. Sarpagantha/Indian Snakeroot (*Rauvolfia serpentina* L.)
19. Senna (*Senna alexandrina* Mill.)
20. Sweet Basil (*Ocimum basilicum* L.)
21. Sweet Flag (*Acorus calamus* L.)
22. Sweet Leaf/Stevia/Charkarai Tulsi (*Stevia rebandiana* Bertoni)
23. Thorn Apple/Devil's Trumpets (*Datura stramonium* L.)
24. Tropical Soda Apple (*Solanum viarum* Dunal)

References **485**

Index **497**

Chapter 1

Fruit Crops

1. Apple (*Malus × domestica* Borkh.)

1. Scab/Black Spot

Symptoms

Leaves

- ☆ Large brown olivaceous spots with wavy margin appear on lower surface of leaves.
- ☆ Later lesions appear on upper surface of leaves.
- ☆ Lesions become velvety with slight depression on upper surface.
- ☆ Lower surface is convex.
- ☆ Infected leaves curved, dwarfed, distorted and fall off.

Fruits

- ☆ Brown or black, circular corky lesions develop on fruits and are surrounded by white halo.
- ☆ Early infection leads to uneven development of fruits and cause cracks in the lesion and the fruits rot due to secondary infection.
- ☆ Market value and keeping quality is reduced.

Fungus

Sexual stage*: Venturia inaequalis* Cooke Wint - saprophytic marsupial stage

Asexual stage*: Spilocaea pomi* Fr.:Fr. - parasitic stage or

Fusicladium dendriticum (Wallr.) Fuckel

The fungus enters a marsupial stage in the form of small black tubercles-pseudothecium (sexual fruiting body), eight haploid ascospores mature two unequal-sized cells. In the presence of moisture, the pathogen ascospores germinated on the leaves, breaking the thin surface epithelium of young leaves.

- ☆ Mycelium is uninucleate grey color, branched septate mycelium.
- ☆ Conidiophores are brown color, continuous or rarely septate, conidiophores from stroma.
- ☆ Conidia are ovate to lanceolate, with a truncate base and pointed apex, and smoky brown color, while arising initially single celled but often they become two celled through septation.

Mode of survival: As overwinters in infected fruit and leaves on the ground as pseudothecia.

Primary spread: From ascospores present in fallen leaves.

Secondary spread: Through conidia spread by air and rain splashes.

Favourable conditions: Free moisture on leaves, high relative humidity of more than 95 per cent.

Management

- ✰ Grow resistant varieties
- ✰ Practice clean cultivation (removal of fallen leaves, twigs and fruits and burning).
- ✰ Antagonistic fungi, *Chaetomium globosum* or *Microsphaeropsis ochracea* is the effective bio-agent working against scab pathogen.
- ✰ Spray urea 5 per cent on infected trees after fruit harvest at the pre-leaf stage and again 2 per cent at before bud-burst stage.
- ✰ Spray dodine or captan or difolatan 0.2 per cent at green tip stage and pink bud stage.

2. Powdery Mildew

Symptoms

- ✰ It is chronic reccurent problem in apple; mostly all parts of the tree are affected.
- ✰ This powdery mildew frequently infects new vegetative growth, causing reduced vigor, leaf malformation, and reduced viability of buds.
- ✰ Small white greyish patches of fungal growth appear of leaves on both the surfaces; leaves turn brown.
- ✰ Affected leaves curl become hard, brittle, crinkled and dry with roughened surface.
- ✰ Powdery growth is also seen on the twigs and buds.
- ✰ Infected flower buds open five to eight days later than healthy buds and exhibit reduced fruit set.
- ✰ Flower petals are distorted and pale yellow or light green.
- ✰ Affected fruits remain small or deformed.

Fungus

Podosphaera leucotricha (Ellis and Everh.) E. S. Salmon

It is an ascomycete fungus in the Erysiphaceae family, asexual conidia are hyaline and contain distinct fibrosin body, Conidia are air-dispersed and do not require free moisture to germinate. Sexually eight ascospores are deliberated from a distinct sac like asci enclosed in chasmothecia (sexual fruiting body).

Mode of survival: As Overwinters in the form of mycelia in dormant buds and fruits.

Primary spread: From conidia formed from dormant mycelia.

Secondary spread: Through air-borne conidia.

Favourable conditions: Temperature of 10-25°C favourable for conidial germination.

Management

- Grow resistant varieties
- Remove infected buds by dormant-season pruning and destroy to reduce the level of primary inoculum.
- Prune and destroy infected shoots during growing season.
- Spray carbendazim 0.1 per cent or dinocap 0.2 per cent.
- Spray the young seedlings with triadimefon 0.1 per cent in the nursery.
- Spray recommended doses of myclobutanil, trifloxystrobin and kresoxim-methyl.
- Horticultural oils, waxes, and biological compounds produced by *Bacillus* strains are also available, but their effectiveness is somewhat inconsistent.

3. Fire Blight

Symptoms

- This is one of the most destructive diseases of apple. It affects flowers, twigs, fruits, branches and trunk.
- It is characterized by sudden browning of leaves as if they are scorched by fire and hence the name fire blight.
- Diseased flowers are water-soaked, shrivel, turn brown to black and exude droplets of bacterial ooze.
- Infected leaves turn brown to black, curl, shrivel, hang downward and usually cling to the curled or blighted twigs.
- Terminal twigs wilt from the tip downward and the tip of the twig hooked and take the shape of Shepherd's crook.
- Brown cankerous lesions are formed on the outer bark of the branch and the pith exhibit a dark brown discoloration.
- Immature fruits are water-soaked, become brown, shrivel and turn black and remain attached to the cluster base.
- Dead fruits cling to the tree for several months.
- Fruits infected through injury develop red, brown or black lesions and exude milky ooze which later turns red to brown.

Bacterium

Erwinia amylovora (Burrill) Winslow *et al.*

It is a rod shaped, Gram-negative measuring 1.0-3.0 × 0.5-1.2 μm are typically motile by peritrichous flagella and grows best at 21°C - 28°C.

Mode of survival: As overwinters in margins of cankers, twigs and dead woods.

Primary spread: From infected twigs.

Secondary spread: Through dissemination of bacterial ooze through honey bees, sucking insects, flies, wasps, birds and rain splashes.

Favourable conditions: Susceptible varieties, rain and temperatures between 18°C and 30°C.

Management

- Avoid growing of susceptible cultivars and rootstocks, during favourable conditions.
- Expose of scions to 43°C for 23 hours kills the bacteria without causing tissue damage to the plants.
- Avoid excess nitrogen fertilizer especially ammonium nitrate.
- Red delicious group of cultivars are inherently resistant.
- Avoid sprinkler irrigation.
- Disinfect pruning tools between cuts with mercuric chloride 0.1 per cent or sodium hypochlorite 10 per cent to reduce spread of bacterium.
- Prune overwintering cankered branches and blighted twigs during dormant season.
- Prune active lesions in the affected branches during crop season at least 30 cm below the visible infection and burn.
- Spray insecticide to control insects and to prevent the spread of bacteria.
- Spray *Erwinia herbicola* or *Pseudomonas fluorescens,* honey bees disseminate the antagonist to flowers more effectively than manual spray.
- Spray Bordeaux mixture (2:6:100) or 1 per cent streptomycin (500-1000 ppm).

4. Crown Gall

Symptoms

- Small tumour like outgrowth (galls) or knobbly swollen galls appears mostly on crown (near soil level) followed by trunk, branches and roots.
- At young stage, the galls are soft, spongy, spherical, and white or flesh coloured.
- The galls vary in size from 7-10 mm in dia.
- Older galls are hard, corky, knobby or knotty on woody stems and become more cleft.
- The affected plants are stunted with chlorotic leaves.

Bacterium

Rhizobium radiobacter (Beijerinck and van Delden)

[Syn. *Agrobacterium tumefaciens* (Smith and Townsend) Conn].

It is a Gram-negative, rod shaped (bacilliform), aerobic, motile bacterium having 1-6 peritrichous flagella. A large extrachromosomal piece of DNA, commonly referred to as a tumor-inducing (Ti) plasmid, is carried by *A. tumefaciens.*

Mode of survival: As free living in sandy loam soil and surface of roots (rhizosphere).

Primary spread: From infested soil.

Secondary spread: Through rainfall, irrigation water, air, insects, tools and plant parts.

Favourable conditions: Wounds made by cultural practices, grafting and insects at soil level in the roots or stems.

Management

- ✫ Field sanitation and summer ploughing.
- ✫ Use of disease free planting materials.
- ✫ Use resistant rootstocks and budded plants than grafts.
- ✫ Avoid wounding in the crown and roots during cultivation.
- ✫ Use the biocontrol agent, *Agrobacterium radiobacter* K84.
- ✫ Eradication of crown galls followed by smearing with copper-based solutions or sodium hypochlorite is transiently effective.

5. Mosaic

Symptoms

- ✫ Leaves exhibit irregular creamy white or pale yellow irregular spots intermingled with green patches.
- ✫ Presence of yellow band along the larger veins on both the sides of leaves.
- ✫ Affected leaves show withering.

Virus

Apple mosaic virus (ApMV).

Primary spread: From natural root grafting, budding and infected tools.

Secondary spread: Through seed and pollen transmission.

Favourable conditions: Winter season.

Management

- ✫ Use of certified virus-tested planting material.

- Inactivate the virus in the seedlings by maintaining at 37°C for 27 days and then at 40°C for 20 days.
- Apical meristem culture of plant material followed by thermotherapy (24 to 32 days at 38°C).

Postharvest Disease of Apple

1. Soft Rot or Blue Mould: *Penicillium expansum*

Symptoms

- Rotted area light in colour.
- Soft and watery texture.
- Rotted area advances, does not sunken.
- Grey-blue cushions fruiting bodies on the diseased skin.

2. Bitter Rot: *Glomerella cingulata*

Symptoms

- Faint, light brown discolouration beneath the skin.
- Discolouration increase and extends in a cone shaped manner.
- Rough, circular lesion becomes depressed.
- Tiny, raised black dot appear beneath the cuticle.
- They give rise to acervuli which break cuticle.
- Entire fruit becomes rotted.
- Rotted fruit drop and mummified.

3. Grey Mould: *Botryis cinerea*

Symptoms

- Translucent watery light brown to dark brown spots.

4. Phytophthora Rot: *Phytophthora cactorum*

Symptoms

- Irregular outlined rotted surface.
- Flesh turns pale to brown.
- Affected portion may become spongy and water soaked.

Management

- Orchard sanitation, pre harvest spraying with Benomyl 500 ppm or 0.25 per cent.
- Precaution in picking, grading and packing.
- Discard injured fruits.

- Adopt proper cold storage 0°C-1.6°C.
- Fruit should be wrapped with wrapper treated with double boiled linseed oil.
- Postharvest dipping in 500 ppm aureofungin for 20 minutes or benomyl 1000 ppm.

2. Banana (*Musa* sp.)

1. Panama Disease/*Fusarium* Wilt

Symptoms

- Yellowing and drooping of lower most leaves and it extends upwards and finally heart leaf alone remains green and upright.
- Leaves break near the base and hang around the pseudostem and dry up.
- Longitudinal splitting at the base of the pseudostem is seen from ground level up to 60 cm.
- Vascular tissues in pseudostem become dark brown in colour.
- Central portion of the affected rhizome is dark brown and purplish streaks radiate from the centre in the cross section.

Fungus

Fusarium oxysporum f. sp. *cubense* (E.F. Smith).

Mycelium is septate, hyaline and branched. Fungus produces microconidia, macroconidia and chlamydospores. Microconidia are single celled hyaline and elliptical or oval. Macroconidia are sickle-shaped, hyaline and 3-septate and they are tapering at both the ends. Chlamydospores are thick walled, spherical to oval and hyaline.

Mode of survival: As chlamydospores in plant debris in soil.

Primary spread: From infected rhizomes or suckers, soil adhering (with chlamydospores of fungus) the rhizome.

Secondary spread: Through contaminated implements and irrigation water.

Favourable conditions: Waterlogged condition, monoculturing, injuries to roots during intercultural operations and injury by burrowing nematode (*Radopholus similis*) and lesion nematode (*Pratylenchus coffeae*).

Management

- Follow crop rotation with paddy or sugarcane.
- Grow resistant varieties such as Poovan, Robusta and Nendran for planting in endemic areas.
- Avoid susceptible varieties such as Rasthali, Monthan, Karpuravalli, Kadali, Rasakadali, Pachanadan, *etc.*

- Plant healthy suckers obtained from healthy plantations which restricts the movement of infected suckers to non-infected areas.
- Paring and pralinage (removal of roots and outer skin of rhizomes) and dipping of the suckers in carbendazim 0.1 per cent solution followed by dipping the sucker in clay slurry and sprinkling 40g carbofuran granules for controlling nematode and wilt problem.
- Apply neem cake 250 kg/ha as basal.
- Provide good drainage especially during rainy season.
- Flood fallowing up to 24 hours is an effective method of reclaiming infested soils.
- Clean the implements after their usage in infested garden before using them in non-infested garden.
- Avoid root injury during intercultural operation.
- Drench carbendazim 0.1 per cent around the pseudostem at bimonthly intervals starting from 3rd MAP (months after planting) or inject the corm with 3 ml of 2.0 per cent carbendazim solution or embedding 60 mg of carbendazim or *Pseudomonas fluorescens* in a capsule in rhizome at 2,4 and 6 MAP using capsule applicator to a depth of 7 cm.
- Apply *Pseudomonas fluorescens* 2.5 kg/ha along with 50 kg of well decomposed FYM or neem cake.
- Apply *Trichoderma viride* and *Pseudomonas fluorescens* at 25 gm/pit for 4 times once at a time of planting in the planting pit and remaining doses at 3, 5 and 7 MAP.
- Uproot infected plants along with sucker and apply lime at 1-2 kg/pit.
- Application of urea + sugarcane trash (250g/pit) followed by lime (1 kg/pit) and neem cake (1-2 kg/pit).
- Application of neem cake at 250 kg/ha was most effective in controlling *Fusarium* wilt in Rasthali cultivar.

2. Anthracnose/Fruit Rot

Symptoms

- It attacks the plants at all the stages of development.
- Dark brown to black diamond shaped spots occur on green fruits which are delimited by pale margin.
- The pulp under the lesion is often affected.
- Infection on the green fruit in the field (latent infection) expresses at the time of ripening during storage.
- On ripening, the fruit develops numerous dark circular sunken spots which enlarge, coalesce.

- ☆ On maturity of lesions, pink spore masses are produce.
- ☆ It is also a postharvest disease causing considerable loss during transport and storage.

Fungus

Gloeosporium musarum Cke. and Mass. (Syn: *Colletotrichum musae* Berk. and Curt.).

Acervuli (asexual fruiting body), conidiophores are hyaline, septate, branched. Conidia are hyaline, aseptate, oval to elliptical or straight cylindrical and 11.0 to 17.0 × 3.0 to 6.0 µm.

Mode of survival: As infected fruits.

Primary spread: From fruits in old plantations.

Secondary spread: Through air-borne conidia and rain splashes.

Favourable conditions: Temperature of 30-35°C and relative humidity of 85 per cent, wounds and bruises to fruits in the field, during harvest, transit and storage.

Management

- ☆ Minimize the injuries to fruit during harvest.
- ☆ Maintain sanitation in plantation and packing house.
- ☆ Pre-harvest spray of carbendazim or thiophanate methyl 0.1 per cent or prochloraz 0.1 per cent or chlorothalonil 0.2 per cent on bunches 15-20 days before harvesting.
- ☆ Postharvest dipping of bunches in carbendazim 400 ppm.
- ☆ Dip the fruits in water extract of *Solanum torvum* 50 per cent conc. effectively reduces the disease and also increases the shelf life up to 26 days.

3. Sigatoka Leaf Spot

Symptoms

- ☆ Disease occurs first on lower older leaves.
- ☆ Elongated dark brown spots appear on leaves parallel to veins.
- ☆ The spots increase in size and show grey centre with brown margin and are surrounded by yellow halos.
- ☆ Numerous spots on leaves coalesce and cause browning and drying of leaves.
- ☆ The affected leaves dry up leaving only the top 3 or 4 younger leaves unaffected.
- ☆ Fructification of the fungus is seen as black dots in the grey centre of the spot.

- ☆ Heavy incidence of disease results in delayed flowering, reduction in size and number of hands and fingers.
- ☆ Bunches harvested from affected orchards ripens prematurely.

Fungus

Sexual stage: *Mycosphaerella musicola* Leach.

Asexual stage: *Pseudocercospora musae* Zimm.

Perithecia (sexual fruiting body) are dark brown to black, ostiolate. Asci are oblong, clavate and measures 28.8 to 36.8 × 8.0 to 10.8 μm. Ascospores are 1-septate, hyaline. Asexual conidia are elongated, narrow and multiseptate and measure 20 to 80 × 2 to 6 μm.

Mode of survival: As perithecia in dried leaves.

Primary spread: From ascospores from diseased leaves.

Secondary spread: Through air-borne conidia and water.

Favourable conditions:

- ☆ Warm temperature of 23-25°C.
- ☆ High relative humidity and rainy or humid weather.
- ☆ Poor and ill-drained soil and shady areas.
- ☆ Closer planting, non-removal of suckers and presence of heavy weeds.
- ☆ Frequent flood irrigation or sprinkler irrigation.
- ☆ Continuous banana cultivation in nearby areas.

Management

- ☆ Adopt crop rotation.
- ☆ Remove severely damaged and dried leaves and burn outside the field.
- ☆ Follow intercultural operation such as timely removal of suckers, weed free cultivation, improving drainage and adopting correct spacing.
- ☆ Spray carbendazim 0.1 per cent or thiophanate methyl 0.1 per cent or propiconozole 0.1 per cent or chlorothalonil 0.2 per cent or mancozeb 0.2 per cent or mineral oil 0.3 per cent or carbendazim 0.1 per cent along with wetting agent.
- ☆ Spray with 1 per cent petroleum oil (banole) + half the dose of recommended fungicides.

4. Freckle Disease/Black Spot

Symptoms

- ☆ Numerous, minute, black, raised spots are seen on the upper leaf surface.
- ☆ Infected leaves become yellow.

- ☆ Spots on fruits are superficial (skin deep) and are produced either singly or in groups.

Fungus

Sexual stage: *Guignardia musae* Racib.

Asexual stage: *Phyllosticta musarum* Cooke.

The fungus produces asexual fruiting body pycnidia and pycnidiospores. Pycnidiospores are needle-shaped, hyaline and multiseptate. Sexually cylindro-clavate asci, and ellipsoid to clavate ascospores were formed.

Mode of survival: As infected plant debris.

Primary spread: From infected plant debris.

Secondary spread: Through air-borne spores.

Favourable conditions: Poor fertilization

Management

- ☆ Remove severely damaged and dried leaves and burn outside the field.
- ☆ Spray copper oxychloride 0.25 per cent or Bordeaux mixture 1 per cent.
- ☆ Spray of mancozeb 0.25 per cent + copper oxychloride 0.3 per cent twice at 21 days interval.

5. Cigar End Rot

Symptoms

- ☆ The damage is caused on the young bunches.
- ☆ A black necrosis spread from the perianth into the tip of immature fingers with powdery conidiophores and conidia.
- ☆ The infection spread slowly along the fingers, grey conidia are formed on the shriveled tip of the fruit *i.e.,* like the ashes at the burnt end of the cigar (cigar end rot).
- ☆ The pulp undergoes a dry rot and becomes mummified.

Fungus

Verticillium theobromae (Turconi) and *Trachysphaera fructigena* Tabor and Bunting.

Conidiophores are single or in small groups. Conidia are hyaline, oblong to cylindrical. Conidia are asexual spores.

Mode of survival: As dormant mycelium in infected plant parts persist in soil.

Primary spread: From dormant mycelium in infected plant parts.

Secondary spread: Through air-borne conidia, soil and dying flower parts (perianth).

Favourable conditions: Rainfall, high humidity (90-92 per cent) and low temp, (18-22°C).

Management

- ☆ Avoid closer planting.
- ☆ Use tolerant varieties such as Grand Naine.
- ☆ Remove pistil and perianth by hand 8-10 days after bunch formation (tipping).
- ☆ Remove and bury infected fingers to prevent spread of fungus.
- ☆ Spray the bunch with mancozeb 0.2 per cent or thiophanate methyl 0.1 per cent or copper oxychloride 0.25 per cent with wetting agent and covered with plastic sleeves.
- ☆ Minimize bruising and follow proper sanitation measures during postharvest handling.

6. Bacterial Wilt/Moko Disease

Symptoms

- ☆ Yellowish discolouration of the inner leaf lamina appears followed by collapse of the leaves and premature ripening of the fruit.
- ☆ Discolouration of the vascular strands is seen from pale yellow to dark brown or bluish dark.
- ☆ Petiole breaks at junction with pseudostem and the plants show rapid wilting.
- ☆ Individual fingers of the bunch turn yellow, distorted and fruit pulp turns dark-brown.
- ☆ Greyish brown bacterial ooze comes out when the pseudostem is cut.
- ☆ Whole plant dies.
- ☆ The most characteristic symptoms appear on the young suckers that have been cut once and begin regrowth. These suckers are blackened and stunted.

Bacterium

Ralstonia solanacearum (Yubucchi *et al.*)

[Syn. *Pseudomonas solanacearum* (Smith), *Burkholderia solanacearum* (Smith)].

The bacterium is rod-shaped, gram negative, lophotrichus bacteria. (more than one polar flagellum), multiplication occurs by bacterial fission.

Mode of survival: As infected sucker and plant parts in soil for 6 months to 2 years and in *Heliconia* sp.

Primary spread: From soil or water and also spreads through infected suckers/ rhizomes.

Secondary spread: Through irrigation water, implements and insects.

Favourable conditions: Soil temp 28-32°C, relative humidity 87-92 per cent, pH slightly acidic to neutral, clay loam, sandy loam soil and poor drained soil.

Management

- Adopt fallowing or rotation with sorghum.
- Follow strict quarantine and phytosanitary measures.
- Use healthy suckers from disease free garden for planting.
- Provide proper drainage and use resistant varieties like Robusta and Grand naine.
- Remove weed hosts like *Heliconia* spp. (alternative host).
- Disinfect the tools with formalin diluted with water at 1:3 as disinfectant since the pruning wounds are the common entry points.
- Remove male buds as soon as the last hand emerges as the insects carry the bacterium on the male flowers.
- Eradicate infected plants and expose soil to direct sunlight.
- Apply copper oxychloride at 3gm per lit and streptocycline at 0.5 gm per lit as a soil drench.
- Use bio-agent like *Pseudomonas fluorescens.*

7. Head Rot/Tip Over/Top Rot

Symptoms

- Newly planted suckers are often affected and exhibit rotting and emit foul odour.
- The tip shrivels, blacken and dry thereby kill the entire sucker.
- In older plants rotting occurs at the collar region and leaf bases.
- In advanced cases, trunk base becomes swollen and split.

Bacterium

Erwinia carotovora sub sp. *carotovora* (Jones).

Erwinia carotovora is a rod shaped gram negative bacterium with peritrichous flagella that was named after the crop of carrots from which it was first isolated.

Mode of survival: As soil borne and enters through wounds and also through leaf sheath of suckers.

Primary spread: From infected sucker to wounded or injured suckers.

Secondary spread: Through irrigation water.

Favourable conditions: Water logged condition and susceptible varieties.

Management

- ☆ Plant disease free suckers.
- ☆ Poovan is resistant and Robusta is susceptible to this disease.
- ☆ Provide good drainage and remove infected plants and burn.
- ☆ Before planting, treat suckers with 0.4 per cent copper oxychloride (4g/L) 30 minutes.
- ☆ Drench methoxy ethyl mercuric chloride (Emisan-6) 0.1 per cent. Avoid drenching after shooting.
- ☆ Dip suckers in copper oxychloride (40g/10 L) + Streptocycline (3g/10 lit) for 30 min before planting.
- ☆ Drench with Sodium hypochlorite 10 per cent or Bleaching powder 20g/ litre/tree immediately irrigate after application and repeat twice at 10 days interval.

8. Bunchy Top

Symptoms

- ☆ Primary symptoms of the disease are seen when infected suckers are planted.
- ☆ Infected suckers put forth narrow, chlorotic leaves with mosaic symptoms.
- ☆ The affected leaves are crowded at the top (bunchy top), erect, rigid, brittle with their margins rolled upwards.
- ☆ Characteristic symptom includes the presence of interrupted dark green streaks along the secondary veins on the under surface of the lamina or midrib or petiole.
- ☆ Affected plants are extremely stunted and do not produce bunch.

Virus

Banana bunchy top virus (BBTV) or *Musa virus-1.* It is ssDNA virus

Mode of survival: As infected plants and suckers.

Primary spread: From Infected suckers.

Secondary spread: Through black aphid vector, *Pentalonia nigronervosa* var. *typica.*

Favourable conditions:

- ☆ Susceptible varieties are hill banana and dwarf cultivars.
- ☆ Abundance of vector population.
- ☆ Disease spread is more during winter in plains and summer in hills because of higher vector population.

Management

- ☆ Select and plant only healthy suckers.
- ☆ Follow strict quarantine regulation to be adopted especially for hill banana in disease free areas.
- ☆ Grow resistant varieties.
- ☆ Use only certified virus free tissue culture plants for new planting.
- ☆ Remove the infected plants with suckers and burn.
- ☆ Inject 4 ml of 2,4-D (prepared with water in the ratio of 1:8. *i.e.*, 125 mg/lit of water) into the psuedostem using pseudostem injector or
- ☆ Follow capsule application of 2,4-D 200 mg/corm to kill the infected plant. Herbicide injected plant collapses and topples in 3 to 5 days of application. This treatment kills the infected plant and vector simultaneously and economically especially in hilly areas.
- ☆ Spray phosphamidon 1ml/lit or methyl demeton 1ml/lit to control insect vector population.
- ☆ Inject dimethoate at 1 ml/plant (1 ml diluted in 4 ml of water) at 45 days interval from 3rd month to flowering stage.
- ☆ Spray Imidachloprid 0.50 ml/lit at monthly intervals when aphid population starts building.

9. Banana Mosaic/Infectious Chlorosis

Symptoms

- ☆ Chlorotic or yellow, linear discontinuous streaks appear on leaves.
- ☆ Leaves are narrow, shorter and show upward curling.
- ☆ Eruption of veins on the leaves, partial unfurling of lamina and occasional rotting of heart leaf are the other symptoms.
- ☆ Diseased plants are stunted and do not produce bunches.

Virus

A strain of *Cucumber mosaic virus* (CMV).

Mode of survival: As infected banana plant, cucumber, pumpkin and weeds.

Primary spread: From infected suckers and from weeds.

Secondary spread: Through aphids (*Aphis gossypii* and *A. maidis*).

Management

- ☆ Avoid growing cucurbitaceous crop as intercrop or in nearby field.
- ☆ Use resistant varieties and keep the field as weed-free.
- ☆ Use disease free suckers.
- ☆ Destroy infected plants along with suckers.

- Dry heat treatment of suckers at 40°C for 1 day has been found effective.
- Spray methyl demeton 1ml/lit.

10. Banana Bract Mosaic/Kokkan Disease

Symptoms

- It is characterized by the presence of spindle shaped pinkish to reddish streaks on pseudostem, midrib and peduncle.
- Typical mosaic and spindle shaped reddish brown streaks appear on bracts of the inflorescence and fingers.
- Suckers exhibit reddish brown streaks at emergence and separation of leaf sheath from central axis.
- In Nendran, clustering of leaves is seen at crown with a 'traveler's palm' appearance.
- In Robusta, fingers fail to develop and give the appearance of 'pencil' (pencil kai).

Virus

Banana bract mosaic virus (BBrMV). It is a flexuous, filamentous virus particle that belongs to the potyvirus group.

Mode of survival: As infected plants

Primary spread: From infected suckers and tissue culture plants.

Secondary spread: Through aphid vectors (*Aphis craccivora, Aphis gossypii* and *Pentalonia nigronervosa*)

Favourable conditions: Nendran, Robusta, Poovan, Ney Poovan, Rasthali, Red Banana, Monthan, Karpuravalli and tissue culture banana are susceptible.

Management

- Strict domestic quarantine is needed to contain the disease spread.
- Use of disease free suckers.
- Avoid growing susceptible varieties and tissue culture banana.
- Remove and destroy the infected plants.
- Spray of methyl demeton 1 ml/lit

11. Banana Streak Mosaic

Symptoms

- Discontinuous yellow streaks appear along the blade or midrib which become necrotic and turn black.
- The leaf lamina may also be narrower, thicker and become brittle.

- ☆ Necrosis is seen on the lower side of the sheath, which leads to drying of leaves.
- ☆ Necrotic streaks also appear on midrib, pseudostem and petiole.
- ☆ Infected plants are stunted.
- ☆ Bunch choking, abortion of bunch and seediness in fruits are seen in infected plants.

Virus

Banana streak virus (BSV). Virus particles are bacilliform. The disease is neither transmitted mechanically nor by insect vector.

Mode of survival: As infected plants.

Primary spread: From infected suckers.

Secondary spread: Through mealy bugs (*Planococcus citri, P. ficus, Dysmicoccus brevipes, Psedococcus, longispirus* and *Ferrisia virgata*).

Favourable conditions: Susceptible varieties (Poovan) and mealy bug infestation.

Management

- ☆ Strict domestic quarantine is needed to contain the disease spread.
- ☆ Plant disease free suckers.
- ☆ Virus indexing is to be done before mass multiplication.
- ☆ Remove and destroy the infected plants.
- ☆ Spray acepate + imidacloprid 1 gm/lit to control insect vectors.

Postharvest Diseases of Banana

1. Anthracnose: *Gloeosporium musarum*

Symptoms

- ☆ Light brown depressed lesion.
- ☆ Coalesce and cover the whole fruit.
- ☆ Severe case pink spore masses on fruit.

2. Botryodiplodia Rot/Black Tip/Finger Tip Rot/Black Rot: *Botrytis cinerea*

Symptoms

- ☆ Downward rotting of bunch.
- ☆ Dropping of fingers.
- ☆ Extensive finger rot is a common symptom from the tip to main stalk end.
- ☆ Dark brown skin, wrinkled with pycnidia.
- ☆ Rotting of pulps.

3. Crown Rot or Pineapple Disease: *Ceratocystis paradoxa*

Symptoms

- ✰ Main stalk decay rapidly.
- ✰ Tissue become blackened.
- ✰ Emit pineapple odour.

Management

- ✰ Reduce inoculum level orchard sanitation, removal of flower erect, tipping or removal of floral remnants.
- ✰ Harvest at correct stage.
- ✰ Careful handling during storage.
- ✰ Postharvest spray with mancozeb 0.2 per cent or carbendazim 0.1 per cent before storage.
- ✰ Store at 12-13°C vary to variety to variety below this temperature cause chilling injury.

3. Ber (Jujube) (*Ziziphus mauritiana* Lam.)

1. Powdery Mildew

Symptoms

- ✰ White powdery growth appears on young leaves and causes them to shrink and defoliate.
- ✰ Small, white powdery growth on young fruits enlarges, coalesce and finally turn brown to dark brown.
- ✰ In severe cases, the whole fruit surface covered with the powdery mass leads to premature dropping of fruits.
- ✰ Infected fruits become corky, cracked, mis-shapened and underdeveloped. Matured fruits turn rusty and unfit for marketing.

Fungus

Oidium erysiphoides f. sp. *zizyphi* Yan and Wang.

Oidium type conidiophore, borne upright single. Conidia are cylindrical, hyaline, measuring about 25.2 - 37.8 × 16.8 - 21.0 μm.

Mode of survival: As bud woods of the host plant.

Primary spread: From infected plant parts.

Secondary spread: Through air-borne conidia.

Favourable conditions: Temperature range from 10-30°C and relative humidity of 32 per cent and above, favour the diseases development.

Management

- ☆ Collect and destroy infected fallen leaves and fruits.
- ☆ Spray of dinocap 0.05 per cent or wettable sulphur 0.25 per cent or carbendazim 0.1 per cent during the fruit attains pea size.

2. Rust

Symptoms

- ☆ Small, irregular, reddish brown uredopustules appear on the lower leaf surface and later they cover the whole area of leaves.
- ☆ The infected leaves dry and defoliate.

Fungus

Phakopsora zizyphi-vulgaris Dietel. or *Uredo ziziphi* Pat.

The pathogen is autoecious in nature.

Mode of survival: As teliospores in the infected leaves.

Primary spread: From teliospores.

Secondary spread: Through air-borne uredospores.

Favourable conditions: Observed during drier months of the year.

Management

- ☆ Spray mancozeb or zineb 0.2 per cent or wettable sulphur 0.3 per cent.
- ☆ Spray with thiophanate methyl 70 per cent WP at 1 gm/lit or propiconazole 25 per cent EC at 0.75 ml/lit or mancozeb 75 per cent WP at 2.5 gm/lit of water.

4. Citrus (*Citrus* sp.)

1. Gummosis

Symptoms

- ☆ The pathogens cause gummosis, foot rot, root rot, crown rot, leaf fall and brown rot diseases.
- ☆ Lesions develop as high as 60 cm from the ground level on the trunk (foot rot)
- ☆ Later it extends below the soil on crown roots as crown rot. Lesions on trunk start oozing gummy substances on the affected bark forming an encrustation (gummosis).
- ☆ Droplets of brownish black gum trickles down the stem.
- ☆ The bark cracks, shreds longitudinally and is destroyed.

- ☆ The feeder roots rot (root rot). If regeneration of feeder root fails, the affected plant shows starvation, less canopy with naked branches, die back and slow decline symptoms.
- ☆ Leaves turn pale and yellow and shed.
- ☆ Affected tree dies in few months.
- ☆ Prior to death, the tree produces more flowers but dies before the fruits mature.
- ☆ Fruits drop and rot and such fallen fruits are covered by whitish cottony fungal growth.

Fungi

Phytophthora palmivora Butler, *P. citrophthora* and *P. parasitica*

It is an oomycete possess aseptate mycelia, zoospore are the asexual spores produced in the sporangium where as oospores are sexual spores or resting/ dormant spores borne in oogonium.

Mode of survival: As survives on fallen fruits, twigs, leaves and in cracks of trees.

Primary spread: From dormant mycelia and oospores present in infected debris and soil.

Secondary spread: Through zoospore spread through soil and irrigation water.

Favourable conditions: Cool weather, temperature 18-22°C, 90-95 per cent RH, high soil moisture, PH 6.0-7.0, heavy and poorly drained soil, use of susceptible root stock, low budding, prolonged wet weather and prolonged contact of water with tree trunk.

Management

- ☆ Select proper planting site with adequate drainage.
- ☆ Use resistant rootstocks (Sour orange, Trifoliate orange).
- ☆ Provision of proper drainage and avoid excess irrigation.
- ☆ Avoid contact of water with the tree trunk by adopting ring method of irrigation are effective.
- ☆ Injuries to crown roots or base of stem during cultural operations should be avoided.
- ☆ Scrap affected parts with sharp knife and paint the trunk with Bordeaux paste up to 1 m height or metalaxyl-MZ or disinfect with mercuric chloride 0.1 per cent or potassium permanganate solution 1 per cent using a swab of cotton before the onset of monsoon as prophylactic measure.
- ☆ Drench the soil either by fosetyl-Al (2.0g/L water) or metalaxyl MZ-72 (2.0 g/L) covering the whole plant canopy and basin of affected plant at 40 days interval after onset of monsoon.

2. *Diplodia* Gummosis

Symptoms

- ☆ Infection starts at cracks.
- ☆ Gum oozes from spots near fork/limbs, dries and become black.
- ☆ The wood tissues become brown in colour.
- ☆ Large limbs are killed and they break at the infected portion.

Fungus

Diplodia natalensis Pole-Evans (*Lasiodiplodia theobromae*)

Diplodia is an asexual stage of fungus, which produces minute globose pycnidia asexually in the infected plant parts.

Mode of survival: As in infected plants.

Primary spread: From infected plants.

Secondary spread: Through air-borne pycnidiospores and rain splashes.

Favourable conditions: Reduced tree vigour, insect damage and old trees.

Management

- ☆ Trees should be kept in vigorous growing condition.
- ☆ Broken limbs should be cut and burnt.
- ☆ Wounds should be scraped and protected with Bordeaux paste.

3. *Ganoderma* Root Rot

Symptoms

- ☆ Infection starts at one or more of the lateral roots of the tree.
- ☆ Later it spreads to crown roots and to the main trunk.
- ☆ Affected wood are discoloured, soft and decomposed.
- ☆ Aerial parts start drying.
- ☆ In advanced stages, woody bracket-like fructification (sporocarp) appears at the base of the trunk.
- ☆ Finally the whole tree dies.

Fungus

Ganoderma lucidum (Curtis) P. Karst.

It is polypore bracket fungus (shelf mushroom) under basidiomycetes. A whitish mat of mycelium can form, eventually turning brown. Underside of the bracket sheds basiodiospores.

Mode of survival: As soil-borne basidiospores.

Primary spread: From basidiospores from sporocarp.

Secondary spread: Through irrigation water and contact of roots of diseased trees.

Favourable conditions: Other susceptible host plants (palm trees, cocoa, coffee, rubber, tea *etc.*), neglected garden, water stress during summer.

Management

- ☆ Incorporate green manure or FYM or oil cakes in the soil.
- ☆ Remove dead and decayed stumps and burn.
- ☆ Maintain the vigour of the trees by proper fertilization.
- ☆ Dig a trench 2.5-3.0 m away from the diseased tree and apply sulphur dust 1 kg/tree.
- ☆ Drench the basin with aureofungin–sol 0.03 per cent or carbendazim 0.1 per cent or Bordeaux mixture 1.0 per cent a day after irrigation.

4. Dry Root Rot

Symptoms

- ☆ Moist decay of the bark is seen in early stages and later bark dries and shreds.
- ☆ The affected roots emit a bad odour.
- ☆ Affected tree defoliates and produces heavy crop of small-sized fruits before death.

Fungus

Macrophomina phaseolina (Tassi) Goid., *Fusarium* sp. and *Diplodia natalensis.*

Mode of survival: As infected plant parts in the soil.

Primary spread: From soil-borne microsclerotia/chlamydospores/spores.

Secondary spread: Through irrigation water and implements.

Favourable conditions: Injuries to roots, heavy soil, insufficient supply of nitrogen and closer planting.

Management

- ☆ Grow and incorporate green manure in the soil.
- ☆ Remove dead and decayed stumps and burn.
- ☆ Apply optimum dose of fertilizers.
- ☆ Drench tree basin with carbendazim 0.1 per cent.

5. Anthracnose and Die back/Wither Tip

Symptoms

- ☆ Affected branches begin to wither from the tip to downward (die back).

- ☆ The leaves wither and drop off.
- ☆ Black dots of acervuli appear on dead twigs.

Fungus

Colletotrichum gloeosporioides Pat and Penz.

[Sexual stage: *Glomerella cingulata* (Stoneman)].

Acervuli (asexual fruiting body) with black setae, conidiophores are hyaline, septate, branched. Conidia are hyaline, aseptate, oval to elliptical or straight cylindrical and 11.0 to 17.0 × 3.0 to 6.0 µm.

Mode of survival: As dead twigs and branches.

Primary spread: From infected twigs and branches.

Secondary spread: Through air-borne conidia from acervulus and rain splashes.

Favourable conditions: Disease is favoured by warm weather at temperature of 30-32°C, RH 80-85 per cent, cloudy weather and presence of susceptible host.

Management

- ☆ Organic manure adequately to improve the plant vigour.
- ☆ Improve drainage facilities.
- ☆ Prune dead branches and apply Bordeaux paste or copper oxychloride paste to cut ends.
- ☆ Prune dead twigs and spray carbendazim 0.1 per cent or captafol 0.2 per cent.

6. Powdery Mildew

Symptoms

- ☆ Whitish powdery growth appears on leaves and young twigs.
- ☆ Tender leaves are highly susceptible.
- ☆ Leaves become yellow, distorted and shed.
- ☆ Young fruits are covered by mildew growth.

Fungus

Oidium tingitaninum J.C. Carter

Septate mycelia, barrel shaped conidia borne in chains, ectophytic, sub epidermal haustoria, external mycelia.

Mode of survival: As dormant mycelium and cleistothecia on leaves and twigs.

Primary spread: From infected leaves and twigs.

Secondary spread: Through air-borne conidia and rain splashes.

Favourable conditions: Damp morning with few hours of sunshine.

Management

- ☆ Remove water shoots and sulphur dust 25 kg/ha in the morning or spray wettable sulphur 0.3 per cent or triadimefon 0.1 per cent at fortnightly intervals.

7. Sooty Mould

Symptoms

- ☆ The disease is common in the orchards where honey dew excreting insects (mealy bug and scale insects) infestation is severe.
- ☆ Black velvety superficial coating on the leaves, twigs and fruits is the characteristic feature of the disease.
- ☆ The photosynthetic activity of the leaves is reduced.
- ☆ Under dry conditions the affected leaves curl and shrivel up.
- ☆ The trees look sickly and fruit yield is reduced.

Fungus

Capnodium citri Berk. and Desm.

Superficial, membranous, composed of branched brown thick-walled hyphae. Sexually produces pseudothecium. Asci are thick-walled, 8-spored. Ascospores arranged biseriately with transverse septa. Asexually pycnidia were formed. Pycnidiospores are small, hyaline and bacillary with transverse septa.

Mode of survival: As dormant mycelium/pseudothecium in infected portions.

Primary spread: From infected plant parts.

Secondary spread: Through air-borne conidia.

Favourable conditions: Lack of timely control of honeydew excreting insects; old and dense orchard with poor light penetration.

Management

- ☆ Spray systemic insecticides like methyl demeton (2 ml/lit) + wettable sulphur (2 g/lit) or copper oxychloride (2.5 g/lit) + gum acacia (3 ml/lit) to control both insects and the disease.
- ☆ Spray starch solution 5 per cent (Mix 1 kg starch/maida/jack seed flour in 5 lit of water. Boil and make up the volume to 20 litres with water). Starch dries and forms flakes over the leaves and are removed along with the fungus when air blows.
- ☆ Spray should cover both the surface of leaves for effective control.

8. Scab/Verucosis

Symptoms

- ☆ Leaves, twigs and fruits of mandarin are attacked.

- ☆ Leaves show irregular, corky, raised, reddish or yellowish lesions mostly on the lower surface leading to severe distortion and wrinkling.
- ☆ The lesions are mostly flat.
- ☆ Dark corky outgrowths are seen on twigs and fruits.
- ☆ Affected fruits are rough, mis-shapened, show prominent warty projections and drop prematurely.

Fungus

Sexual stage: *Elsinoe fawcettii* Bitancourt and A.E. Jenkins

Asexual stage: *Sphaceloma fawcettii* var. *fawcettii* A.E. Jenkins

- ☆ Common scab or sour orange scab - *Elsinoe fawcettii*
- ☆ Sweet orange scab – *E. australis*
- ☆ Tryone scab – *Sphaceloma fawcettii* var. *scabiosa*

Sexually pseudothecium were formed as globose, dark, pseudo-parenchymatous, and multilocular. Asci are subglobose or ovoid, bitunicate, eight-spored. Ascospores hyaline, ellipsoidal, with two to four cells, usually constricted at the central septum. Asexually acervuli were formed with pale-brown phialidic conidiophores produces hyaline, unicellular, ellipsoid conidia.

Mode of survival: As pseudothecium on leaves and fruits.

Primary spread: From ascospores in infected plant parts.

Secondary spread: Through air-borne conidia.

Favourable conditions: Low temperature, high humidity and presence of moisture on leaves and fruits. Young leaves are more susceptible.

Management

- ☆ Collect and burn diseased leaves, twigs and fruits
- ☆ Spray carbendazim 0.1 per cent or Bordeaux mixture 1 per cent or copper oxychloride 0.25 per cent or difolatan 0.2 per cent at 15 days interval.

9. Pink Disease

Symptoms

- ☆ The affected branches are covered with fine silvery white mycelium (cob-web).
- ☆ Later, characteristic pink colour is produced by the fungus on the twigs and branches and hence the name pink disease.
- ☆ Leaves turn yellow and dry.
- ☆ Oozing of gummy substances from the branches.
- ☆ In severely infected branches, bark shreds and the wood is exposed.

Fungus

Erythricium salmonicolor (Berk. and Broome) Dastur

[Syn. *Corticium salmonicolor* (Berk. and Broome)]

Asexual stage: *Necator decretus* Mass.

Erythricium salmonicolor is a basidiomycetous fungus, produces the silvery white, later pink and finally dirty septate mycelium. Basidial fructification with scattered hymenia sparingly produced basidia bearing basidiospores.

Mode of survival: In susceptible host plants (cashew, jack, mango, rubber, coffee, *etc.*)

Primary spread: From Spores in infected plant parts.

Secondary spread: Through Air-borne spores and rain splashes.

Favourable conditions: Overcrowding, bad drainage, high humidity, warm and moist climatic conditions are conducive. Disease appears during or just after monsoon rain.

Management

- Prune all the diseased portions and paint cut ends with Bordeaux paste.
- Spray Bordeaux mixture 1 per cent mixed with crude oil emulsion or oil based copper fungicides to destroy the mycelium in the cracks.
- Spray copper fungicides or triadimefon for effective management.

10. Canker

Symptoms

- Acid lime, lemon and grapefruits are mostly affected.
- The disease attacks all the plant parts *viz.*, leaves, twigs, thorns, older branches and fruits.
- The spots on the leaves are raised, circular, 2-10 mm dia, brown, corky outgrowth with yellow halo.
- Older cankerous spots are irregular and often aggregated.
- The cankerous spots are with raised margin and a sunken centre (crater-like).
- On the twigs, irregular and raised cankers are formed, causing death of twigs above the point of attack.
- On the fruit surface round to irregular, raised cankerous spots without yellow halos are seen. The spots are superficial on fruits and such fruits fetch reduced price.

Bacterium

Xanthomonas axonopodis pv. *citri* (Hasse) (Syn. *Xanthomonas campestris* pv. *citri.*). The bacterium is rod-shaped, aerobic, Gram negative, motile by polar flagellum and non-spore former.

Mode of survival: As infected leaves (viable up to 6 months), twigs (up to 76 months) and fruits in the trees; also in fallen leaves and fruits.

Primary spread: From infected plant parts.

Secondary spread: Through air splashed rain, diseased planting materials and leaf miner attack.

Favourable conditions: It can be a serious disease where rainfall and warm temperatures are frequent during periods of shoot emergence and early fruit development. Free moisture on the host surface and temperature between 20°C and 35°C with evenly distributed rain. Leaf miner infestation predisposes the leaves to canker.

Management

- ☆ Collect and destroy diseased leaves and fruits.
- ☆ Plant air breaks to reduce the speed of air and the spread by air blown rain splashes.
- ☆ Maintain vigour by proper manuring and irrigation.
- ☆ Spray neem cake solution 5 per cent is effective in minimizing leaf miner damage as well as bacterial canker.
- ☆ Prune and destroy badly infected twigs before onset of monsoon and swab the cut ends with Bordeaux paste, followed by 3-4 sprays with streptomycin sulphate + tetracycline combination (streptocycline) 100 ppm + copper oxychloride 0.25 per cent with a sticking agent.

11. Citrus Greening/Huanglongbing (HLB)

Symptoms

- ☆ Leaves become small, yellow or yellow with green veins.
- ☆ Leaf mottle is one of the best diagnostic symptoms of greening.
- ☆ Leaf blade dwarfed, thickened, become upright and leathery.
- ☆ Twigs show short internodes, multiple bud formation and are upright and produce small leaves.
- ☆ Infected trees are stunted and may not bear fruits.
- ☆ Symptoms are often seen on part of the canopy or even only in a branch or a twig.
- ☆ Fruits are small, lopsided with aborted seeds and drop prematurely.
- ☆ Fruits which remain on the tree show orange colour on the sides exposed to sunlight whereas the shaded side remains green and hence the name "greening".
- ☆ Fruits are low in soluble solids but high in acid.

Bacterium

Candidatus Liberibacter asiaticus (Phloem limited Fastidious vascular bacterium): It is an obligate, Gram-negative bacterium.

Primary spread: From infected bud wood.

Secondary spread: Through insect vector (Citrus psylla, *Diaphorina citri*).

Favourable conditions: Susceptible varieties and abundance of insect vector population.

Management

- ☆ Remove affected and unproductive trees.
- ☆ Through indexing, select greening-free parent trees for bud wood.
- ☆ Plant disease free budded plants raised on improved rootstock.
- ☆ Spray tetracycline 500 ppm at fortnightly interval which inhibits the multiplication of the pathogen.
- ☆ Spray Imidacloprid (1ml/lit) or apply dimethoate 10 per cent granules to soil around the plant basin to control nymphs and adults of insect vector.

12. Tristeza/Quick Decline/Stem Pitting Disease

Symptoms

- ☆ Lime and sweet orange are highly susceptible.
- ☆ Leaves lose their dark green lustre, become dull.
- ☆ Leaves exhibit large number of vein flecks (chlorosis along midrib and lateral veins).
- ☆ Soon after discolouration, the leaves begin to fall and the branches become exposed and gradually begin to show die back symptom.
- ☆ Characteristic pits or depressions on wood (stem pitting) can be seen on its stem and branches after removal of bark.
- ☆ Reduces yield of fruits in severe cases.
- ☆ Stunting and sudden collapse or abrupt wilting that justifies the name quick decline.

Virus

Citrus tristeza virus (CTV). It is a flexuous rod having dsRNA.

Primary spread: From infected bud wood.

Secondary spread: Through black citrus aphid (*Toxoptera citricidus*) and dodder (*Cuscuta reflexa*).

Favourable conditions: Susceptible *Citrus* spp. and abundance of vector population.

Management

- ☆ Good cultural practices, increasing the fertility levels of the soil and good drainage are useful in reducing the disease.
- ☆ Use virus free materials on tolerant rootstock.
- ☆ Remove and burn diseased trees.
- ☆ For acid lime, pre-immunization (cross protection) with mild strain gives effective control (Cross protection: Use of bud wood from mild strain trees for grafting to protect the graft against the infection by severe strains of the same virus).
- ☆ Spray imidacloprid 0.2 ml/lit to control insect vector.

13. Red Rust/Algal Spot/Green Scurf

Symptoms

- ☆ Reddish brown circular to irregular spots develop on the leaves.
- ☆ When the twigs are attacked the bark is thickened and develops cracks leading to chlorosis of foliage and death of the affected twigs.
- ☆ Reddish spots are seen on fruits.

Alga

Cephaleuros virescens Kunze ex E.M. Fries.

This organism is not a parasite and grows only superficially, but can damage trees. Under moist conditions motile zoospoores are released from the sporangia, penetrate the host cuticle, and generate disc-like algal thalli with thread like algal filaments.

Mode of survival: As on different plant parts in living trees.

Primary spread: From other hosts.

Secondary spread: Through air-blown rain splashes.

Favourable conditions: Closer planting and frequent rain and wet condition.

Management

- ☆ Provide proper fertilization and irrigation.
- ☆ Proper pruning to enhance air circulation within the canopy and sunlight penetration, managing weeds and wider tree spacing.
- ☆ Avoid closer planting.
- ☆ Adopt balanced application of fertilizers.
- ☆ Managing insect, mite and other foliar diseases increases tree vigor and lessens susceptibility to this disease.
- ☆ Spray with Bordeaux mixture 1 per cent or Copper oxychloride 0.25 per cent.

14. Dodder (Phanerogamic Parasite)

Symptoms

- Small masses of branched, thread-like, cream yellow or orange coloured, leafless stems, devoid of green pigments twine around the stem or leaves of the host.
- It penetrates the host through haustoria, by which it draws nutrients from the host.
- Affected trees become weak and unproductive and some trees may die.

Parasite

Cuscuta subinclusa Durand and Hilg., *Cuscuta asiatica* Pall.

It is a flowering parasitic plant, which totally depend on the host plant for its nutrition hence named as total stem parasite. It is yellowish twining vine. It produces tiny, white, pink or yellowish flowers in clusters. It produces nearly 3000 seeds/plant.

Mode of survival: It perpetuates through seeds in the soil remain viable until favourable season return.

Primary spread: From Seeds of the parasite.

Secondary spread: Through animals, implements, irrigation water and human beings.

Favourable conditions: Continuous cultivation of susceptible hosts in the same field, closer planting.

Management

- The use of dodder-free planting seed
- Clean machinery before moving from an infested area to a non-infested area
- Burning reduces a dodder infestation
- Crop rotation with non-host crops (except onion, linseed, alfalfa, weeds, *Morinda* sp., *etc.*)
- Remove and destroy of the parasite.
- Apply pendimethalin 0.5-1.5 kg/ha as pre-emergence.

Postharvest Diseases of Citrus

1. Blue Mould: *Penicillium italicum*; Green Mould: *Penicillium digitatum*

Symptoms

- Soft watery spot.

- ☆ Initially white mycelial growth.
- ☆ Later develop blue or green mold.
- ☆ Spore masses are powdery or velvetty.
- ☆ Fruits shrink and mummified.
- ☆ In case green mold when fruits are wrapped the wrappers stick to the fruit.
- ☆ This will not happen in Blue mold.

3. Brown Rot: *Phytophthora* spp.

Symptoms

- ☆ Fruits near ground level mostly affected.
- ☆ Dull white or brown colour discolouration.
- ☆ Affected rinds are leathery and gradually soften.
- ☆ Fungus restricted to rind due to acidic.
- ☆ Emit aromatic odour.

4. Stem End Rot: *Phomopsis citri*

Symptoms

- ☆ Leathery rind softening of the pulp.
- ☆ Disease start from stem end or from side.
- ☆ Affected rind turn brown or black colour.
- ☆ Immature fruits are more susceptible.

5. Anthracnose: *Colletotrichum gloeosporioides*

Symptoms

- ☆ Red spots on the skin.
- ☆ Subsequently rotting occurs due to latent infection.

Management

- ☆ Good sanitation.
- ☆ Pruning branches 3 feet above ground level.
- ☆ Pre harvest treatment with copper fungicide.
- ☆ Hot water treatment for 2-4 min at 115-120°F, 36 hour after picking (or)
- ☆ Soaking in warm solution (43°C) or soak in Borax or sodium carbonate solution.
- ☆ Wrapping fruits paper impregnated with fungicidal solution.

5. Custard Apple/Sugar Apple (*Annona squamosa* L.)

1. Anthracnose

Symptoms

- ☆ Flower petals exhibit dark-brown lesions that enlarge, become black and cause flowers to shed, thereby reducing fruit set.
- ☆ Infections on young fruit cause either rotting and mummification.
- ☆ Infected branches shows witches broom appearance.
- ☆ Floral infection leads to shedding of flowers and young berries.
- ☆ Stem lesions develop as cankers and do not exude gum.

Fungus

Sexual stage: *Glomerella cingulata* (Stoneman)

[Asexual stage: *Colletotrichum gloeosporioides* Pat and Penz.].

Other sp: *C. anonicola* Ciferri; *Gloeosporium anonae* (Batista)

Acervuli (asexual fruiting body) with black setae, conidiophores are hyaline, septate, branched. Conidia are hyaline, aseptate, oval to elliptical or straight cylindrical and 11.0 to 17.0 × 3.0 to 6.0 µm.

Mode of survival: As older fruits left on the trees provide inoculum for disease spread.

Primary spread: From affected fruits and leaves.

Secondary spread: Through air-borne conidia.

Favourable conditions: Wet weather and heavy wind.

Management

- ☆ Grow resistant varieties.
- ☆ Spray carbendazim 0.2 per cent or mancozeb 0.2 per cent from flower initiation to harvest.

2. Cylindrocladium Leaf and Fruit Spot

Symptoms

- ☆ Symptoms on leaves and fruit begin as black spots.
- ☆ Spots on fruit are skin deep, not sunken and eventually dry out and crack.
- ☆ Spots enlarge; coalesce to form large, irregular, brownish-black lesions.

Fungus

Cylindrocladium colhounii Peerally

The pathogen produces three to four celled macroconidia. Macroconidiophores are septate pedicellate arranged. Perithecia were produced rarely.

Mode of survival: As soils of infected field.

Primary spread: From affected plant debris and soil.

Secondary spread: Through soil-borne conidia.

Favourable conditions: Dense planting.

Management

- ☆ Low-lying branches should be removed to avoid fruit and foliage contact with soil and increase air circulation.
- ☆ Mulch under trees and permanent sod in the middles.

3. Alternaria Leaf Spot

Symptoms

- ☆ Small yellowish spots first appear along the leaf margins, which gradually enlarge and turn into brownish patches with concentric rings.
- ☆ Severe infection leads to drying and defoliation.
- ☆ Dark brown-purplish patches appear on the infected fruits andrachis just below its attachment with the shoots.

Fungus

Alternaria alternata (Fr.) Keissl.

It has elongated dark brown color conidiophores. In favourable conditions conidia are produced (growing as buds from the conidiophores) from the conidiophores asexually.

Mode of survival: As infected plant debris.

Primary spread: From affected plant debris.

Secondary spread: Through air-borne conidia.

Favourable conditions: Moist wet weather.

Management

- ☆ Spray thiram or captan or mancozeb 0.2 per cent at 10 days interval

4. Diplodia Rot

Symptoms

- ☆ Purplish to black spots or blotches on surface of the fruit and eventually covered with white mycelia and black pycnidia.
- ☆ Diplodia rot is distinguished by its dark internal discolouration and the extensive corky rotting produces.

Fungi

Botryodiplodia theobromae Patouillard; *Diplodia theobromae* (Pat.) W. Nowell

Survival and spread:

- ✰ Survives in infected plant parts which serve as source of inoculums.

Favourable conditions:

- ✰ Optimum temperature is 25.9°C to 31.5°C and relative humidity is 80 per cent.

Management

- ✰ Use mulching under trees to reduce soil splash.
- ✰ Remove dead twigs and mummified fruits.
- ✰ Maintain proper plant canopy to reduce the high humidity by appropriate pruning and training.
- ✰ Prune trees 50 cm above the ground.

5. Black Canker

Symptoms

- ✰ Irregularly shaped spots ranging from small specks to large blotches
- ✰ Spots have an indistinct 'feathered' edge.
- ✰ Tissue damage under the spots is no more than 10 mm deep.

Fungi

Phomopsis annonacearum (Bond) Mont.

Survival: As mummified fruit and leaves.

Spread: From infected plant parts

Favourable conditions

- ✰ Optimum temperature is 25.9°C to 31.5°C and relative humidity is 80 per cent.

Management

- ✰ Use mulching under trees to reduce soil splash.
- ✰ Regularly monitor the diseases and remove dead twigs and mummified fruits.
- ✰ Follow proper plant canopy by appropriate pruning and training.
- ✰ Prune trees 50 cm above the ground.

6. Grapevine (*Vitis vinifera* L.)

1. Powdery Mildew

Symptoms

- ✰ White powdery coating appears on upper surface of leaf either in patches or in full.

- ✰ Leaves turn chlorotic then brown, malformed and fall.
- ✰ Stem turns dark brown.
- ✰ Floral infection leads to shedding of flowers and young berries.
- ✰ Berries are covered by white growth, show cracking, shriveling and mis-shapening.

Fungus

Uncinula necator (Schwein.) Burrill. (syn. *Erysiphe necator*)

Asexual stage: *Oidium tuckeri* Berk.

Mycelium is external, septate and hyaline. Conidiophores short and arise from ectophytic mycelium. Conidia are single celled, hyaline, barrel-shaped and in chain (*Oidium* type). Sexual fruiting body (circinoid cleistothecium) contains several asci and ascospores.

Mode of survival: As dormant mycelia present in infected shoots and buds.

Primary spread: From dormant mycelia from infected plant parts.

Secondary spread: Through air-borne conidia.

Favourable conditions: Sultry warm conditions with dull cloudy weather.

Management

- ✰ Grow resistant varieties.
- ✰ Avoid overcrowding and dense growth of vines by proper pruning.
- ✰ Spray wettable sulphur 0.2 per cent or dinocap 0.2 per cent or carbendazim 0.1 per cent.

2. Downy Mildew

Symptoms

- ✰ Leaves show irregular yellowish spots on the upper surface.
- ✰ Corresponding lower leaf surface shows white cushiony downy growth.
- ✰ In advanced stages, affected leaves are totally covered with downy growth, become yellow, brown, dry and fall prematurely.
- ✰ Tender shoots are dwarfed.
- ✰ Brown sunken lesions appear on the stem.
- ✰ Developed berries turn reddish brown and result in soft-rot.
- ✰ Berries become leathery, shrivel, dry and mummified.

Chromista

Plasmopara viticola (Berk. and M.A. Curtis).

It is a heterothallic oomycete. Mycelium is coenocytic and intercellular. Sporangiophores branch at right angles to the main axis and the tips of branches are blunt. Sporangia are thin walled, lemon-shaped, single celled and hyaline and produce zoospores. Oospores are thick walled.

Mode of survival: As oospores in infected plant parts.

Primary spread: From oospores from infected debris in soil.

Secondary spread: Through air-borne sporangia.

Favourable conditions: Optimum temperature of 20-22°C and relative humidity of 80-100 per cent and dew formation.

Management

- ✰ Prune and burn diseased leaves, shoots, flowers and berries
- ✰ Plant with proper spacing and training
- ✰ Spray Bordeaux mixture 1.0 per cent or difolatan 0.2 per cent or chlorothalonil 0.2 per cent or metalaxyl + mancozeb 0.2 per cent and repeat at weekly intervals.

3. Anthracnose/Bird's Eye Spot

Symptoms

- ✰ Young shoots are more susceptible than leaves.
- ✰ Circular spots surrounded by yellow halo on the leaves.
- ✰ The centre of the spot later become grey, sunken and fall off giving a shot hole symptom.
- ✰ Black elliptical sunken lesions on young shoot cause girdling and death.
- ✰ Infection on stalk of bunches and berries result in shedding of bunches and berries respectively.
- ✰ Sunken pots with ashy grey centre and dark brown margin resembling bird's eye appear on berries.
- ✰ Severe infection leads to shedding of berries.

Fungus

Elsinoe ampelina Shear

Asexual stage*: Gloeosporium ampelophagum* (Pass.) Sacc.

Elsinoe produces pseudothecia as a sexual structure, sclerotia as a resting structure and Acervuli as a asexual structure.

Mode of survival: As acervuli and sclerotia in the infected stem.

Primary spread: From ascospores from infected plant parts and infected seed vines.

Secondary spread: Through air-borne conidia.

Favourable conditions: Warm wet weather, heavy rain after pruning and badly drained soils.

Management

- ☆ Prune diseased leaves and twigs.
- ☆ Spray Bordeaux mixture 1.0 per cent or copper oxychloride 0.25 per cent or carbendazim 0.1 per cent or mancozeb 0.2 per cent or difolatan 0.2 per cent or chlorothalonil 0.2 per cent or bitertenol 0.1 per cent.

4. Black Rot

Symptoms

- ☆ Circular red spots with black margin appear on the leaves.
- ☆ Fruiting bodies are seen as minute black dots in concentric rings on the spots.
- ☆ Light brown circular spots on the berries and they increase in size.
- ☆ Infected berries are later discoloured and decayed and transformed into hard, black shriveled mummies.

Fungus

Guignardia bidwellii (Ellis)

Sexually produces pseudothecia and asexually produces pycnidia as conidiomata, spermogonia or ascomata primordial.

Mode of survival: As perithecia in infected berries.

Primary spread: From ascospores present on mummified fruits.

Secondary spread: Through air-borne pycnidiospores.

Favourable conditions: Temperature 25°C, leaf wetness of 6 hours favours the disease. Ascospores are released shortly after bud break until about 2 weeks after bloom and are dispersed by air and rain.

Management

- ☆ Collect and destroy diseased berries and leaves.
- ☆ Spray Bordeaux mixture 1.0 per cent or chlorothalonil 0.2 per cent or Pyraclostrobin 0.2 per cent.

Postharvest Disease of Grapes

1. Grey Mould Rot: *Botrytis cinerea*

Symptoms

- ☆ Skin condensed from flesh and gets separated.
- ☆ Skin become brown.
- ☆ Flesh becomes soft and watery.
- ☆ Decay and foul smell noticed.

Management

- Fumigate the basket with sodium sulphide.
- Transport in refrigerator van (4°C).
- Store the fruit at 0°C.

7. Guava (*Psidium guajava* L.)

1. *Fusarium* Wilt

Symptoms

- It is severe at the time of fruit bearing stage though it affects at any stage of the crop.
- Leaves of affected branches show yellowing, curling and drying of terminal branches but they do not shed for some time.
- Later leaves turn red followed by premature shedding.
- The twigs become bare and fail to put forth new leaves or flowers and eventually dry up.
- Barks of stem turn light brown with severe splitting. Light brown discoloration is also noticed in vascular tissues.
- Fruits are underdeveloped, hard and stony.
- Roots show rotting and the tree dies within a year of infection.

Fungus

Fusarium oxysporum f. sp. *psidii* Prasad (Asexual).

It is a saprophytic fungus, produces septate, hyaline to pinkish branched mycelium, upon maturity it produces microconidia, macroconidia and chlamydospores.

- Microconidia are small, one or two celled, more abundant and frequently produced by the fungus in the vessels of plant under all conditions.
- Macroconidia are larger conidia, which are three to five celled, gradually pointed and curved toward the ends. These spores are produced by sporodochia, commonly found on the surface of plants killed by this pathogen.
- Chlamydospores possess one or two cells, are round, thick-walled spores, produced either terminally or intercalary on older mycelium. Remarkably, they can remain dormant in soil and infect other hosts for as long as 30 years.

Mode of survival: As saprophyte and chlamydospores in soil.

Primary spread: From soil-borne pathogen.

Secondary spread: Through irrigation water and contaminated implements.

Favourable conditions: Monsoon period (Sep-Oct); alkaline soil with soil moisture of 60-80 per cent ; sandy loam and clayey loam soils; root injuries during cultural operations; nematode infestation.

Management

- ☆ Avoid waterlogged condition.
- ☆ Treat the pits with formalin and keep it covered for 3 days by polythene sheet and take up transplanting after two weeks.
- ☆ Maintain proper tree vigour through timely application of fertilizers and irrigation.
- ☆ Apply gypsum or lime to make the soil pH 6.0 to 6.5.
- ☆ Apply oil cakes, green manures and organic amendments.
- ☆ Uproot infected trees and burn.
- ☆ Spot drenching with carbendazim 0.1 per cent or thiophanate-methyl 0.1 per cent.

2. Anthracnose/Fruit Rot

Symptoms

- ☆ Dark brown to black, sunken, circular spots appear on the fully grown unripe fruits.
- ☆ Numerous, minute, black acervuli are seen in the centre of the lesions.
- ☆ Several spots coalesce to form bigger lesions.
- ☆ Diseased portions are comparatively harder and develop cracks.
- ☆ Unopened flowers and buds are attacked and shed.
- ☆ Necrotic ashy grey lesions occur on leaf tips and margins.
- ☆ Main branches show die back symptoms.

Fungus

Gloeosporium psidii Delacr/*Colletotrichum psidii* Curzi.

(Sexual stage*: Glomerella psidii* (Del.) sheld.).

Acervuli are dark brown to black. Conidiophore is cylindrical and tapering towards apex. It is hyaline and septate with single terminal phialide. Conidia are hyaline in single and salmon color in mass, aseptate, oval to elliptical or straight, cylindrical, flattened at base. Sexually perithecia were formed.

Mode of survival: As dormant propagules in unripe fruits and mummified fruits.

Primary spread: From infected fruits.

Secondary spread: Through air-borne conidia from dead leaves, twigs and mummified fruits.

Favourable conditions: Closer planting, poor canopy management, lack of timely harvest of fruits. In moist weather, acervuli are produced in abundance on dead twigs with pinkish spore masses.

Management

- ☆ Apple Guava (light red fleshed) is moderately resistant.
- ☆ Spray Bordeaux mixture 1.0 per cent or copper oxychloride 0.25 per cent before monsoon sets.

3. Grey Blight and Fruit Canker/Scab

Symptoms

- ☆ The disease generally occurs on green fruits and rarely on leaves.
- ☆ Minute, brown or rust coloured, circular, necrotic spots appear on fruits.
- ☆ The lesions tear open the epidermis and the margin is elevated with a depressed centre (crater like appearance).
- ☆ The cankerous lesion is very shallow without penetrating deep into the flesh.
- ☆ In severe cases, fruits break open to expose seeds.
- ☆ The infected fruits remain underdeveloped, become hard, malformed and mummified and drop.
- ☆ On the leaves, small rusty grey spots with brown margin appear with acervuli at the centre.

Fungus

Pestalotiopsis psidii Pat.

It is a coelomycetous pigmented fungi, characterized by spores having four-euseptate and pigmented median cells with two to four apical hairs like appendages.

Mode of survival: As acervuli in infected fruits and leaves.

Primary spread: From infected fruits and leaves.

Secondary spread: Through air-borne conidia.

Favourable conditions: Appear to be weak parasites infecting readily the fruits and leaves of guava injured by pinpricks. Naturally *Helopeltis* bug made punctures cause infection.

Management

- ☆ Spray Bordeaux mixture 1.0 per cent or copper oxychloride 0.25 per cent.

4. Sooty Mould

Symptoms

- ☆ Black sooty growth is found on the upper surface of leaves, branches and fruits.

- Sooty growth (mycelium, conidiophores and conidia) is associated with the infestation of honey dew excreting insects like scales and mealy bugs.

Fungi

Meliola sp. and *Capnodium* sp.

Mode of survival: As in mango.

Management

- Spray wettable sulphur + chloropyriphos + gum acacia (0.2 + 0.1 + 0.3 per cent).

5. Red Rust/Algal Spot

Symptoms

- Brown or red, circular or irregular rough spots appear on the upper leaf surface.
- The spots become faded after release of sporangia.
- Fruit lesions are small and dark green to black in colour.

Alga

Cephaleuros virescens Kunze ex E.M.Fries.

Management

- Spray Bordeaux mixture 1.0 per cent or copper oxychloride 0.25 per cent.

6. Phanerogamic Parasite - Giant Mistletoe

Symptoms

- The flowering parasite attacks young branches of trees in neglected gardens.
- Branches above the point of attack appear sickly or dry.

Parasite

Dendrophthoe falcata (L. f.) Ettingsh. and *D. longiflora* (Desr.) Ettingsh.

This is a partial stem parasite with dull green leaves and profuse flowering.

Favourable condition: Trees in poorly maintained or neglected plantations are highly susceptible.

Management

- Infected branches should be cut 15 cm below the point of attachment of the parasite.
- Cut ends protected by painting with Bordeaux paste.
- Spray 30-40 per cent diesel oil emulsion in soap water to eradicate the parasite.

8. Indian Gooseberry/Aonla (*Phyllanthus emblica* L.)

1. Rust

Symptoms

- Reddish brown, circular, 2-5 mm dia, uredo pustules appear on both surfaces of the leaflets.
- The uredosori rupture the epidermis expose uredospores.
- On fruits, pustules are black at first and later they develop into rings.
- Later the pustules join together and cover a big area of the fruit.
- The black spores get exposed by rupturing a paper covering and the fruit gives a dirty appearance.

Fungus

Ravenelia emblicae Syd. and P. Syd. and *Phakopsora phyllanthi* Dietel.

Mode of survival: As infected leaves and fruits.

Primary spread: From teliospores by air or splashing rain.

Secondary spread: Through air-borne uredospores.

Favourable conditions: Cool conditions with long periods of wetness.

Management

- Spray wettable sulphur 0.3 per cent or zineb 0.2 per cent.

2. Anthracnose

Symptoms

- Circular, brown to grey spots with yellow margin studded with dot-like fruiting bodies are seen on the leaflets.
- On the fruits, lesions are depressed and dark at the centre.
- Acervuli arranged in rings are found on the fruit spots.
- The fruits become shriveled and rot.

Fungus

Sexual stage: *Glomerella cingulata* (Stoneman) [*Colletotrichum gloeosporioides* Pat and Penz.].

Acervuli (asexual fruiting body) with black setae, conidiophores are hyaline, septate, branched. Conidia are hyaline, aseptate, oval to elliptical or straight cylindrical and 11.0 to 17.0 × 3.0 to 6.0 μm.

Mode of survival: As infected fruits.

Primary spread: From affected fruits and leaves.

Secondary spread: Through air-borne conidia.

Favourable conditions: Hot and humid weather.

Management

- Keep the orchard clean.
- Collect and destroy the affected fruits from the orchard.
- Spray carbendazim 0.1 per cent before harvesting of fruits.

3. Sooty Mould

Symptoms

- Sooty mould causes velvety covering of black fungal growth on the surface of leaves, twigs and flower.
- Infection is restricted only to the surface and do not penetrate into leaves.

Fungus

Capnodium sp.

Management

- Spray starch 2 per cent + lambda cyhalothrin 0.05 per cent + wettable sulphur 0.2 per cent.

9. Jackfruit (*Artocarpus heterophyllus* L.)

1. Pink Disease

Symptoms

- Pinkish outgrowth is seen on the surface of young branches.
- The fungus penetrates the bark and grows inside and subsequently enters cortex.
- Due to interruption of flow of nutrients, leaves become yellow.
- Leaves crinkle, shed and twigs dry.

Fungus

Pellicularia salmonicolor (Berk. and Br.) Dastur

Primary spread: From infected plant parts.

Secondary spread: Through air-borne conidia.

Mode of survival: As dormant mycelium

Management

- Cut the affected branches and wound dress with Bordeaux paste on the cut end.

2. *Rhizopus* Fruit Rot

Symptoms

- ✰ Young fruits and male inflorescences are infected.
- ✰ In young fruit, rotting begins in the stalk and covered by the blackish mycelium and sporangia.
- ✰ The fruits are mummified and drop prematurely.

Fungus

Rhizopus atrocarpi Vuillemin

Rhizopus is a saprophytic zygomycetes fungus, produces filamentous aseptate dark color hyphae. In asexual reproduction, sporangiospores are produced inside the sporangium. Sporangia are supported by columella. In sexual reproduction, zygospore is produced.

Mode of survival: As survives in fallen fruits in soil.

Primary spread: From zygospores from infected fallen fruits in soil.

Secondary spread: Through air-borne sporangia.

Favourable conditions: Poorly maintained plantation and rainy season.

Management

- ✰ Improve sanitation.
- ✰ Spray Bordeaux mixture 1.0 per cent or copper oxychloride 0.25 per cent or captan 0.2 per cent.

3. Leaf Spot

Symptoms

- ✰ Circular dark brown to reddish brown spots with 10-30 mm size with black margin are produced on the leaf lamina.
- ✰ In extreme case defoliation occurs because of coalescence of several spots on the foliage.

Fungus

Colletotrichum gloeosporioides Pat and Penz.

Mode of survival: As acervuli in infected plant debris.

Primary spread: From infected fallen leaves.

Secondary spread: Through air-borne conidia.

Management

- ✰ Collect and burn the infected fallen leaves.
- ✰ Spray mancozeb 0.2 per cent or carbendazim 0.1 per cent

10. Mango (*Mangifera indica* L.)

1. Powdery Mildew

Symptoms

- ☆ Whitish powdery growth is seen on the inflorescence, leaves, stalk of inflorescence and young fruits resulting in its shedding.
- ☆ The sepals are more susceptible than petals. Affected flowers fail to open, blighted and drop prematurely.
- ☆ Young fruits are covered by the whitish powdery growth. Epidermis of the infected maturing fruits cracks and corky tissues are formed.
- ☆ Affected fruits do not develop in size due to powdery growth and drop before attaining pea size.
- ☆ Infection on young leaves is noticed on both the sides as white powdery patches when their colour changes from metallic brown to light green.
- ☆ The pathogen is more restricted to the area of the central and lateral veins of the infected leaf and the leaves twist curl and become distorted.

Fungus

Oidium mangiferae Bert. (Syn. *Acrosporium mangiferae* (Berthet) Subram.).

Mycelium is hyaline, ectophytic, conidiophores non-septate, short and hyaline, conidia hyaline, single celled, barrel-shaped and produced in chains.

Mode of survival: As infected plant parts.

Primary spread: From diseased older leaves and dried inflorescences.

Secondary spread: Through air-borne conidia.

Favourable conditions: Cool night accompanied with heavy morning mist during flowering; temperature of 15-30°C and relative humidity of 70-85 per cent ; cloudy weather and high air velocity for 3-4 days; closer planting and non-adoption of canopy management.

Management

- ☆ Adopt canopy management by systematic pruning.
- ☆ Collect and destroy diseased and fallen young fruits.
- ☆ Spray wettable sulphur 0.2 per cent or carbendazim 0.1 per cent dinocap 0.1 per cent or dust sulphur 25 kg/ha before flowering and repeat if necessary.

2. Anthracnose

Symptoms

- ☆ It infects leaves, twigs, stems, blossoms and fruits.

- On leaves round to irregular and dark brown spots appear and the affected portion dries and falls off exhibiting 'shot hole' symptom (leaf spot).
- Elongated black necrotic areas appear on twigs and the young branches dry from tip downwards (die back or wither tip).
- Black spots appear on panicles and flowers which enlarge and coalesce to cause drying of flowers. The infected flowers fall-off (blossom blight).
- Infection on fruits is more common during transit and storage. Black round sunken spots of more than 5 mm seen on matured fruits which later crack. Spots increase in size rapidly, coalesces covering major fruit surface. Fruit skin becomes thin, black with pinkish growth (fructification) and the pulp is rotten (fruit rot).
- Latent infection on unripe fruit is carried from the field and develops further in storage causing rotting of fruits in store houses.

Fungus

Colletotrichum gloeosporioides Pat and Penz.

[Sexual stage: *Glomerella cingulata* (Stoneman)].

Acervuli (asexual fruiting body) with black setae, conidiophores are hyaline, septate, branched. Conidia are hyaline, aseptate, oval to elliptical or straight cylindrical and 11.0 to 17.0 × 3.0 to 6.0 µm.

Mode of survival: As dried fallen leaves, defoliated dead branches and diseased twigs attached to trees (survive for 14 months).

Primary spread: From diseased twigs and leaves.

Secondary spread: Through air-borne spores, rain drops and contact with diseased fruit during transport and storage and latent infection in matured green fruits.

Favourable conditions: Temperature of 24-30°C and high relative humidity of 95-97 per cent ; frequent rain and free moisture on the foliage; closer planting and non-adoption of canopy management; unhygienic orchard maintenance.

Management

- Diseased twigs, leaves and fruits should be collected and destroyed.
- Prune and burn infected twigs.
- Spray carbendazim 0.1 per cent or thiophanate methyl 0.1 per cent or chlorothalonil 0.2 per cent or copper oxychloride 0.25 per cent before flower opening and during fruit formation.
- Spray hexaconazole 0.1 per cent or carbendazim 0.1 per cent at 15 days interval prior to harvest. Last spray should be 15 days prior to harvest.
- Dip the fruits in thiabendazole 1000 ppm for 5 min or carbendazim 500 ppm before storage.

- ☆ Treat the matured fruits in hot water at 52° C ± 1°C for 20 min. Reduce the duration to 10 min by adding carbendazim or prochloraz 0.05 per cent in hot water.

3. Grey Blight

Symptoms

- ☆ Elongated spots on leaves are seen at the tip or margin or any part of leaf.
- ☆ Spots are light brown with a dark brown margin.
- ☆ Several spots coalesce to cover major leaf area with pinhead like fructifications (acervuli) and cause drying of foliage.
- ☆ Elliptical lesions on stem girdle the stem.
- ☆ Water soaked lesions on fruits enlarge rapidly and cause rotting particularly in storage.

Fungus

Pestalotiopsis mangiferae (Henn.) (Syn. *Pestalotia mangiferae* Henn).

It is a coelomycetous pigmented fungus, produces septate mycelia and acervulus as an asexual fruiting body. Conidia are 5-celled, oblong to clavate or clavate to fusiform, colored cells are 15-16 micro meter long, upper two of them slightly darker than the lowest olivaceaus colored cells, septa and walls sometimes black and will have long pedicel.

Mode of survival: As infected plant parts.

Primary spread: From diseased leaves to healthy leaves by rain splashes.

Secondary spread: Through air-borne conidia.

Favourable conditions: During young stage of plant, pathogen reside as latent infection under favourable conditions *viz.*, monsoon period with temperature of 20-25°C resumes as active infection.

Management

- ☆ Remove and destroy infected plant parts.
- ☆ Spray copper oxychloride 0.25 per cent or Bordeaux mixture 1 per cent.

4. Sooty Mould

Symptoms

- ☆ The disease is associated with honey dew secreting insects like scales, mealy bugs and mango hoppers.
- ☆ The honey secreted by the insects on the leaves and twigs attracts the fungus, which multiplies rapidly, spreading on the plant surface.
- ☆ Black fungal encrustation is seen on tender and old leaves, stem, inflorescence and fruits.

- Because of the production of superficial mycelium and masses of black spores, which stick to the leaf surface due to the sticky honey dew, the foliage appears black and ugly and hence the name sooty mould.
- Photosynthetic activity, fruit set and yield are reduced.
- Black coloured fruits fetch low market value.

Fungi

Melioa mangiferae Earle., *Capnodium ramosum* Cke., *C. mangiferae* Cke. and Brown and *Trichospermum acerinum* (Syd). Speg.

Mode of survival: As dormant mycelium in infected leaves, inflorescence and twigs.

Primary spread: From infected plant parts.

Secondary spread: Through air-borne conidia.

Favourable conditions: Lack of timely control of honeydew excreting insects; high incidence of leaf hoppers; old and dense orchard where light penetration is low; Continuous and heavy rainfall washes down these substances but high humidity proved congenial for growth of the fungus.

Management

- Spray systemic insecticides like methyl demeton 0.05 per cent + wettable sulphur 0.2 per cent or copper oxychloride 0.25 per cent or Bordeaux mixture 1 per cent to control both insects and the disease.
- Spray starch solution 5 per cent (Mix 1 kg starch/maida/jack seed flour in 5 lit of water. Boil and make up the volume to 20 litres with water). Starch dries and forms flakes over the leaves and are removed along with the fungus when air blows.
- Spray wettable sulphur 0.2 per cent + methyl demeton 0.05 per cent + gum acacia 0.3 per cent.
- Spray should cover both the surface of leaves for effective control.

5. *Diplodia* Stem-end Rot

Symptoms

- It is an important postharvest disease of ripe mango fruits.
- Initially circular brown spots appear near the stem end which gradually enlarges into dark brown or black area towards the lower portion of the fruit.
- The whole fruit completely turns black and rots within two to three days.
- The pulp of the diseased fruit becomes brown.
- The infected fruits lose ascorbic acid rapidly and they have low content of non- reducing sugars in the pulp.

Fungus

Diplodia natalensis Pole-Evans (*Lasiodiplodia theobromae* Pat.)

Diplodia is an asexual stageic fungus, which produces minute globose pycnidia asexually in the infected plant parts. Pycnidiospores are hyaline and thin walled, becoming thick walled, and dark brown and one septate.

Mode of survival: As infected plant parts.

Primary spread: From fungus is a wound parasite. Dead twigs and bark of the trees harbour the fungus. The spores are spread through rain splashes.

Secondary spread: Through air-borne pycnidiospores and contact with diseased fruits in the storage.

Favourable conditions: Temperature of 26-32°C and relative humidity above 80 per cent.

Management

- ✰ Spray carbendazim or thiophanate methyl 0.1 per cent 15 days before harvesting control the postharvest rot
- ✰ Harvest the fruits along with the stalk of 5 cm length to check the infection from stem end. If fruits are harvested without stalk, the opening should be closed with wax.
- ✰ Dip mangoes in borax solution 6 per cent at 43°C for three min or in hot water 52±1°C with 0.05 per cent carbendazim for 5 min.
- ✰ Fruits should be covered in cellophane paper and sent to the ripening houses immediately after harvest.
- ✰ Injuries to fruits should be avoided at all stages of handling.

6. Bacterial Leaf Spot (Black Spot) and Canker

Symptoms

- ✰ It attacks leaves, petioles, peduncle, twigs, stem, branches and fruits.
- ✰ Small water-soaked lesions appear in groups towards the tip of the leaf.
- ✰ They increase in size, turn black, angular and are surrounded by a distinct yellow halo.
- ✰ Cankerous lesions appear on petioles and branches.
- ✰ On young fruits, water-soaked, raised brown or black lesions appear and become cankerous later.
- ✰ Infected fruits exhibit cracks on the cankered area with gummy exudation.
- ✰ Badly affected fruits drop prematurely.

Bacterium

Xanthomonas axonopodis pv. *mangiferae-indicae* Ah-You *et al.*

The bacterium is Gram negative, rod-shaped, motile by means of a polar flagellum.

Mode of survival: As infected plant parts in the tree

Primary spread: From infected planting materials and infected mango stones.

Secondary spread: Through air splashed rain.

Favourable conditions: Susceptible varieties; temperature of 25-30°C and relative humidity of more than 90 per cent ; high air velocity and lack of growing air breaks; weevil damage

Management

- Maintain orchard sanitation and seedling/graft certification.
- Use mango stones from healthy fruits for raising seedlings (root stock).
- Raise air-breaks to reduce bruising/wounding and subsequent infection.
- Spray streptomycin sulphate + oxytetracycline combination (streptocycline) at 200- 250 ppm or copper oxychloride 0.25-0.3 per cent at 10 days interval.

7. Phanerogamic Parasite - Giant Mistletoe

Symptoms

- The stem of flowering parasite creeps along the stem of mango and it attacks by peg-like growth at several places. It is a partial stem parasite.
- It forms a bushy growth with dark green leaves.
- At the point of attachment haustorium gives a bulged appearance and it penetrates through haustoria and absorbs food materials.
- Affected branches are weak, lose the foliage and dry up.
- Trees gradually deteriorate and die in 2-4 years.
- The parasite also attacks *Citrus* spp., jack, guava, sapota, neem, *etc.*

Parasite

Dendrophthoe falcata (L.f.) Ettingsh, *D. ampullaceus, D. capillatus* and *D. elasticus* (Syn. *Loranthus longiflorus*). Leaves of the parasites are narrow, thick, leathery and green. Fruits are fleshy with sticky mucilaginous contents.

Mode of survival: As parasite survives on host trees.

Primary spread: From other hosts.

Secondary spread: Through seeds of parasite through birds.

Favourable conditions: Poorly maintained unhygienic orchard and other host plants near the orchard.

Management

- ☆ Cut the infected branches atleast 30 cm below the point of attachment of the parasite. If small bit of haustorium is left in the host, it re-grows with renewed vigour and soon starts damaging the host as before.
- ☆ Protect the cut ends by wound dressing with Bordeaux paste or copper oxychloride.
- ☆ The plant growth regulator ethephon used as directed by the label to control.
- ☆ Spray 30-40 per cent diesel oil emulsion in soap water to eradicate the parasite.

8. Red Rust/Algal Spot

Symptoms

- ☆ Rusty-red, circular or irregular, elevated spots occur mostly on leaves, veins, petioles, young twigs and fruits are infected.
- ☆ Algal spots grow and merge with each other.
- ☆ The spots are reddish brown initially and turn greenish grey after of shedding of spores.
- ☆ The disease is more common on closely planted orchard.
- ☆ Photosynthesis reduced leading to poor fruit yield.

Causal Organism

Alga, *Cephaleuros virescens* Kunze ex E.M. Fries.

It is an algal parasite. In the asexual reproduction it produces thallus consisting of setae (sterile hairs, sporangiophores and sporangia).

In sexual reproduction is by isogametic copulation. Large, sessile, flask shaped gametangia are produced in the thallus. They are formed in each microsporangium as a result of meiosis. These microzoospores rarely infect the host.

Mode of survival: The alga survives on different plant parts in living trees.

Primary spread: From zoospores from collatral hosts.

Secondary spread: Through air-blown rain splashes.

Favourable conditions: Closer planting and rrainy season.

Management

- ☆ Avoid closer planting.
- ☆ Follow balanced application of fertilizers.
- ☆ Spray with Bordeaux mixture 1 per cent or copper oxychloride 0.25 per cent.

9. Malformation

Symptoms

Vegetative malformation: It occurs mostly on young plants. Witches, broom-like growth of shoot is known as vegetative malformation or bunchy top. These infected branches do not bear fruits.

Floral malformation: The flower buds are transformed into vegetative buds. The flower buds seldom open and remain dull green and appear in clusters. Abnormal development of inflorescence is known as floral malformation.

Causal Organism

There are a number of schools of thoughts and accordingly virus, fungus, (*Fusarium moniliforme* var. *subglutinans*), mites, nutrient deficiency, and role of C : N ratios are responsible.

Management

- ☆ Prune diseased parts along with basal 15-20 cm apparently healthy portions followed by spray carbendazim 0.1 per cent or captafol 0.2 per cent along with insecticides.
- ☆ Spray chelated Zinc 100 ppm and Copper 40 ppm during flower bud differentiation and flowering stage combined with spray of carbendazim 0.1 per cent.
- ☆ Spray naphthalene acetic acid (NAA) 200 ppm or Planofix (90 ml/20 L) during first week of October (Before bud differentiation stages) followed by deblossoming spray in late December or January (bud burst stage).

10. Black-tip

(*Physiological Disorder*)

Symptoms

- ☆ Occur in orchards located in close proximity to a brick kiln. The gases like carbon monoxide, sulphur dioxide and ethylene constituting the fumes of brick kiln are known to damage growing tip of fruits and give rise to the symptoms of black tip.
- ☆ The infection of fruits is initiated at marble stage, with a characteristic yellowing of tissues at the distal end.
- ☆ Gradually, the colour intensifies into brown and finally black extends toward the upper part of the fruit.
- ☆ Growth and development of the fruit is retarded.
- ☆ Fruits exposed on outer side of the tree are more prone than those unexposed.

Cause: Boron deficiency. Trees exposed to brick kiln fumes are deficient in boron.

Favourable conditions: Irrigation, vigour of the tree and management practices.

Management

- ☆ Establish mango orchards 5-6 km away from the brick kilns.
- ☆ Spray borax 1.0 per cent at pea stage and repeat two or three times at 15 days interval.

Postharvest Diseases of Mango

1. Stem End Rot or Stalk End Rot: *Botryodiplodia theobromae*

Symptoms

- ☆ Light brown spots near pedicel.
- ☆ Spot enlarge form brownish black patches.
- ☆ Under favourable conditions pycnidia form on the skin.

2. Anthracnose: *Colletotrichum gloeosporioides*

Symptoms

- ☆ Black round, irregular sunken spots on the skin.
- ☆ Extend to whole surface during ripening.
- ☆ Produce acervuli.

3. Aspergillus Rot: *Aspergillus niger*

Symptoms

- ☆ Occur during mechanical injury.
- ☆ Sunken rotting areas on fruits
- ☆ Advanced stage conidial head seen.

4. Rhizopus Rot: *Rhizopus* spp.

Symptoms

- ☆ Soft watery rot with strands of mycelium
- ☆ Crack develops with slight pressure.

Management

- ☆ Regular pre harvest spray schedule mancozeb 0.25 per cent
- ☆ Hot water treatment 50-55°C for 15 minutes
- ☆ Dipping fruits in suitable fungicide as postharvest treatment.

11. Papaya (*Carica papaya* L.)

1. Damping Off

Symptoms

- Young seedlings are killed before they come out of soil.
- Germinated seedlings topple over at any time after emergence.

Chromista

Pythium aphanidermatum (Edson) Fitzp.

Mycelium is well developed, coenocytic, hyphae 2.8-7.5 µm in diameter. Sporangia are loculate, toruloid and vesicles formed at the time of germination, zoospores, 30-40 in number, are present within the vesicle and are borne on spherical oogonia. They germinate by producing a germ tube. The sexual spores are the oospores, borne singly which are moderately thick walled.

Mode of survival: As oospores in the soil.

Primary spread: From dormant mycelium in infected plant materials and and oospores in soil.

Secondary spread: Through zoospores spread through rain and irrigation water.

Favourable conditions: High soil moisture, poor drainage and crowded seedlings.

Management

- Treat seeds with *Trichoderma viride* 4g/kg or *Pseudomonas fluorescens* 10g/kg or captan 4g/kg.
- Raise the seedlings in well-drained soil.
- Drench Bordeaux mixture 1.0 per cent or captan 0.2 per cent or ccopper oxychloride 0.25 per cent or metalaxyl 0.1 per cent.

2. Foot Rot/Stem Rot

Symptoms

- Water-soaked lesions appear on the stem bark near the ground level and enlarge.
- Affected tissues turn dark brown or black and the infection girdle the stem at the base. The stem becomes watery and shrinks followed by death of the plant.
- Terminal leaves become yellow, droop and wilt.
- Fruits shrivel, malformed and drop.
- Roots disintegrate and decay.

Chromista

Pythium aphanidermatum (Edson) Fitzp. and Fungi *Rhizoctonia solani* Kuhn.

Mode of survival: As oospores and sclerotia in infected plant debris in soil.

Primary spread: From oospores and sclerotia.

Secondary spread: Through dormant mycelia, sclerotia and zoospores in irrigation water.

Management

- ☆ Raise the seedlings in well-drained soil and avoid waterlogging condition.
- ☆ Remove diseased seedlings carefully and burn.
- ☆ Select area for new planting with drainage facilities.
- ☆ Apply *Trichoderma viride* (15g/plant) mixed with well-decomposed FYM around the root zone of the plants at the time of planting.
- ☆ Adopt ring method of irrigation.
- ☆ Drench the stem base with Bordeaux mixture 1.0 per cent or captan 0.2 per cent or copper oxychloride 0.25 per cent or metalaxyl 0.1 per cent.

3. Powdery Mildew

Symptoms

- ☆ It attacks seedlings and matured trees.
- ☆ White powdery growth appears on upper surface of leaves and fruits.
- ☆ Young infected leaves dry up prematurely and drop down.
- ☆ Infected fruits are small and malformed.

Fungus

Oidium caricae-papayae J.M. Yen.

The pathogen is an obligate parasite. Mycelium is hyaline, septate, ectophytic and creeping. The asexual spore is conidia which are borne in long chains at 2-4 hyaline, elliptical and are barrel shaped. The sexual spores are the ascospores which are globose, and gregarious. Each ascocarp contains one ascus.

Mode of survival: As cleistothecia in infected plant parts.

Primary spread: From ascospores from infected plant parts.

Secondary spread: Through air-borne conidia.

Favourable conditions: Temperature of 24-26°C and relative humidity of 80-85 per cent.

Management

- Spray wettable sulphur 0.3 per cent or carbendazim 0.1 per cent or triadimefon 0.2 per cent.

4. Anthracnose

Symptoms

- The spots on green immature fruits are brown, circular and sunken with 1-3 cm in dia and they coalesce later.
- Sparse mycelial growth appears on the margins of spots on fruits.
- The centers of these spots later turn black and then pink when the fungus produces spores.
- The flesh beneath the spots becomes soft and watery, which spreads to the entire fruit.
- It attacks leaves, petioles and stems and produces water soaked irregular spots where acervuli are seen in concentric rings at the centre.
- It infects the fruits in storage also (postharvest pathogen).

Fungus

Colletotrichum gloeosporioides Pat and Penz.

The fungus enters through pores of green fruits where it is still green and develops further inside the flesh during ripening period. The pathogen produces ascospores in the senescing petioles which become airborne, lodge on fruit surface, germinate and produce appressoria. The pathogen can cause latent infection in mature fruits though lenticels. The fungus grows in fruit flesh hydrolyzing sucrose during the course of infection.

Mode of survival: As infected fallen fruits.

Primary spread: From fungus survives as dormant mycelia in the fallen leaves.

Secondary spread: Through air and rain water splash borne conidia produced in the acervuli.

Favourable conditions: The pathogen is more severe at 25°-30°C and relative humidity of 85-90 per cent. The conidia of pathogen are disseminated by raindrop splashes and insects.

Management

- Harvest the fruits as soon as they mature.
- Remove and destroy the affected fruits
- Spray carbendazim 0.1 per cent or chlorothalonil 0.2 per cent or mancozeb 0.2 per cent or thiophanate methyl 0.1 per cent.
- Treat the fruits in hot water or fungicidal wax after harvest to control postharvest damage.

5. Dry Root Rot/Charcoal Rot

Symptoms

- Leaves exhibit yellowing, drooping and later drying.
- Basal portion of the stem and roots exhibit decay.
- Bark of the trees at the basal region splits into shreds and a large number of sclerotial bodies are seen on affected root tissues.

Fungus

Macrophomina phaseolina (Tassi) Goid. (Pycnidial stage)

Rhizoctonia bataticola (Taub.) Butler (Sclerotial stage)

It is a soil borne asexual fungi, characterized by the production of both pycnidia and sclerotia in host tissues.

Mode of survival: As microsclerotia in soil and plant debris.

Primary spread: From pycnidiospores from plant debris.

Secondary spread: Through sclerotia and pycnidiospores through irrigation water.

Favourable conditions: High temperature of above 30°C and low soil moisture.

Management

- Remove and burn affected plants.
- Spot drench with carbendazim 0.1 per cent.

6. Mosaic

Symptoms

- The disease attacks the papaya plants of all age groups, but is most serious on young plants.
- Typical mosaic symptoms showing chlorosis with dark green blisters alternated with yellowish green patches on top young leaves.
- The petiole is reduced in length and the top leaves are upright in position.
- Leaf lobes are thin and are increased in number and show shoe-string symptoms.
- Fruits show circular, water-soaked lesions with white spots in the centre. Such fruits are elongated and reduced in size.

Virus

Papaya mosaic virus (PapMV) Potx virus.

Primary spread: From infected seedlings.

Secondary spread: Through aphids (*Myzus persicae, Aphis malvae, A. medicaginis, A. gossypii* and *Macrosiphum sonchi*).

Favourable conditions: Susceptible hosts and abundance of insect vector

Management

- ☆ Use healthy seedlings for planting.
- ☆ Rogue out diseased plants at periodical intervals
- ☆ Application of Carbofuran (1 kg a.i./ha) at the time of sowing followed by 2-3 foliar sprays of Phosphamidon (0.05 per cent) at an interval of 10 days starting from 15-20 days after sowing effectively checks the population of aphids.

7. Leaf Curl

Symptoms

- ☆ Mostly the young leaves are affected.
- ☆ Leaves show severe curling, crinkling and distortion accompanied by vein clearing with reduced leaf size.
- ☆ Inverted cups are formed due to inward and downward curling of leaves.
- ☆ Leaves become leathery and brittle; veins become thick and dark green.
- ☆ Petioles are twisted into zig-zag manner.
- ☆ Infected plants are stunted.
- ☆ Affected plants do not flower or bear few and small fruits in advanced stages.

Virus

Tobacco leaf curl virus (TLCV).

Mode of survival: As infected trees and several other hosts.

Primary spread: From infected seedlings.

Secondary spread: Through insect vector, whitefly (*Bemisia tabaci*).

Management

- ☆ Use healthy seedlings.
- ☆ Uproot affected plants and burn.
- ☆ Avoid growing collateral hosts (tomato, tobacco, chilli, *etc.*) near papaya plantation.
- ☆ Spray dimethoate or methyl demeton 0.05 per cent.

8. Ring Spot

Symptoms

- ☆ Yellowing and clearing of veins in young leaves.

- ☆ Later leaves become mottled or with severe blistering and distortion.
- ☆ Dark-green streaks and rings also appear in the leaf stalks and stems.
- ☆ Dark green concentric rings and spots or C-shaped markings develop on the green fruits and hence the ring spot.
- ☆ Symptoms persist on the ripe fruit as darker orange-brown rings.
- ☆ Virus induces shoe-string symptom and stunting of plants.
- ☆ Fruits are smaller, deeply lobed and have low sugar and poor latex content.

Virus

Papaya ring spot virus (PRSV) Poty virus.

Mode of survival: As infected trees and other hosts.

Primary spread: From infected seedlings.

Secondary spread: Through aphids (*Aphis gossypii* and *Myzus persicae*).

Management

- ☆ Remove and destroy diseased plants.
- ☆ Spray systemic insecticide to control insect vectors.

12. Peach (*Prunus persica* L.)

1. Leaf Curl

Symptoms

- ☆ The disease first appears in the early spring when the leaves begin to unfold.
- ☆ Leaf blade thickens and puckers along the midrib and curls and veins do not develop properly.
- ☆ Young leaves gradually change to yellow and finally to a reddish purple.
- ☆ The reddish velvety surface of the lamina is covered with a whitish grey bloom of the fungus on the upper surface.
- ☆ Both the leaves and petioles may curl and the affected leaves die and drop prematurely.
- ☆ Twigs become pale green to yellow, swollen, stunted and exude gummy material.
- ☆ Fruit set is poor. Affected flowers and fruits drop prematurely.

Fungus

Taphrina deformans (Berk.)

Mycelia are intercellular and produce naked asci (Open ascus) individually. Each ascus bears eight ascospores of 3 to 7 µm diameter.

Mode of survival: As conidia and ascospores in infected twigs and buds.

Primary spread: From ascospores and sprout conidia from barks and buds.

Secondary spread: Through ascospores and sprout conidia.

Favourable conditions: Cool and wet weather and young and unfolded leaves. The expanded leaves are resistant to infection

Management

- ✰ Grow resistant varieties.
- ✰ Remove and burn infected twigs and fallen leaves.
- ✰ Spray chlorothalonil 0.2 per cent or copper oxycloride 0.25 per cent before bud burst and full bloom.

2. Rust

Symptoms

- ✰ Pale yellow spots appear on both the leaf surfaces.
- ✰ On the under surface brown rusty pustules are seen.
- ✰ When the leaves are severely infected defoliation occurs.
- ✰ Fruit infected with rust is unsaleable. The infections penetrate several millimetres into the flesh.

Fungus

Puccinia pruni- spinosae Pers. (Syn.*Tranzschelia discolor* f. sp. *persicae* Fuckel).

Tranzschelia is a macrocyclic, heteroceous rust, the spore stages are single celled, rusty brown urediospore, dark brown teliospore, basidiospore, and aeciospore.

Mode of survival: As The fungus persists as perennial mycelium, in the under ground stems of *Anemone coronaria*.

Primary spread: From urediospore.

Secondary spread: Through aeciospore from alternate host *Anemone coronaria.*

Management

- ✰ Spray mancozeb 0.2 per cent or carbendazim 0.1 per cent.

3. Powdery Mildew

Symptoms

- ✰ The young leaves are coated on the upper surface with a thick layer of mycelium.
- ✰ Terminal portion of the growing shoot is covered with a white powdery layer.

- ☆ White, round spots develop on fruits and turn the fruits into pink to dark brown.

Fungus

Sphaerotheca pannosa var. *persicae* [Syn: *Podosphaera pannosa* (Wallr.) de Bary.]

Mycelium is septate, hyaline branched, and produces aseptate conidiophore bears chains of globose shape conidia. Sexually it produces myceloid type of Cleistothecium.

Mode of survival: As cleistothecia in infected plant parts.

Primary spread: From ascospores are then released and spread by air, insects, and rain.

Secondary spread: Through air-borne conidia.

Management

- ☆ Spray carbendazim 0.1 per cent or wettable sulphur 0.3 per cent.
- ☆ Spray recommeted doses of myclobutanil, azoxystrobin, triadimefon and thiophanate-methyl formulations.

Postharvest Diseases of Papaya

1. Rhizopus Rot: *Rhizopus stolonifer*

Symptoms

- ☆ Irregular, water soaked lesion.
- ☆ Gradually enlarge and get covered.
- ☆ White and dark brown fungal growth.
- ☆ Finally fruit collapse.

2. Anthracnose: *Colletotrichum* spp.

Symptoms

- ☆ Brown superficial discoloration on the skin later develops into circular slightly sunken area.
- ☆ Lesion coalesce and white mycelial growth appear.
- ☆ Salmon pink spore masses on the fruit.

3. Botryodiplodia Fruit Rot: *Botryodiplodia theobromae*

Symptoms

- ☆ Initially produce water soaked spots.
- ☆ Black and have a rough surface later.
- ☆ Erumpent, confluent arrangement of pycnidia.

Management

- ✰ Orchard sanitation.
- ✰ Preharvest spray with 0.2 per cent mancozeb or bordeaux mixture 1 per cent.
- ✰ Storing fruit at 10°C.
- ✰ Hot water treatment at 49°C for 20 minutes.

13. Pear (*Pyrus communis* L.)

1. Scab

Symptoms

- ✰ Olive green circular spots appear which later become grey and necrotic.
- ✰ Spots on fruits are black and affected fruits are deformed and small in size.
- ✰ Fruits develop cracks, become corky.

Fungus

Venturia pirina Aderh.

Venturia is a heterothallic, hemibiotrophic fungus. *V. pyrina* develops sexual stage (ascospores) in leaf litter, where it overwinters as a saprotroph during the dormant season and asexual (conidia) is formed on infected plants.

Mode of survival: As overwinters in fallen leaves and infected twigs.

Primary spread: From ascospores in fallen leaves.

Secondary spread: Through conidia by air and rain splashes.

Favourable conditions: Free moisture on leaves, high relative humidity of more than 95 per cent.

Management

- ✰ Grow resistant varieties.
- ✰ Practice clean cultivation (removal of fallen leaves, twigs and fruits and burning).
- ✰ Spray dodine or captan or difolatan 0.2 per cent.

2. Fire Blight

Fire blight is most serious disease in pear than apple.

Symptoms, Pathogen, Epidemiology and Management: As in apple.

14. Pineapple (*Ananas comosus* L.)

1. Butt Rot/Black Rot/Stem End Rot

Symptoms

- ☆ Symptoms are seen only on crowns, slips and suckers before or immediately after planting.
- ☆ Grey to black rot of the soft butt tissue develops, leaving stringy fibers and a cavity at the base of the stem.
- ☆ If affected material is planted, partial decay of the butt severely reduces plant growth
- ☆ When butt decay is severe, plants fail to establish, wilt rapidly and leaf tissue dies.
- ☆ Unlike *Phytophthora* heart rot, the young leaves remain firmly attached to the top of the stem.
- ☆ Infected plants can easily be broken at ground level.

Fungus

Sexual stage*: Ceratocystis paradoxa* (Dade) C. Moreau

Asexual stage*: Thielaviopsis paradoxa* (De Seynes).

Mode of survival: As chlamydospores in soil and decaying pineapple residues

Primary spread: From chlamydospores in soil.

Secondary spread: Through air-borne spores.

Favourable conditions: Warm wet soil, prolonged wet weather and wounds.

Management

- ☆ Improve soil drainage.
- ☆ Destroy diseased plants.
- ☆ Avoid suckers from the infected area for planting.
- ☆ Avoid planting during wet weather.
- ☆ Dip setts in Captafol 0.2 per cent
- ☆ Dip fruits after harvest in Thiabendazole 0.1 per cent.

2. Heart Rot

Symptoms

- ☆ Plants of all ages are attacked, but 3-4 month old crown plantings are most susceptible. Fruiting plants or suckers on ratoon plants are affected.
- ☆ The colour of the heart leaves changes to yellow or light coppery brown.
- ☆ Later, the heart leaves wilt (causing the leaf edges to roll under), turn brown and eventually die.

- Young leaves are easily pulled from the plant.
- The basal white leaf tissue becomes water- soaked and rotten with a foul smell due to the invasion of secondary organisms.
- The growing point of the stem becomes yellowish brown with a dark line between healthy and diseased areas.

Chromista

Phytophthora cinnamomi Rands and *P. parasitica*

Mode of survival: As chlamydospores in soil and in infected plant debris;

Primary spread: From chlamydospores from soil.

Secondary spread: Through irrigation water and farm implements

Favourable conditions: High rainfall and poor drainage

Management

- Avoid deep planting and prevent soil entering the hearts during planting.
- Avoid soil spilling over heart of plant.
- Provide drainage facilities.
- Remove and burn diseased plant.
- Treat with Bordeaux mixture 1 per cent or metalaxyl+mancozeb 0.2 per cent or copper oxychloride 0.25 per cent as prophylactic measure.

15. Plum (*Prunus domestica* L.)

1. Leaf Curl

Symptoms

- The shoot tips appear swollen and are often twisted and curled.
- Distortion of the leaves also occur.
- The fruit frequently appears swollen, misshapen, and bladder like with thick spongy flesh.
- The centers of infected fruit are hollow due to the failure of the young seed to develop.

Fungus

Taphrina deformans (Berk.)

Mycelia are intercellular and produce naked asci (Open ascus) individually. Each ascus bears eight ascospores of 3 to 7 μm diameter.

Mode of survival: As conidia and ascospores in infected twigs and buds.

Primary spread: From ascospores and sprout conidia from barks and buds.

Secondary spread: Through ascospores and sprout conidia.

Favourable conditions: Cool and wet weather and young and unfolded leaves.

Management

- ☆ Spray copper oxychloride 0.25 per cent or zineb 0.2 per cent

16. Pomegranate (*Punica granatum* L.)

1. Leaf Spot and Fruit Spot

Symptoms

- ☆ Small, dull violet spots with yellow halo appear on the leaves.
- ☆ Infected leaves turn yellow and drop off.
- ☆ Dark brown, depressed spots are seen on fruits with minute, black dots (acervuli).
- ☆ Infected fruits crack and become unmarketable.

Fungus

Colletotrichum gloeosporioides Pat and Penz.

Acervuli (asexual fruiting body), conidiophores are hyaline, septate, branched. Conidia are hyaline, aseptate, oval to elliptical or straight cylindrical and 11.0 to 17.0 × 3.0 to 6.0 µm.

Mode of survival: As infected plant parts

Primary spread: From infected plant debris.

Secondary spread: Through air-borne conidia

Favourable conditions: Temperature of 20-27°C and high relative humidity

Management

- ☆ Prune and destroy diseased twigs.
- ☆ Spray thiophanate methyl 0.1 per cent or chlorothalonil 0.2 per cent or mancozeb 0.2 per cent

2. *Cercospora* Leaf Spot

Symptoms

- ☆ Light brown spots appear on leaves and fruits and the spots on fruits are bigger.
- ☆ Black and elliptic spots appear on the twigs.
- ☆ Infected twigs are flattened, depressed with raised edge and dry up.
- ☆ In severe infection the whole plant dies.

Fungus

Cercospora punicae Henn.

It is a leaf spot causing fungi, produces filiform (thread like) shape conidia from the conidiomata.

Mode of survival: As infected leaves and twigs.

Primary spread: From infected debris.

Secondary spread: Through air-borne conidia.

Management

- ✰ Prune affected twigs and remove severely infected and dried fruits.
- ✰ Spray thiophanate methyl at 0.1 per cent or mancozeb or chlorothalonil at 0.2 per cent.

3. Bacterial Blight

Symptoms

- ✰ Spots on leaves are dark brown, irregular, necrotic which are surrounded by prominent, yellow or water-soaked area and these spots coalesce to form large patches.
- ✰ Severely infected leaves are often distorted and malformed and fall off prematurely.
- ✰ On the twigs and branches, brown to black spots appear around the nodes, girdle the stem and cause the branches to break.
- ✰ On fruits water soaked, oily, brown to black raised spots appear on the skin surface.
- ✰ Spots enlarge and merge with each other to cover large area and the infection is restricted to the rind.
- ✰ In severe cases, small cracks appear on the spots and the fruit splits open along the lesions.
- ✰ The spots covered with thin shining white encrustation consisting of bacteria.

Bacterium

Xanthomonas axonopodis pv. *punicae* (Hingorani and Singh) Vauterin.

It a Gram negative, rod-shaped, motile with single polar flagellum.

Mode of survival: As plant debris (8 months) and infected stem cankers (for several years) and dormant buds plant parts in living trees.

Primary spread: From infected planting materials.

Secondary spread: Through air splashed rain, irrigation water and pruning tools.

Favourable conditions: Rainy season, temperature of 25-35°C and relative humidity above 50 per cent.

Management

- Plant disease-free planting materials
- Follow clean cultivation procedures
- Spray Bordeaux mixture 1.0 per cent or streptomycin+tetracycline (streptocycline) 100 ppm or copper oxychloride or copper hydroxide 0.25 per cent.

17. Sapota (*Manilkara zapota* L.)

1. Leaf Spot

Symptoms

- Small, round and pinkish to dark brown spots with grey centre on leaves.
- They are either distributed or grouped on the leaf.
- The spots coalesce and cause premature leaf fall and reduction fruit yield.

Fungus

Phavophleospora indica Chinnappa and *Pestalotiopsis versicolor*

Pestalotiopsis asexual fruiting bodies of the fungus is Acervulus.

Mode of survival: As infected plant parts.

Primary spread: From infected plant debris.

Secondary spread: Through air-borne conidia.

Management

- Spray zineb or mancozeb 0.2 per cent or copper oxychloride 0.25 per cent.

2. Flat Limb

Symptoms

- It is a sporadic disease in India.
- Branches are typically flat, twisted with thin, yellow, hollow leaves formed in bunches.
- Bunchy appearance of leaves at the tip of the limb
- Fruits are small, hard and dry.

Fungus

Lasiodiplodia theobromae Pat.

[*Botryodiplodia theobromae* (Pat.) Griff. and Maubl].

The asexual fruiting body of the fungus is pycndia. The pycnidium produces and releases two-celled, dark brown, striated conidia.

Management

- ☆ Cut the affected branches and protect the cut ends with Bordeaux paste.

3. Sooty Mould

Symptoms

- ☆ Black sooty growth develops on the upper surface of the leaves which reduces vigour and the photosynthetic ability of the trees.

Fungus

Capnodium sp.

Mode of survival: As in mango.

Management

As in mango.

Chapter 2

Vegetable Crops

1. Bean (*Phaseolus vulgaris* L.)

1. Anthracnose

Symptoms

- ☆ The pathogen can cause infection on all above-ground parts
- ☆ Symptoms on the underside of the leaf appear as dark brown-black, slender lesions that follow the leaf veins.
- ☆ Older lesions are sunken and occur on both sides of the leaves.
- ☆ Brown to black, slender to oval-shaped lesion appears on petioles and stem. The margins of the lesions are slightly raised, well-defined black ring surrounded by a red-brown halo. The centre of the lesions appear light to dark brown.
- ☆ Lesions on pods are circular, sunken with reddish-brown margins and reddish centres. The fruiting bodies on the lesions are with orange-pink spores.
- ☆ Infected seeds are shriveled and discoloured with very poor germination and the seedlings from them die at an early stage.

Fungus

Colletotrichum lindemuthianum (Sacc. and Magnus)

[Sexual stage: *Glomerella lindemuthianum* Shear]

It is a hemibiotrophic pathogen, it spends part of its infection cycle as a biotroph, and the other part as a necrotroph, and upon infection light pink color conidia were produced. Matured conidia produced conidial anastomosis tube in nature.

Mode of survival: In seeds (up to 5 years) and in the crop refuse (up to 2 years).

Primary spread: From infected seeds and plant refuses.

Secondary spread: Through air-borne conidia and air-driven rain splashes.

Favourable conditions: Temperature of 17°C, relative humidity of 92 per cent, presence of free moisture on plant surface and disease intensity increases with age of the plant.

Management

- ☆ Follow deep summer ploughing to incorporate the infected debris.
- ☆ Use resistant varieties and adopt crop rotation.
- ☆ Collect and use seeds from disease free pods/certified seeds for sowing.
- ☆ Collect and destroy infected plants.

- ✰ Treat the seeds with carbendazim 2g/kg.
- ✰ In field, the spread of the disease can be limited by application of fungicides like mancozeb (0.25 per cent) or combination of mancozeb (0.25 per cent) and carbendazim (0.05 per cent) or tricyclazole (0.03 per cent) and repeat at 10 -14 days interval.

2. Rust

Symptoms

- ✰ Reddish brown pustules are mostly seen on the lower surface of the leaves.
- ✰ The corresponding upper surface of pustules becomes chlorotic.
- ✰ Black telia are formed later.
- ✰ The severely infected leaves turn brown, dry and fall off prematurely.

Fungus

Uromyces phaseoli-typica Arthur

It is a macrocyclic rust, produces five spores *viz.*, spermatiospores, aeciospores, urediospores, teliopores and basidiospores.

Mode of survival: As teliospores in plant debris and volunteer plants.

Primary spread: From teliospores from infected plant debris.

Secondary spread: Through air-borne basidiospores and uredospores.

Favourable conditions: Free moisture on leaves, temperature of 17-27°C, relative humidity of more than 95 per cent and cloudy and humid weather with night dew.

Management

- ✰ Use resistant varieties
- ✰ Adopt crop rotation with non-host crops.
- ✰ Collect and destroy severely diseased leaves and volunteer bean plants.
- ✰ Spray mancozeb or chlorothalonil 0.2 per cent.

3. Powdery Mildew

Symptoms

- ✰ Initial symptoms appear as small and white talc-like spots on the upper surface of leaves. These spots increase in size and form a whitish, powdery growth spreading to cover a large area of the leaves.
- ✰ Infected leaves gradually curl downward, pale yellow or brown, die and fall off.

Fungus

Erysiphe polygoni DC.

It produces amphigenous white septate mycelium, lobed single or paired appressoria, oidium type conidiophore, solitary cylindrical conidia, brown, septate myceloid Cleistothecia with 3-10 asci, containing 3-4 spores.

Mode of survival: As cleistothecia in the plant debris in soil

Primary spread: From infected plant debris.

Secondary spread: Through air-borne conidia.

Favourable conditions: Shady conditions, high N fertilization, high relative humidity and moderate temperature.

Management

- ☆ Grow resistant varieties.
- ☆ Remove and destroy infected plant materials.
- ☆ Spray wettable sulphur 0.3 per cent.

4. Dry Root Rot

Symptoms

- ☆ It attacks seedlings and grown up plants.
- ☆ Black sunken lesions appear near the base of the cotyledons and it spreads to the stem and first pair of unfolded leaves.
- ☆ Infection on stem of grown up plants shows sunken and reddish brown lesions which spread both upward and downward causing death of plants.
- ☆ Affected tissues show dry rot and large number of fruiting bodies (Sclerotia on root barks or pycnidia on stem).

Fungus

Macrophomina phaseolina (Tassi) Goid. (Pycnidial stage)

Rhizoctonia bataticola (Taub.) Butler (Sclerotial stage)

It is a soil borne monocyclic fungus possesses two asexual phases, a saprophytic phase (*R. bataticola*) that forms microsclerotia and mycelia and a pathogenic phase (*M. phaseolina*) present in host tissues that forms microsclerotia, mycelia and pycnidia.

Mode of survival: As sclerotia in infected plant refuse in soil.

Primary spread: From soil-borne sclerotia.

Secondary spread: Through irrigation water and implements.

Favourable conditions: Dry weather, high temperature and low soil moisture.

Management

- Adopt long crop rotation.
- Rogue out and destroy infected plant debris.
- Treat the seeds with carbendazim 2g/kg or *Trichoderma viride* 4g/kg or *Pseudomonas fluorescens* 10g/kg.
- Spot drench with carbendazim 0.1 per cent.

5. Angular Leaf Spot

Symptoms

- The symptoms appear as circular spots on cotyledonary leaves while, dark grayish in colour on upper surface while light gray on the lower surface.
- In severe infections, leaves show upward curling and defoliate prematurely.
- Elongated dark brown lesions also appear on stem and petioles.
- Yellowish brown discolouration is also observed on infected seeds located underneath the pod lesions.

Fungus

Phaeoisariopsis griseola (Sacc.) Ferr. (Syn. *Isariopsis griseola*)

Columnar hyphae producing conidiophores are laterally joined together in the form of synnemata or coremia, which are darker either curved, flexuous or septate. Conidia are light grey to hyaline, cylindrical to spindle shaped and straight to slightly curved, 1-3 (rarely 4-6) septate which are borne singly on the tips of conidiophore.

Mode of survival: As in infected plant refuse in soil.

Primary spread: From rain splashes water and wind.

Favourable conditions: Dry weather, high temperature and low soil moisture.

Management

- Collect and destroy the infected plant debris.
- Follow at least 2-3 years crop rotation with non-leguminous crops.
- Use healthy seed and treat them with carbendazim (0.2 per cent).
- Spray the crop with mancozeb (0.25 per cent) or combination of mancozeb (0.25 per cent) and carbendazim/benomyl/thiophanate methyl (0.05 per cent) or hexaconazole (0.05 per cent) and repeat at 10-14 days interval.

5. Floury Leaf Spot

Symptoms

- The disease appears in the form of white tufty growth of the fungus on the lower leaf surface appearing as sprinkled with coarse flour; hence the disease is named as floury leaf spot.

- The corresponding upper side of the leaf turns yellow, then brown and finally dry up. Young leaves are less susceptible.

Fungus

Mycovellosiella phaseoli (Drummond) Deighton

(Syn.: *Ramularia phaseolina* Petrak)

It is an obligate parasite. Conidiophores emerge in tuft, simple or irregularly dichotomous, a few short, erect, single conidiophores. Conidia are ellipsoid or ovoid, stright hyaline, aseptate, rarely one septate.

Mode of survival: As infected plant refuse in soil.

Primary spread: From Through rain splashes water and wind.

Favourable conditions: Dry weather, high temperature and low soil moisture.

Management

- Collect and destroy the infected plant debris.
- Follow at least 2-3 years crop rotation with non-leguminous crops.
- Collect and destroy the infected plant debris.
- Follow at least three years of crop rotation with non-leguminous crops.
- Spray the crop with carbendazim (0.1 per cent) and repeat at 10-14 days interval.

5. Common Bacterial Blight

Symptoms

- Small, angular, light green, water-soaked or translucent spots appear on the leaves.
- The spots rapidly enlarge and merge.
- As they develop the centre of the spot becomes brown and surrounded by a distinct yellow zone.
- In susceptible varieties, the lesions expand and coalesce. Such leaves appear scorched, wither and fall off.
- Pod lesions are round and water-soaked that enlarge merge and form sunken, irregular reddish brown blotches.
- Infected pods are shriveled and dry.
- The seeds in severely affected pods do not develop properly but get shriveled.

Bacterium

Xanthomonas campestris pv. *phaseoli* (Smith) Dye.

Mode of survival: In seed (more than two years) and in the diseased plant debris.

Primary spread: From infected seeds.

Secondary spread: Through air-blown rain and irrigation water.

Favourable conditions: Optimum temperature of 26- 32°C, wounds by air-blown soil particles and insects.

Management

- Follow crop rotation.
- Grow resistant/tolerant varieties.
- Sow only disease-free healthy seeds.
- Soak the seeds in streptocycline at 100 ppm before sowing.
- Avoid water stagnation and maintain proper drainage.
- Rogue out and burn severely infected plants and crop debris.
- Spray copper oxychloride 0.25 per cent.

6. Bean Common Mosaic

Symptoms

- Leaves show irregular-shaped, light-yellow and dark-green areas in a mosaic-like pattern.
- Infected leaves show puckering, stunting, malformation and downward curling.
- Infected leaves are narrower and longer than normal ones.
- Early infected plants are yellowish and dwarfed.
- Systemic infection causes necrosis roots, stems, leaf veins and pods.
- Infected pods are chlorotic, undersized with fewer seeds.

Virus

Bean common mosaic virus (BCMV)

Mode of survival: In seed (seed-borne) and weed hosts.

Primary spread: From infected seeds.

Secondary spread: Through insect vector, aphids (*Myzus persicae, Acyrthiosiphon pisum* and *Aphis fabae*).

Favourable conditions: Temperature of 20-25°C, monocropping, abundance of aphid vector and weed hosts.

Management

- Grow tolerant varieties.
- Collect seeds from healthy plants.
- Remove and destroy infected plants.

☆ Intercropping of maize with bean also reduces the incidence of this disease.

☆ Control insect vectors by spray methyl demeton 2ml/lit or other systemic insecticides.

2. Beetroot (*Beta vulgaris* sub sp. *vulgaris* L.)

1. *Cercospora* Leaf Spot

Symptoms

☆ This is a commonly occurring disease on foliage of beetroots.

☆ Numerous small circular spots appear on the leaf surface, which increase in size, becoming brownish or purplish in colour. The centre of the spot dries up and fall leaving a shot-hole symptom.

☆ Several spots coalesce to form larger areas of dead tissue.

☆ In case of severe infection spots cover the entire leaf surface resulting in pre-mature death and dropping of the leaves.

☆ As leaves die, the crown becomes cone-shaped with a rosette of dead leaves at the base. Severe incidence causes reduction in root size and yield.

Fungus

Cercospora beticola Sacc.

Mycelium is septate and dark in colour. Conidia are borne singly at the tip of conidiophores. They are hyaline, elongate, filiform and multiseptate.

Mode of survival: As infected crop debris and seed-borne.

Primary spread: From infected seeds and from crop debris.

Secondary spread: Through air-borne conidia

Favourable conditions: Temperature of 77-95°F, high relative humidity, free moisture on foliage.

Management

☆ Adopt deep summer ploughing to bury crop debris.

☆ Follow three year crop rotation.

☆ Sow disease free seeds.

☆ Treat the seeds with captan or thiram 4g/kg.

☆ Spray mancozeb 0.2 per cent or carbendazim 0.1 per cent.

2. *Sclerotium* Root Rot

Symptoms

- The most important sign is the presence of white strands of fungal mycelium at collar region of the plant.
- Numerous brown mustard seed-like sclerotia on the affected parts are seen.
- Yellowing of leaves and wilting of plants.
- The fleshy roots are completely or partially decay.

Fungus

Sclerotium rolfsii Sacc. [Sexual stage*: Athelia rolfsii* (Curzi)].

It is a soil living saprophytic pathogen, produces ribbon like hyphae with clamp connections. Basidia are club-shaped, bearing four smooth, ellipsoid basidiospores. Small, brownish sclerotia (hyphal propagules) are also formed.

Mode of survival: In soil as sclerotia on crop refuse.

Primary spread: From soil-borne sclerotia.

Secondary spread: Through irrigation water.

Favourable conditions: Temperature above 28°C and high soil moisture.

Management

- Remove and destroy the diseased plants along with the mycelia and sclerotia.
- Treat the seeds with *Trichoderma viride* 4g/kg or carbendazim 2g/kg.
- Apply carbendazim 0.1 per cent as soil drench.

3. Curly Top

Symptoms

- It appears at any stage of the crop but damage was more at seedling stage.
- It affects leaves, stems, flowers, fruits or roots.
- Infected plant parts are distorted through curling, twisting, rolling, stunting, *etc.*
- Leaves become thickened and leathery.
- The disease impairs both yield and quality of the roots.

Virus

Beet curly top virus (BCTV)

Mode of survival: In other crop hosts like capsicum, bean, potato and tomato and weed hosts.

Primary spread: From infected crop hosts and volunteer plants.

Secondary spread: Through insect vector, leaf hopper (*Circulifer tenellus*).

Favourable conditions: Abundance of leaf hopper population.

Management

- ☆ Grow resistant varieties.
- ☆ Adopt sanitary measures like eradication of susceptible weeds and volunteer plants from a previous planting.
- ☆ Regulate the time of planting in order to avoid the flights of leafhopper.
- ☆ Use of barrier crops.
- ☆ Remove and destroy infected plants.
- ☆ Spray malathion 2ml/lit to control insect vector.

3. Bhendi [*Abelmoschus esculentus* (L.) Moench]

1. Powdery Mildew

Symptoms

- ☆ The disease is found mainly on the older leaves and stems of plants.
- ☆ Powdery white patches appear on the upper surface of leaves covering entire leaf surface later.
- ☆ Fungal growth changes brown and the affected leaves dry and fall.
- ☆ Infection on stem is also common.

Fungus

Erysiphe cichoracearum DC.

Mode of survival: As cleistothecia in plant debris

Primary spread: From infected plant debris.

Secondary spread: Through air-borne conidia

Favourable conditions: Cool and humid weather.

Management

- ☆ Spray wettable sulphur at or dinocap at 0.2 per cent or carbendazim 0.1 per cent.

2. *Cercospora* Leaf Spot

Symptoms

- ☆ Irregular, brown spots with grey centre and dark coloured margins appear on leaves.
- ☆ Severely affected leaves fall off prematurely.

Fungus

Cercospora abelmoschi (Ellis and Everh.) Deighton

Mode of survival: In crop refuse.

Primary spread: From infected crop debris.

Secondary spread: Through air-borne conidia.

Favourable conditions: Temperature of 25-29°C.

Management

- ✰ Remove and destroy the infected plant parts.
- ✰ Spray copper oxychloride 0.25 per cent or Bordeaux mixture 1.0 per cent or zineb or captan or difolatan 0.2 per cent or carbendazim 0.1 per cent.

3. *Alternaria* Leaf Spot

Symptoms

- ✰ Brown, sub-circular spots mostly on the lower leaves increase in size and show concentric rings.
- ✰ Severely affected leaves fall off.

Fungus

Alternaria hibiscinum (Thüm).

Mode of survival: In crop refuse.

Primary spread: From infected crop debris.

Secondary spread: Through air-borne conidia.

Favourable conditions: Temperature of 25-30°C.

Management

- ✰ Spray mancozeb or difolatan 0.2 per cent.

4. Yellow Vein Mosaic/Vein Clearing

Symptoms

- ✰ The green colour of main and lateral veins is bleached and the veins and veinlets exhibit yellow network (vein clearing). Veins are thickened.
- ✰ The interveinal portions exhibit small portions of green colour.
- ✰ As the disease develops, these green portions turn chlorotic and finally the entire leaves become yellow and papery.
- ✰ The affected plant is severely stunted with smaller leaves and shortened internodes.
- ✰ Fruits are smaller, malformed, thick, fibrous, hard and yellow and unfit for consumption.

Virus

Bhendi yellow vein mosaic virus (BYVMV).

Mode of survival: In weed hosts.

Primary spread: From weed hosts and volunteer plants.

Secondary spread: Through insect vector, whitefly (*Bemisia tabaci*).

Favourable conditions: Susceptible varieties, weed hosts and abundance of insect vector.

Management

- ☆ Grow resistant cultivars (Arka Anamika, Arka Abhay, Parbani Kranti).
- ☆ Rogue out affected plants and weed hosts
- ☆ Raise barrier crops of sorghum, pearlmillet, maize with 5-6 rows before raising bhendi.
- ☆ Install yellow sticky traps
- ☆ Treat the seeds with imidacloprid at 5 ml/kg
- ☆ Apply carbofuran to soil 1.5 kg ai/ha at sowing followed by spray of dimethoate or methyl demeton 2ml/lit.

4. Brinjal (*Solanum melongena* L.)

1. Damping Off

Symptoms

- ☆ Pre-emergence of damping off results in seed and seedling rot before emergence from soil.
- ☆ Post-emergence of damping off characterized by water-soaked, brown discolouration at collar region followed by softening of tissues and collapse at soil level.

Chromista

Pythium aphanidermatum (Edson) Fitzp.

It produces hyaline, aseptate mycelium, oospores may form a germ tube directly to infect the plant or they may form sporangia. The sporangia produce zoospores, which are the motile form of the fungus.

Mode of survival: As oospores in infected plant parts in soil.

Primary spread: From oospores in the soil.

Secondary spread: Through irrigation water.

Favourable conditions: High soil moisture and moderate temperature along with high relative humidity especially in the rainy season

Management

- ✰ Sow healthy seeds.
- ✰ Treat seeds with thiram or captan 4g/kg or *Trichoderma viride* at 4g/kg or *Pseudomonas fluorescens* 10g/kg.
- ✰ Avoid raising nursery continuously in the same area.
- ✰ Drench nursery soil with copper oxychloride 0.25 per cent or Bordeaux mixture 1 per cent at 4 lit/m^2.
- ✰ Solarize the soil by spreading 25 μm transparent polythene sheet over the seed bed for 40-45 days during summer.
- ✰ Application of *Trichoderma viride* or *Pseudomonas fluorescens* at 40 g/ sq.m along with FYM.

2. *Fusarium* Wilt

Symptoms

- ✰ It attacks seedlings as well as matured plants.
- ✰ Leaves become chlorotic, flaccid and hang down.
- ✰ Vascular brown streaks are seen in the longitudinal section of the stem.
- ✰ Roots exhibit rotting.
- ✰ Whitish fungal growth seen on the basal stem and roots.

Fungus

Fusarium solani (Mart.) Sacc.

[Sexual stage: *Nectria haematococca* Berk. and Broome]

Mode of survival: In crop debris and soil as chlamydospores.

Primary spread: From soil and diseased seedlings.

Secondary spread: Through irrigation water

Favourable conditions: Root-knot nematode predisposes the plant to disease incidence.

Management

- ✰ Treat the seeds with *Trichoderma viride* at 4g/kg or *Pseudomonas fluorescens* 10g/kg or carbendazim 2g/kg.
- ✰ Soil application of *Trichoderma viride* or *Pseudomonas fluorescens* at 2.5 kg + 50 kg FYM/ha on 30 DAP in the main field.
- ✰ Remove and destroy infected plants and spot drench with carbendazim 0.1 per cent.

3. *Verticillium* Wilt

Symptoms

- ☆ The disease attacks the young plants as well as mature plants.
- ☆ The leaves show presence of irregularly scattered necrotic pale yellow spots which later coalesce resulting in complete wilting of leaves.
- ☆ Drying of branches followed by wilting and collapse of plant.
- ☆ The longitudinal section of the root shows a characteristic dark brown discoloration on the xylem vessels.

Fungus

Verticillium dahliae Kleb.

The mycelium is hyaline, septate, and multinucleate. Conidia are ovoid or ellipsoid and usually single-celled. They are borne on phialides, which are specialized hyphae produced in a whorl around each conidiophore. *Verticillium* is named for this verticillate (=whorled) arrangement of the phialides on the conidiophore.

Mode of survival: As microsclerotia in soil.

Primary spread: From soil-borne inoculum.

Secondary spread: Through irrigation water

Favourable conditions: Temperature of 19-21°C and other crop hosts.

Management

- ☆ Follow crop rotation avoiding bhendi, tomato, cotton and potato.
- ☆ Drench the soil in root region of plants with carbendazim 0.1 per cent.
- ☆ Treat the seeds with *Trichoderma viride* at 4g/kg or *Pseudomonas fluorescens* 10g/kg or carbendazim 2g/kg.
- ☆ Follow soil application of *Trichoderma viride* or *Pseudomonas fluorescens* at 2.5 kg + FYM 50 kg of/ha.

4. *Alternaria* Leaf Spot

Symptoms

- ☆ Irregular, brown and necrotic spots with concentric rings appear on leaves.
- ☆ Leaves dry due to bigger necrotic patches and then fall down prematurely.
- ☆ Spots on fruits are dark brown, sunken, turn yellow and cause fruit to drop.

Fungus

Alternaria solani Sorauer

It is a polycyclic necrotrophic pathogen, produces dark colored, multinucleate muriform and beaked conidia arise singly in simple conidiophore.

Mode of survival: In seed and soil (Seed-borne and soil-borne).

Primary spread: From infected seeds and plant debris.

Secondary spread: Through air-borne conidia

Management

- Use disease free seeds.
- Remove and destroy infected plant debris.
- Spray mancozeb 0.2 per cent or chlorothalanil 0.2 per cent.

5. *Cercospora* Leaf Spot

Symptoms

- Large irregular or circular, brown spots with grey centre are seen on leaves.
- Spots are also seen on fruits.

Fungus

Cercospora melongenae Welles

Conidiophores appear in fascicles (clusters) of 3 to 12. They are pale brown, occasionally septate, and unbranched, and they bear hyaline, mildly curved conidia.

Mode of survival: In infected plant debris in soil upto one year and seed.

Primary spread: From infected plant debris and seed.

Secondary spread: Through air-borne conidia

Favourable conditions: Leaf wetness and high relative humidity.

Management

- Use disease free seeds.
- Remove and destroy affected leaves.
- Spray mancozeb 0.2 per cent or difolatan 0.2 per cent.

6. Rust

Symptoms

- Pycnial and aecial stages of pearlmillet rust are formed on leaves of brinjal (alternate host).
- Upper surface of leaf becomes depressed and the corresponding lower surface is convex.
- Pycnia are formed as orange yellow pustules on the upper leaf surface.

- ☆ Aecial cups are formed in groups on the lower surface of leaves.
- ☆ Severe infection of rust causes drying of leaves.

Fungus

Puccinia substriata var. *penicillariae* (speg.).

It is a macrocyclic heterocious with uredial and telial stage on *Pennisetum* spp. and spermogonial and aecial stages on *Solanum* spp.

Mode of survival and spread: Primary host (pearlmillet)

Favourable conditions: Availability of pearlmillet crop in nearby fields.

Management

- ☆ Spray mancozeb or difolatan 0.2 per cent or oxycarboxin 0.1 per cent.

7. *Phomopsis* Blight and Fruit Rot

Symptoms

- ☆ The pathogen infects the crop from seedling to harvest.
- ☆ On young seedlings, brown or dark sunken lesions appear on stem slightly above the soil surface results in collapse and death (seedling infection).
- ☆ Leaves coming in contact with soil are affected.
- ☆ Circular or irregular, grey to brown spots with light coloured centre embedded with pycnidia appear on leaves and the affected leaves turn yellow and drop prematurely (leaf infection).
- ☆ Stem lesions appear at basal part near nodal region as grey dry rot with constriction.
- ☆ The infected branch dries and giving a partial wilting symptom (stem infection).
- ☆ The fruit is attacked while in the plant.
- ☆ Pale, sunken, oval spots on fruits enlarge covers the entire fruit at the end. Internal portion of fruits rots and the whole fruit is mummified and remain attached with the plant (fruit infection).
- ☆ Diseased fruits are unmarketable.

Fungus

Sexual stage: *Diaporthe vexans* (Sacc. and Syd.) Gratz.

Asexual stage: *Phomopsis vexans* (Sacc. and Syd.).

The mycelium was hyaline and septate; the conidiophores (phialides) within the pycnidium were hyaline, simple, or septate. Two types of conidia (alpha and beta) were observed. Alpha conidia were hyaline, single celled, biguttulate and subcylindrical (4.1–6.5 × 1.2–1.9 µm). Beta conidia were filiform, curved, hyaline and septate (6.2–7.6 × 0.5–0.8 µm).

Mode of survival: In plant debris, soil and seed.

Primary spread: From plant debris and seed.

Secondary spread: Through rain splashes, insects, contaminated equipment and air.

Favourable conditions: Temperature of 21- 32°C and hot and wet weather.

Management

- ✰ Follow crop rotation with cereals.
- ✰ Use disease free seeds/tolerant varieties.
- ✰ Plant disease-free seedlings.
- ✰ Collect and burn infected crop residue and fallen mummified fruits.
- ✰ Seed treatment with carbendazim 2g/kg or thiram 4g/kg or hot water at 50°C for 30 min.
- ✰ Spray with mancozeb 0.2 per cent in the nursery.
- ✰ Spray difolatan 0.2 per cent or carbendazim 0.1 per cent in main field.

8. *Sclerotium* Collar Rot

Symptoms

- ✰ Necrosis of stem tissues is seen near the soil line.
- ✰ White, cottony and silvery mycelial growth is visible on the affected portion.
- ✰ Fungal growth below the soil surface is also common.
- ✰ White to light brown mustard like sclerotia is observed on the infected collar region and soil.
- ✰ Progressive drooping and wilting of the entire plant is observed.

Fungus

Sclerotium rolfsii Sacc. [Sexual stage*: Athelia rolfsii* (Curzi)].

It is a soil living saprophytic pathogen, produces ribbon like hyphae with clamp connections. Basidia are club-shaped, bearing four smooth, ellipsoid basidiospores. Small, brownish sclerotia (hyphal propagules) are also formed.

Mode of survival: As sclerotia in plant debris.

Primary spread: From soil-borne sclerotia.

Secondary spread: Through irrigation water.

Favourable conditions: High soil moisture, ill drained soil, dense canopy and several cultivated crop hosts and weeds.

Management

- ✰ Follow crop rotation with cereals.

- ☆ Removal of weed hosts.
- ☆ Deep summer ploughing.
- ☆ Avoid use of infected infected FYM, compost, soil, *etc.*
- ☆ Use disease free seeds/tolerant varieties.
- ☆ Apply green manure followed by *Trichoderma viride* or *Pseudomonas fluorescens* 2.5 kg/ha in soil.
- ☆ Avoid closer spacing and maintain proper aeration and sunlight near the collar region.
- ☆ Spot drench carbendazim 0.1 per cent or copper oxychloride 0.25 per cent.

9. Bacterial Wilt

Symptoms

- ☆ The characteristic symptoms are wilting of foliage and sudden death of plant.
- ☆ The vascular browning of stem followed by rotting of roots and rootlets are the other symptoms.

Bacterium

Ralstonia solanacearum (Yubucchi *et al.*)

[Syn. *Pseudomonas solanacearum* (Smith), *Burkholderia solanacearum* (Smith)].

The bacterium is rod-shaped, gram negative, lophotrichus bacteria. (more than one polar flagellum), multiplication occurs by bacterial fission.

Mode of survival: In soil.

Primary spread: From soil.

Secondary spread: Through irrigation water.

Favourable conditions: High soil moisture, acidic pH, intercultural wounds and nematode injury.

Management

- ☆ Follow deep summer ploughing.
- ☆ Grow disease resistant varieties.
- ☆ Follow crop rotation avoiding bhendi, tomato and potato.
- ☆ Avoid water stagnation in the field.
- ☆ Soak seeds in a solution of streptocycline 25 ppm (2.5g/100 litres of water) for 30 min before sowing.
- ☆ Remove and destroy affected plant parts.

10. Little Leaf

Symptoms

- The infected plants exhibit general chlorosis and reduction in plant height.
- Leaf size is abnormally reduced.
- Axillary buds are induced to grow to produce many, small narrow and thin leaves.
- Reduction in internodal length leads to clustering and overcrowding of leaves giving bushy appearance.
- Flowers are phylloid and fruits very rare.
- Early infected plants are completely sterile.

Bacterium

Candidatus Phytoplasma trifolii.

Mode of survival: In weed hosts (*Argemone mexicana, Datura metel, Physalis minima, Solanum indicum, Solanum trilobatum, Catharanthus roseus* and *Withania somnifera*).

Primary spread: From weed hosts.

Secondary spread: Through insect vector, leafhopper (*Hishimonas phycitis*).

Favourable conditions: Susceptible varieties, presence of other crop hosts and weeds and abundance of vector.

Management

- Eradicate solanaceous and other weed hosts.
- Roguee out infected plants.
- Grow moderately resistant varieties *viz.*, Arka Sheel, Banara Giant, Pusa Purple Cluster.
- Spray dimethoate or methyl demeton 2ml/lit to control insect vector, leafhoppers.
- Spray tetracycline 100 ppm to suppress multiplication of phytoplasma.

CRUCIFERS

5. Cabbage and Cauliflower (*Brassica oleracea* L.)

1. Damping Off

Symptoms

- Damping-off kills seedlings before or soon after they emerge.
- When infection occurs before the seedling emergence causes seed rot which results in poor germination.

Chromista

Pythium debaryanum R. Hesse

Mode of survival: As oospores in soil and plant debris.

Primary spread: From infested soil.

Secondary spread: Through irrigation water

Favourable conditions: Cool moist cloudy weather with heavy rain, high humidity, wet soils, overcrowded seedlings in the nursery and compact soil.

Management

- Choose the site with good drainage for preparing seed beds.
- Avoid excessive irrigation.
- Treat the seeds with *Trichoderma viride* 4g/kg or thiram 4g/kg and soil drenching with copper oxychloride 0.25 per cent.
- Spray copper oxychloride 0.25 per cent or Bordeaux mixture 1 per cent.

2. Wire Stem

Symptoms

- It is more serious in nursery beds.
- Reddish brown discolouration of the stem appears near the ground level on young seedlings.
- The affected portion gets constricted and the plants bend or twist without breaking.
- The lesion is sunken and the stem resembles a wire, hence the name 'wire stem'.
- The girdled seedling eventually dies.

Fungus

Rhizoctonia solani Kuhn.

The vegetative mycelium is brown colored at mature. The mycelium consists of septate hyphae partitioned by dough-nut shaped pore. It does not produce asexual spores and rarely produce basidiospores.

Mode of survival: In soil and plant debris as sclerotia.

Primary spread: From infested soil.

Secondary spread: Through irrigation water and farm implements.

Favourable conditions: Cool, cloudy weather, high humidity, wet and compact soil and overcrowding.

Management

- Adopt raised seed bed.

- Avoid excessive irrigation to reduce humidity around the plants.
- Discard the diseased seedlings at transplanting.
- Treat the seeds with *Trichoderma viride* or thiram 4g/kg.
- Soil drench with carbendazim 0.1 per cent.

3. Club Root

Symptoms

- Infected plants show yellowing of leaves and wilting of plants in the middle of hot, sunny days, which recover during night.
- Young plants are killed while older plants remain alive but stunted and fail to produce marketable heads.
- Infected roots enlarge to form galls that differ in size and shape depending on host plant.
- The roots are with spindle-shaped or club-shaped swellings due to hypertrophy and hyperplasia.
- Infected roots are unable to absorb water and nutrients hence the growth of the plant is stunted.
- Clubbed roots disintegrate at the end of the season due to invasion by bacteria and other saprophytic organisms.

Fungus

Plasmodiophora brassicae Woronin

The pathogen has a complex life cycle comprising three stages: survival in the soil as resting spores, root hair infection and, finally, cortical infection.

Mode of survival: As resting spores in soil and debris up to 20 years.

Primary spread: From primary zoospores from infested soil.

Secondary spread: Through soil and by farm implements, manure from cattle fed on diseased crops, flood water, infected seedlings, air-blown soil and animals.

Favourable conditions: Low lying areas, ill-drained soils with a soil temperature of 15-25°C, acidic soils and continuous cultivation of cruciferous crops.

Management

- Maintain a soil pH of 7.2 by liming (Field with pH 5.0 needs approximately 2.5 t of lime/ha). Finely ground lime alters the pH more quickly than coarse granules.
- Follow crop rotation with non-cruciferous crops for 7 year.
- Apply formalin 2 per cent at 10 lit/sq. m. of seed bed or fumigate with methyl bromide.
- Seed treatment with *Pseudomonas fluorescens* 10g/kg followed by seedling dip of *Pseudomonas fluorescens* 5g/lit and soil application of *Pseudomonas*

fluorescens 2.5 kg/ha + 50 kg FYM before planting or carbendazim 2g/kg followed by seedling dip in carbendazim solution at 2g/lit for 20 min.

- Improve drainage facilities.
- Use disease free transplants.
- Remove brassica weeds.
- Avoid moving infected transplants and/or infested soil on farm equipment to clean fields.
- Clean and disinfect all machinery before moving it from infested to non-infested land.
- Do not use irrigation water that has been contaminated with resting spores by run-off from infested fields into irrigation ponds.
- Spot drenching with carbendazim at 0.1 per cent.

4. Black Leg

Symptoms

- The fungus kills seedlings or produces sunken, black cankerous lesions at the stem base near ground level and extends below the soil surface causing black rot of lower stem and roots.
- The fungus interferes with water conduction in tissues, wilting and debilitating plants.
- Severely infected plants are stunted and wilt.
- As plants mature, they fall sideways from lack of root anchorage.
- Tiny, black bodies (pycnidia) are seen on the stem cankers and from which millions of spores ooze out and spread during wet weather.

Fungus

Sexual stage: *Leptosphaeria maculans* (Sowerby) P. Karst

Asexual stage: *Phoma lingam* (Tode) Desmaz

Two types of pycnidia can be found; one on infected, live plants and the other on crop residues. Pycnidia are globose to subglobose, yellow brown to brownish black. Pycnidiospores are hyaline, cylindrical to ellipsoidal, sometimes slightly curved, with rounded ends. Pseudothecia produces ascospores.

Mode of survival: In infected seeds, cruciferous weeds and residues of cruciferous crops remaining in or on the soil.

Primary spread: From infected seeds.

Secondary spread: Through pycnidiospores or ascospores are dispersed by splashing rain or irrigation water.

Favourable conditions: Continues drizzling, wet weather.

Management

- Follow deep summer ploughing to incorporate crop debris promptly after harvest to hasten decay.
- Practice four year rotation in seedbeds and fields.
- Use certified/disease free seeds.
- Treat seeds with hot water as it spreads mainly through seed.
- Rogue out diseased plants from seedbeds.
- Improve soil drainage and air circulation.
- Avoid working in the fields when wet.
- Dress seeds with Imidacloprid + Metalaxyl + Carbendazim based products.

5. Cabbage Yellows/*Fusarium* Wilt

Symptoms

- The disease attacks the seedlings in nursery stage, however plants exhibit symptoms 2 to 4 weeks after transplanting.
- Initial symptom appears as the development of yellowish green colour on one side of the plant. A lateral warping or curling of the stem and leaves occurs.
- Lower leaves turn yellow and later the upper leaves are also affected. With time, the yellow leaves turn brown and the affected tissue become dry and brittle.
- Seedlings may die rapidly after transplanting, whereas surviving plants become deformed and stunted due to root decay.

Fungus

Fusarium oxysporum f. sp *conglutinans* (Wollenw.) Snyder and Hansen.

Mode of survival: In the soil for many years on the remains of crops or on the roots of weeds.

Primary spread: From infected seedlings in the nursery and also machinery.

Secondary spread: Through spores are spread in soil and water.

Favourable conditions: Soil moisture, less pH and the temperature about 27°C.

Management

- Grow resistant varieties.
- The conventional controls such as rotation, seed treatment, fungicide sprays, and destruction of crop refuse are of little value once the fungus has established itself on a farm or in a specific field. Therefore, the use of resistant varieties is the only control.
- However, the vulnerable stage of the young seedlings to the infection can be avoided by very early sowing of cabbage.

6. Downy Mildew

Symptoms

- ☆ The disease appears from nursery to curd formation stage.
- ☆ Infection at the seedling stage causes complete death of seedlings.
- ☆ On matured plants, it infects leaves, stems and inflorescence and the initial infection starts from lower most leaves.
- ☆ Greyish white downy fungal growth appears on the under surface of the leaves and the corresponding upper surface shows yellowish spots.
- ☆ Infected leaves dry up and fall.
- ☆ Floral parts are deformed.
- ☆ Cauliflower curds look brownish at the top.
- ☆ Dark brown and depressed lesions on stem later develop with downy growth of the fungus.

Chromista

Peronospora parasitica (Pers.) de Bary;

Hyaloperonospora brassicae (Pers.) Constant.

The pathogen is a polycyclic, oomycetes obligate parasite produce intercellular mycelium with large finger shaped intracellular haustoria. The sporangiophores are dichotomously branched with sterigmata bear the single sporangium at the tip. Sexually produces globose shape oospores in soil and asexually sporangia releases flagellated zoospores.

Mode of survival: As oospores in soil and plant debris and cruciferous weeds.

Primary spread: From oospores from soil and plant debris.

Secondary spread: Through air and irrigation water

Favourable conditions: Cool moist conditions, high humidity, fog, drizzling rains and heavy dew.

Management

- ☆ Follow crop rotation with non-cruciferous crops.
- ☆ Grow tolerant varieties.
- ☆ Treat seeds with metalaxyl 6g/kg.
- ☆ Remove cruciferous weeds in the field.
- ☆ Spray metalaxyl + mancozeb or mancozeb 0.2 per cent.

7. Black Leaf Spot/*Alternaria* Leaf Spot and Blight

Symptoms

- ☆ It usually appears in early stage of plant growth in cauliflower while in later stage in cabbage.

- Small yellow spots appear on the leaf surface.
- Spots become brown, enlarge to circular areas with concentric rings surrounded by yellow haloes.
- Individual spots coalesce into large necrotic area and the necrotic tissues fall off giving a shot-hole symptom.
- The spots become black due to sporulation during humid weather.
- Affected leaves turn yellow and fall off.
- Lesions on stems and petioles are linear and in severe cases the entire plant defoliates.
- Inflorescence and siliqua are severely infected in seed crop.
- The curd of cauliflower infected as brown discoloration of individual florets and flower clusters.

Fungi

Alternaria brassicae (Berk.) Sacc. and *A. brassicicola* (Schwein.)

It is a necrotrophic fungus, produces muriform dark colored large beaked conidia from the simple erect conidiophores. Mycelium is immersed, hyphae are branched, septate and hyaline.

Mode of survival: As internally and externally seed-borne and diseased crop debris.

Primary spread: From infected seeds.

Secondary spread: Through air-borne conidia

Favourable conditions: Moist atmosphere and a relative humidity of more than 90 per cent.

Management

- Follow crop rotation.
- Use disease free seeds/certified seeds.
- Treat seeds with hot water at 50°C for 30 min or with thiram or captan 4g/kg.
- Detach and burn all the infected lower leaves.
- Spray copper oxycloride 0.25 per cent or mancozeb or chlorothalonil 0.2 per cent.

8. White Rust

Symptoms

- Prominent white or creamy yellow blisters of 1 to 2 mm dia are found on the lower surface of the leaves and the corresponding upper surface exhibits yellow patches.

- ☆ The blisters are sub-epidermal and due to the formation of sporangia the epidermal layer ruptures and the sporangia are released.
- ☆ Blisters appear on the stem and inflorescence also.
- ☆ Systemic infection of seedlings or stems and inflorescence causes hypertrophy and hyperplasia of floral tissues which leads to swellings and distortions.
- ☆ Floral organs become partly or wholly swollen, fleshy, green or violet in colour and mostly sterile.
- ☆ Petals and stamens become green and leaf-like. Ovules and pollen grain are atrophied. Ovary is sterile.
- ☆ Early infection causes stunting of plants.

Chromista

Albugo candida (Pers.) Kuntze.

It is an obligate parasite. Mycelium is hyaline, coenocytic and intercellular. Sporangiophores are single celled, club-shaped and hyaline. Sporangia spherical, hyaline and are arranged in chain. Sporangia release zoospores which are reniform and biflagellate. Oospores are produced produced in the infected tissues.

Mode of survival: As oospores in the plant debris in soil.

Primary spread: From soil-borne oospores.

Secondary spread: Through irrigation water

Favourable conditions: Moist cool weather and a thin film of water on the leaves.

Management

- ☆ Remove and destroy infected plant debris.
- ☆ Treat the seeds in hot water at 52°C for 20 min or metalaxyl 4-6 g/kg.
- ☆ Remove cruciferous weeds in the field.
- ☆ Spray Bordeaux mixture 1.0 per cent or mancozeb or metalaxyl + mancozeb 0.2 per cent.

9. Black Rot

Symptoms

- ☆ Plants are infected at all growth stages and the symptoms resemble nutritional deficiencies.
- ☆ Infected seedlings become yellow and die.
- ☆ The classic symptom of black rot is the development of "V" shaped chlorotic to yellow lesions from the leaf margin which turns brown and dies.

- These "V" shaped areas enlarge as the disease progresses and the point of V follows a vein towards the base of the leaf.
- Veins and veinlets of the leaves, petioles and stem turn black. When the veins are cut, sticky yellowish liquid containing bacteria oozes out.
- As infection continues leaves turn black, wilt and die.
- Heads of cauliflower are invaded and discoloured and remain small with reduced quality making it unfit for marketing.

Bacterium

Xanthomonas campestris pv. *campestris* (Pammel) Dowson.

It is a Gram-negative and rod-shaped, aerobic, non-spore forming, motile with single polar flagellum.

Mode of survival: As overwinters in diseased plant refuse and in seed.

Primary spread: From infected plant debris, seed and seedlings.

Secondary spread: Through air-blown rain and implements.

Favourable conditions: Hot and humid conditions, insect injury and hailstorm.

Management

- Use tolerant varieties.
- Treat the seeds in hot water at 50°C for 30 min or with streptocycline 100 ppm for 30 min.
- Grow seedlings on raised beds in non-infested area or frequently change the nursery site.
- Fields cultivated with crucifers should be kept fallow for 2 year.
- Raise intercrops like blackgram or greengram as mulches to reduce the spread of pathogen by rain splashes.
- Remove and destroy diseased plants.
- Spray copper oxychloride 2.5 g/lit + streptocycline at 100 ppm or Kasugamycin 0.2 per cent.

10. Mosaic

Symptoms

- Initially vein clearing or banding symptoms are produced on the leaves ultimately leading to mosaic pattern.
- Early infection may cause reduction in plant growth leading to stunting.

Virus

Cauliflower mosaic virus (CaMV)

Mode of survival: In volunteer other *Brassica* sp.

Transmission: Through aphids (*Brevicorne brassicae* and *Myzus persicae*)

Management

- ☆ Use tolerant varieties.
- ☆ Control of aphid vectors through insecticides like Metasystox (0.1 per cent) at an appropriate stage.

11. Whiptail

Symptoms

- ☆ The deficiency of molybdenum generally occurs in acid soils.
- ☆ In young plants, the deficiency symptom appears as chlorosis of leaf margins and then the whole leaf turns white.
- ☆ The leaf blades do not develop properly.
- ☆ When the deficiency is severe, only the midribs develop resembling whip-like appearance and hence the name 'Whiptail'.
- ☆ The growing point of the plant is also deformed which prevents the curd development.

Cause

Molybdenum deficiency.

Favourable conditions: Acid soil with a pH below 5.5.

Management

- ☆ Apply lime in acidic soils to increase the availability of molybdenum. The quantity of lime is determined by initially measuring the pH of the soil. Alternately, soil application of sodium molybdate (10-15 kg/ha) effectively controls the deficiency symptoms.

6. Carrot [*Daucus carota* sub sp. *sativus* (Hoffm.)]

1. *Cercospora* Leaf Blight

Symptoms

- ☆ At first the disease appears along the margins of the leaves, causing them to curl.
- ☆ Spots in the leaf are small, roughly circular and tan or grey to brown colour with a dead centre.
- ☆ As the lesions increase in number and size, the entire leaflet withers and dies.
- ☆ In heavily infested fields, infection is seen on older and younger leaves, petioles and stems.
- ☆ The lesions may merge and girdle the stem, causing the leaves to die.

Fungus

Cercospora carotae (Pass.) Solheim

Conidiophores emerging from the lesions were simple and straight to subflexuous with a bulbous base. Conidia were solitary, cylindrical to narrowly-obclavate, and hyaline to subhyaline with 2 to 6 septa.

Mode of survival: In diseased crop debris and seed.

Primary spread: From infected seeds.

Secondary spread: Through air-borne conidia.

Favourable conditions: Optimum temperature 19°C - 20°C and wet weather.

Management

- Use of disease-free seeds.
- Adopt crop rotation.
- Spray difolatan 0.2 per cent or copper oxychloride 0.25 per cent.

2. *Alternaria* Blight

Symptoms

- First disease appear at the margin of the leaflets as dark brown to black and irregular spots.
- Older leaves are more prone to infection.
- As the disease progresses entire leaflets may shrivel die and appear as if scorched by fire.
- Lesions on petioles and stems are dark brown and they coalesce and girdle the stems.

Fungus

Alternaria dauci (J.G. Kuhn).

Mode of survival: In plant debris in soil and seed.

Primary spread: From infected seeds.

Secondary spread: Through air-borne conidia.

Favourable conditions: Warm temperatures and extended periods of leaf wetness due to rainfall, dew, or sprinkler irrigation.

Management

- Adopt crop rotation.
- Provide proper drainage.
- Treat the seeds in hot water at 50°C for 15 min or thiram 4g/kg.
- Destroy infected plant material in the field.
- Spray difolatan 0.2 per cent or copper oxychloride 0.25 per cent.

3. Powdery Mildew

Symptoms

- ✰ White powdery growth is seen on leaves and petioles causing them to turn brown and dry.

Fungus

Erysiphe polygoni DC.

Mode of survival: In diseased plant parts.

Primary spread: From ascospores.

Secondary spread: Through air-borne conidia.

Management

- ✰ Spray carbendazim 0.1 per cent.

4. Watery Soft Rot/White Rot

Symptoms

- ✰ The disease is present in field and in storage.
- ✰ It appears as whitish fungal growth with black sclerotia on the stalk, stump and collar region and crown of infected carrot plants.
- ✰ In storage, a soft, watery rot with white mold and black sclerotia are found on carrot.

Fungus

Sclerotinia sclerotiorum (Lib.) de bary.

It produces fluffy white mycelium on and in infected plant parts. This mycelium aggregates itself into sclerotia. Sclerotia buried in soil can survive and remain infective for up to 5 years. Apothecial (carpogenous) germination of sclerotia was observed. Ascospores are hyaline (clear or non-pigmented), unicellular and thin-walled.

Mode of survival: As sclerotia in soil.

Primary spread: From soil-borne sclerotia.

Secondary spread: Through irrigation water.

Favourable conditions: High soil moisture.

Management

- ✰ Adopt crop rotation with rice or other cereals.
- ✰ Follow field sanitation by uprooting and burning of infected plants and weeds.

- ☆ Adopt summer fallowing.
- ☆ Apply green manure followed by soil application of *T. viride* 2.5 kg/ha.
- ☆ Adopt ridges and furrows.
- ☆ Treat the seeds with *T. viride* 4g/kg.
- ☆ Provide rapid cooling prior to storage and sanitation of all storage components to reduce losses from this disease.

5. Collar Rot

Symptoms

- ☆ Wet and soft rot of white, cottony fungal growth on carrot roots are observed at initial stage.
- ☆ Rotting covers entire root along with stump and leaves of the plant near collar region.
- ☆ Later this hyphal mat is converted into small, mustard seed-like sclerotia.

Fungus

Sclerotium rolfsii Sacc. [Sexual stage: *Athelia rolfsii* (Curzi)].

It is a soil living saprophytic pathogen, produces ribbon like hyphae with clamp connections. Basidia are club-shaped, bearing four smooth, ellipsoid basidiospores. Small, brownish sclerotia (hyphal propagules) are also formed.

Mode of survival: As sclerotia in soil.

Primary spread: From soil-borne sclerotia.

Secondary spread: Through irrigation water.

Favourable conditions: High soil moisture.

Management

- ☆ Adopt crop rotation.
- ☆ Remove and burn infected plants along with mycelia mat and sclerotia and weeds.
- ☆ Irrigate at regular intervals to maintain required moisture level.
- ☆ Seed treatment and soil application of *Trichoderma viride* reduces disease incidence.

6. Bacterial Soft Rot

Symptoms

- ☆ Occur when soil conditions are wet or storage conditions are poor.
- ☆ Water-soaked, irregular lesions are seen superficially on the roots and soon spread to cover the inner tissues.

- ☆ The foliage remains green until the disease on the tuber advances considerably.
- ☆ The entire plant wilts after complete rotting of the tuber.
- ☆ A foul odour emits from the decayed roots.

Bacterium

Pectobacterium carotovorum sub *sp. carotovorum*

[Syn. *Erwinia carotovora* pv. *carotovora* (Jones) Bergey *et al.*]

The bacterium is Gram-negative, rod-shaped and motile with 1-6 peritrichous flagella.

Mode of survival: In infected tubers left over in the soil.

Primary spread: From soil-borne inoculum.

Secondary spread: Through rain splashes and irrigation.

Favourable conditions: Abundant moisture on the root surface.

Management

- ☆ Plant seeds on raised beds.
- ☆ Handle carefully the tubers during harvest and grading to avoid injuries and provide adequate sanitation to reduce soft rot.
- ☆ Do not store diseased tubers along with the healthy materials.

7. Cassava/Tapioca (*Manihot esculenta* Crantz)

1. Brown Leaf Spot

Symptoms

- ☆ Usually lower canopy are infected first and progresses to upper leaves. In susceptible cultivars both lower and upper leaves are equally infected.
- ☆ Spots are seen on both the sides of the leaves but more pronounced on upper surface.
- ☆ The spots are circular and brown with a distinct dark brown border.
- ☆ The spots increase in size and become irregular or angular and restricted by leaf margin or veinlets and measure 3 to 12 mm in dia.
- ☆ In the centre of brown spots, greyish growth of conidiophores and conidia are seen during humid condition.
- ☆ Indefinite halo around each spot or blighted area is seen.
- ☆ Leaves turn yellow, dry and drop off.

Fungus

Cercosporidium henningsii (Allesch.) Deighton. (Syn. *Cercospora henningsii, Cercospora manihotis;* Sexual stage*: Mycosphaerella manihotis* Ghesq. and Henrard).

The conidiophores are pale olivaceous brown, unbranched, geniculate, and rounded at the tip. The amphigenous conidia produced singly at the apex of each conidiophore are cylindrical, pale olivaceous. Black perithecia scattered in the necrotic tissues. The asci are elongate-clavate, 8-spored, and sub-sessile. The ascospores are ovoid, uniseptate, constricted at the septum.

Favourable conditions: Temperature of 27°C, relative humidity of more than 90 per cent, Aug-Sept. months and excessive N application.

Mode of survival: In diseased plant debris.

Primary spread: From self-sown cassava and from regrowth in stored stems.

Secondary spread: Through air-borne conidia

Management

- ✰ Remove and destroy affected leaves.
- ✰ Adopt balanced fertilization.
- ✰ Spray copper oxychloride 0.25 per cent or Bordeaux mixture 1.0 per cent or carbendazim 0.1 per cent or thiophanate methyl 0.1 per cent at monthly interval.

2. Sett Rot/Stem Rot

Symptoms

- ✰ It is found in stored setts and in new plantings in dry areas.
- ✰ Rotting of the stem cuttings (setts) occur which leads to failure in germination.
- ✰ Epidermal blisters are seen on the stem. Black masses of pycnidia are seen on the stem when blister ruptures.
- ✰ Vascular discolouration (black) and necrosis leads to stem rot.

Fungus

Diplodia natalensis Pole-Evans.

Favourable conditions: Dry climate.

Mode of survival: In diseased and dried stems and plant debris.

Primary spread: From infected setts.

Secondary spread: Through air-borne conidia.

Management

- ✰ Select disease-free healthy setts for planting.
- ✰ Avoid wounds in stem cuttings, stored for planting.
- ✰ Dip setts in captan + carbendazim, chlorothalonil + mancozeb 0.1 per cent, carboxin 0.1 per cent or thiram 0.2 per cent for 15 min.

3. Cassava Mosaic

Symptoms

- ☆ The mosaic symptoms are first seen on younger leaves.
- ☆ Infected leaves exhibit chlorotic areas in the leaves and distortion of lobes.
- ☆ Leaves are reduced in size, misshapen and twisted with dark green areas separated by normal green tissue.
- ☆ Infected plants are stunted. The stems are thin and soft.
- ☆ Tubers are small and show longitudinal splitting.
- ☆ Tubers contain high fibre and unfit for consumption or starch manufacture.
- ☆ The disease reduces the quality and quantity in tubers and seed-stems.
- ☆ The plants infected soon after germination of setts do not grow further and dies soon.

Virus

Indian cassava mosaic virus (ICMV).

Mode of survival: In setts, volunteer plants.

Primary spread: From infected setts from diseased plants.

Secondary spread: Through insect vector, whitefly (*Bemisia tabaci*)

Management

- ☆ Plant resistant varieties (MnGa-1) or hybrids
- ☆ Plant virus-free planting materials obtained from meristem tip culture or setts from healthy plants.
- ☆ Rogue out and destroy infected plants
- ☆ Place yellow sticky traps at 12/ha to attract and reduce insect vector population.
- ☆ Spray methyl demeton or dimethoate 2ml/lit.

8. Celery (*Apium graveolens* L.)

1. Damping Off

Symptoms

- ☆ Pre-emergence damping off of seed decay or death of seedlings before emergence from soil.
- ☆ Water soaked reddish brown spots on the stem at the soil level or on underground parts causing the stem to girdle and the plants to wilt and fall over and die.

Chromista: *Pythium debaryanum* R.Hesse., *P. ultimum* Trow.

Fungi: *Rhizoctonia solani* Kuhn.

Mode of survival: As oospores or sclerotia in soil.

Primary spread: From oospores or *sclerotia.*

Secondary spread: Through irrigation water.

Favourable condition: High soil moisture, poorly drained soil, high seed rate.

Management

- ☆ Use optimum seed rate.
- ☆ Provide better drainage.
- ☆ Treat seeds with captan or thiram 4 g/kg.

2. Pink Rot (White Mold)

Symptoms

- ☆ Seedlings, grown up plants and stored celery are attacked.
- ☆ Stem of seedlings at ground level is attacked and soft rot develops and causes seedlings to collapse.
- ☆ Affected tissues are covered with a fluffy white fungal growth and bleach sclerotia.
- ☆ Celery plants in the field are attacked at the soil level leading to watery soft rot of lower portion of the head.
- ☆ Diseased tissues become pink in colour (hence the name pink rot) and covered with white mould and black sclerotia.
- ☆ Affected celery heads wilt an collapse.
- ☆ Celery with latent pink rot infection during storage causes water-soaked light brown lesions with pinkish border but no odour occurs. The tissues rapidly turn soft and decay and are covered with white mycelium and sclerotia.

Fungus

Sclerotinia sclerotiorum (Lib.) de Bary.

Mode of spread and survival: As mycelia and sclerotia.

Primary spread: From sclerotia in infected plant debris.

Secondary spread: Through irrigation water, farm equipments, overhead irrigation.

Favourable conditions: Cool, moist weather, high humidity and poor ventilation.

Management

- ☆ Follow deep summer ploughing.
- ☆ Three year crop rotation with non-hosts like onion, corn or cereal or beet.
- ☆ Collect and remove infected parts and plant debris.
- ☆ Control weeds to reduce microclimate favourable to the disease.
- ☆ Spray chlorothalonil 0.2 per cent.

3. *Fusarium* Yellows

Symptoms

- ☆ Mild or late infection causes slight stunting and stiffening of outer stalks or petioles and brown vascular discolouration
- ☆ Leaves become brittle and curl upward with rougher texture
- ☆ In severe infection outer leaves become yellow first and yellowing spreads to older leaves through vascular tissues of roots and crown
- ☆ Later foliage turns brown and dies.
- ☆ Roots are discoloured brown.

Fungus

Fusarium oxysporum f. sp. *apii* (R. Nelson and Sherb.)

Mode of survival: As chlamydospores

Primary spread: From infected seedlings

Secondary spread: Through farm equipment, workers and irrigation water, wind-blown soil.

Management

- ☆ Grow tolerant or resistant cultivars.
- ☆ Plant disease free seedlings only.
- ☆ Avoid continuous celery cultivation and rotate with onion or lettuce for 2-3 years. Avoid carrot or sweet corn.
- ☆ Collect and burn celery trimmings.
- ☆ Control weeds which support multiplication of the fungus in root zone.
- ☆ Drench the soil with *Trichoderma viride* or *T. harzianum*.

4. Basal Stalk Rot

Symptoms

- ☆ Sunken, reddish brown spots occur on the underside of lower stalks
- ☆ Brown spots increase in size and rot
- ☆ White brown mycelium grow on rotten portion

Fungus

Rhizoctonia solani Kuhn.

Mode of survival: As sclerotia in infected plant debris in soil

Primary spread: From soil-borne sclerotia

Secondary spread: Through irrigation water, farm equipment

Management

- ✰ Follow rotation of crops avoiding alfalfa.
- ✰ Plant disease free seedlings.
- ✰ Irrigate field to required moisture level.
- ✰ Protective spray with chlorothalonil.

5. Early Blight

Symptoms

- ✰ Initially older and outer leaves are affected.
- ✰ Light brown, circular or angular, greasy 0.6-1.8 mm in dia with or without yellow halo appear on leaves.
- ✰ Spots increase in size and gradually darker and develop a grey fuzzy fungus growth in the centre.
- ✰ Infection on stalk causes large elongated brown to grey lesions which may cover entire leaf stalk.
- ✰ Leaf spots dry and become papery and may even split and crack.
- ✰ Damage to stalks or leaves reduce market value.

Fungus

Cercospora apii Fresen

Mode of survival: As conidia in infected seeds and in infected plant parts.

Primary spread: From infected seeds, volunteer plants, diseased seedlings.

Secondary spread: Through wind-borne conidia, rain and irrigation water, farm equipment.

Favourable conditions: Warm (15-30°C) and wet conditions with prolonged heavy dew.

Management

- ✰ Grow resistant/tolerant varieties.
- ✰ Treat the seeds with hot water (48°C for 30 min.) (Do not treat seeds of more than 3 years old and hot water treatment may reduce germination).
- ✰ Use certified disease free seedlings.

- ☆ Follow wider spacing to improve air circulation for the leaves to dry out rapidly.
- ☆ Collect and destroy diseased plant parts.
- ☆ Avoid overhead irrigation.
- ☆ Spray aerially with chlorothalonil 0.2 per cent or mancozeb 0.2 per cent or azoxystrobin 0.1 per cent or propiconazole 0.2 per cent.

6. Late Blight

Symptoms

- ☆ Small circular, water soaked spots appear on leaves.
- ☆ Small yellow spots which later turn brown or grey on older outer leaves.
- ☆ Black fruiting bodies (pycnidia) develops within the centre of the spots.
- ☆ Lesions increase in size and coalesce together.
- ☆ Infection spreads to petiole as elongated and brown spots without well-defined margin but with pycnidia.
- ☆ Diseased leaf and stalk fetches low value in the market.

Fungus

Septoria apiicola Speg.

Mode of survival: As infected seeds and plant debris in soil.

Primary spread: From infected seeds and seedlings.

Secondary spread: Through wind, rain, irrigation water, animals, farm workers and machinery.

Favourable conditions: Cool weather (below 24°C).

Management

- ☆ Follow crop rotation with crops like onion protective.
- ☆ Sow disease-free seeds and transplant disease-free seedlings.
- ☆ Soak seeds in warm thiram solution for 24 h or in hot water at 48°C for 30 min.
- ☆ Restrict entry into celery field when it is wet.
- ☆ Avoid excess N application.
- ☆ Avoid overhead irrigation.
- ☆ Collect and destroy diseased plant parts or plough under after harvest.
- ☆ Protective spray with propiconazole 0.2 per cent or azoxystrobin 0.2 per cent or copper oxychloride 0.25 per cent.

7. Bacterial Blight

Symptoms

- Small, angular, deep red spots, water soaked appearance with sharply delineated border occur on leaves.
- More number of spots cause blighting.
- Brown, long discolouration on petiole during wet seasons causing petiole necrosis. Discolouration may be to entire petiole length.

Bacterium

Pseudomonas cichorii (Swingle 1925) Stapp 1928

Mode of survival: As bacterial cells in infected plant debris.

Primary spread: From infected seedlings.

Secondary spread: Through rain and irrigation water, farm implements.

Favourable conditions: Warm and wet weather.

Management

- Plant disease free seedlings.
- Minimize working in fields when plants are wet.
- Spray copper oxychloride 0.25 per cent.

8. Celery Mosaic (CeMV)

Symptoms

- Distinct mosaic or mottling symptoms on leaves.
- Distorted leaf growth with rosette formation.
- Crinkling, vein clearing and malformation of leaves.
- Older leaves exhibit chlorotic/necrotic spots.
- Petioles are flattened.
- Plants are stunted.

Virus

Celery mosaic virus (CeMV) or Apium virus 1

Transmission: Through aphids and sap transmission and not seed transmitted.

Management

- Grow resistant cultivars.
- Plant disease free seedlings.
- Clear out the weeds from crop especially umbelliferous plants.
- Plant new crops far away from mature crops.

- ☆ Plough in old crops and crop debris after harvest.
- ☆ Break of 2-3 months between 2 celery crops.
- ☆ Use of reflective mulches.
- ☆ Spray insecticide to control aphid vectors.

9. Heart Mosaic

Symptoms

- ☆ Mottling of leaves and yellowing of veins.
- ☆ Deformation and crinkling of leaves.
- ☆ Petioles at the heart are curved downward giving the plant an open appearance.
- ☆ Characteristic elongate, brown, translucent, slightly sunken spots appear on the petiole.
- ☆ Spots on entire length of petiole reduces market value.

Virus

Cucumber mosaic virus

Mode of survival: As volunteer celery plants and weed hosts.

Primary spread: From volunteer plants and weed hosts.

Secondary spread: Through aphids.

Management

- ☆ Eradicate weed hosts in the field and its edges.
- ☆ Destroy infected plants after harvest.
- ☆ Spray insecticide to control aphid vector.

9. Curcurbits (*Cucurbita* L.)

1. *Fusarium* Wilt

Symptoms

- ☆ Seedlings and grown up plants are attacked.
- ☆ Seeds decay in the soil. Young seedlings wilt, collapse (damping off) and die.
- ☆ Older plants wilt, wither and die.
- ☆ In wet weather, dead stems show white or pinkish mass of fungal growth.
- ☆ Browning of water conducting vessel (xylem).
- ☆ The roots appear normal at first, but later they turn reddish brown and finally die.
- ☆ Fruits if formed are small and unmarketable.

Fungus

Fusarium oxysporum f. sp. *raphani* J.B. Kendr. and W.C. Snyder

[Syn: *F. oxysporum* f. sp. *conglutinans* (Wollenw.) Snyder and Hansen]

Mode of survival: In infected vines in soil and seed (seed-borne and soil-borne).

Primary spread: From seed and soil.

Secondary spread: Through irrigation water and implements.

Favourable conditions: Temperature of 23-28°C and many susceptible crop hosts.

Management

- ✰ Use resistant varieties.
- ✰ Use disease free seeds.
- ✰ Provide proper drainage.
- ✰ Adopt crop rotation.
- ✰ Keep plants vigorously by means of a good fertility program.
- ✰ Treat the seeds with carbendazim 2g/kg.
- ✰ Prevent the spread of infested soil carried on equipment, tools, feet and irrigation water.
- ✰ Remove and destroy the infected plants. Do not put diseased vines in compost or manure piles.
- ✰ Spot drench soil in root region of plants around diseased plants with carbendazim 1g/lit.

2. Anthracnose

Symptoms

- ✰ All the above ground plant parts are attacked.
- ✰ Drying and death of young seedlings occur
- ✰ Spots on leaves of older plants are water-soaked, yellowish brown angular to roughly circular with yellow halos.
- ✰ In severe cases, diseased and necrotic portion falls off leading to "shot hole".
- ✰ Number of spots coalesce and leaves dry up subsequently.
- ✰ Water soaked, brown, sunken lesions appear on stem, which later girdle the stem and cause death of shoots.
- ✰ Young fruits shrivel with dark colour if pedicels are infected. Circular, black, sunken lesions of varying size appear on matured fruits.
- ✰ On older lesions, black acervuli are arranged in concentric rings.

Fungus

Colletotrichum dematium (Pers.) Grove

The fungus produces black stromata, bearing black setae and hyaline conidiophores. Conidia are produced on conidiophores by budding and are one celled, hyaline, oblong to ovoid.

Mode of survival: In seed and infected plant debris.

Primary spread: From infected seeds.

Secondary spread: Through air-bone conidia, irrigation water and rain splashes.

Favourable conditions: Temperature of 22-27°C and frequent rain.

Management

- ☆ Collect seeds from healthy crop.
- ☆ Maintain proper drainage in the field.
- ☆ Treat the seeds with carbendazim 2g/kg.
- ☆ Removee and burn infected plants.
- ☆ Spray mancozeb 0.2 per cent or chlorothalonil 0.2 per cent or carbendazim 0.1 per cent.

3. Downy Mildew

Symptoms

- ☆ Disease mostly occur on older leaves and spreads to younger leaves.
- ☆ Irregular white patches of downy growth are seen on the under surface of leaves and the corresponding upper surface shows yellow patches.
- ☆ The yellow areas are angular and bounded by veins.
- ☆ Such spots merge to form large brown areas on the leaves.
- ☆ Entire leaf dries up quickly.
- ☆ Fruits are small, few with poor taste.

Chromista

Pseudoperonospora cubensis (Berk. and Curt.) Rustow.

The mycelium is hyaline, aseptate intracellular. Sporagiophores developes in groups of 1-5, bulbous at base with dichotomously branched. Sporagia occur singly and are pale grey to purple in color. Sporangia germinate and produce zoospores. Oospores are produced rarely and are spherical.

Mode of survival: In plant debris as oospores.

Primary spread: From infected debris.

Secondary spread: Through air-borne sporangia, rain splashes and irrigation water.

Favourable conditions: Temperature of 15-22°C with high humidity and heavy dew.

Management

- ☆ Maintain wider spacing and provide drainage facilities.
- ☆ Destroy weed hosts and severely diseased and dried plants.
- ☆ Treat the seeds with metalaxyl 4-6 g/kg.
- ☆ Spray mancozeb 0.2 per cent or metalaxyl + mancozeb or chlorothalonil 0.2 per cent.
- ☆ Avoid copper fungicides which are phytotoxic to cucurbits.

4. Powdery Mildew

Symptoms

- ☆ White or dirty grey powdery growth seen superficially on the upper surface of leaves in patches.
- ☆ Later they cover the entire leaf surface and the affected leaves turn brown and dry.
- ☆ Premature defoliation and death of vines occur.
- ☆ Few flowers are produced and the fruits are undersized, deformed with reduced quality.

Fungus

Erysiphe cichoracearum DC and *Sphaerotheca fuligena* Schlecht. ex. Fr.

Erysiphe – The mycelium is ectophytic, floury white color produce simple germ tube followed by appressoria, barrel to cylindrical shaped conidia. Cleistotheccium produce 10-25 asci.

Sphaerotheca – The mycelium is endophytic, rusty brown color produce forked germ tube, ellipsoidal conidia with fibrosin bodies. Cleistotheccium produce single asci.

Mode of survival: In collateral hosts and as dormant mycelia and cleistothecia in infected crop debris.

Primary spread: From infected crop debris.

Secondary spread: Through air-borne conidia.

Favourable conditions: Dry weather with low humidity.

Management

- ☆ Destroy affected plant parts.
- ☆ Spray thiophanate methyl 0.1 per cent or carbendazim 0.1 per cent or dinocap 0.2 per cent.
- ☆ Avoid sulphur fungicides which are phytotoxic to cucurbits.

5. *Cercospora* Leaf Spot

Symptoms

- ☆ Spots are small, circular to irregular with greyish centre with black margin and coalesce to form large blotches.
- ☆ Infected leaves dry and die.
- ☆ Young succulent stems and petioles are also attacked.

Fungus

Cercospora cruciferarum Ellis and Everh., *Cercospora atrogrisea* Ellis and Everh.

Mode of survival: As In crop debris and collateral hosts.

Primary spread: From infected crop debris.

Secondary spread: Through Air-borne conidia.

Favourable conditions: The disease organism is favoured by cool temperatures (55°F to 65°F) and free moisture on the leaves.

Management

- ☆ Follow field sanitation.
- ☆ Select healthy seeds.
- ☆ Adopt crop rotation.
- ☆ Spray mancozeb or zineb 0.2 per cent or hexaconazole 0.1 per cent.

6. *Alternaria* Blight

Symptoms

- ☆ Small, circular, and somewhat water-soaked or transparent spots appear on the leaves.
- ☆ They become dark brown to black, increase in size and show concentric rings on the upper leaf surface.
- ☆ Spots coalesce and leaves dry up.

Fungus

Alternaria raphani J.W. Groves and Skolko

Mode of survival: As dormant mycelium in diseased crop refuse and weeds.

Primary spread: From infected crop refuse.

Secondary spread: Through air-borne conidia, rain splashes, irrigation water and implements

Favourable conditions: Warm and dry conditions, weakened plants due to poor nutrition and drought.

Management

- ☆ Plough the field immediately after harvesting.
- ☆ Follow crop rotation with other vegetables, flowers and fruits.
- ☆ Spray mancozeb or chlorothalonil 0.2 per cent.

7. Bacterial Wilt

Symptoms

- ☆ It affects the leaves, lateral shoots and entire plant.
- ☆ Initial symptom appears on leaves as drooping and wilting.
- ☆ Leaf symptom is associated with chewing injury made by cucumber beetles [Striped cucumber beetle (*Acalymma vittata*) and spotted cucumber beetle (*Diabrotica undecimpunctata*)].
- ☆ Wilting starts from infected leaf to lateral shoot and then the entire plant.
- ☆ When wilted stems are cross-sectioned, viscid and sticky bacterial ooze exudes from the vascular bundles (Diagnostic method).
- ☆ When older plants are infected, only one shoot wilts initially and at fruiting stage entire plant collapses.

Bacterium

Erwinia tracheiphila (Smith).

It is Gram-negative, rod-shaped, motile bacterium with 4-8 peritrichous flagella.

Mode of survival: As overwinters in cucumber beetles (insect vector). Infected beetle remain infective for 3 weeks. The pathogen is not seed-transmitted and does not survive in soil.

Primary spread: Through beetles while feeding on young leaves or cotyledons.

Favourable conditions: Temperature of 25-30°C, high relative humidity and presence of moisture on the foliage. Winter period with warm temperature is more favourable for overwintering beetles which in turn cause primary infection.

Management

- ☆ Adopt summer ploughing to expose all the stages of beetles.
- ☆ Treat the soil with carbofuran 33 kg/ha or neem cake (150-250 kg/ha) at the time of planting to control beetles.
- ☆ Roguee out diseased plants
- ☆ Spray insecticides to control beetles transmitting the disease.
- ☆ Spray streptocycline 500 ppm.

8. Angular Leaf Spot

Symptoms

- ☆ Disease occurs on leaves, young green stems and fruits.
- ☆ Water-soaked, irregular or angular lesions appear on the leaves.
- ☆ In wet weather, bacterial ooze is seen on the water soaked areas which later dries to form white crusts.
- ☆ The spots enlarge and turn brown.
- ☆ The necrotic tissues dry and fall off leaving irregular holes.
- ☆ Circular spots appear superficially on fruits.

Bacterium

Pseudomonas syringae pv. *lachrymans* (Smith and Bryan). Bacterium is rod-shaped, Gram-negative and motile by 1 to 5 polar flagella.

Mode of survival: As infected crop debris.

Primary spread: From infected seeds and crop debris.

Secondary spread: Through irrigation water and rain splashes.

Favourable conditions: Optimum temperature for disease development is 20-24°C and high moisture is conducive.

Management

- ☆ Soak the seeds in streptocycline (100 ppm) for 30 min.
- ☆ Spray streptocycline 100 ppm or copper oxychloride (0.3 per cent).

9. Cucumber Mosaic

Symptoms

- ☆ Leaves show clear mosaic symptoms of green and yellow patches.
- ☆ Affected leaves are puckered and deformed, curled downward and attain only half the size of the normal leaves.
- ☆ Internodes are shortened with excessive branching.
- ☆ Proliferation of leaves, twigs and petioles gives the plant a bushy appearance.
- ☆ Flowers are dwarfed and few in number.
- ☆ Fruits are misshapened, small with yellowish green mottling and warty.

Virus

Cucumber mosaic virus (CMV).

Mode of survival: In weeds and cultivated crop hosts.

Primary spread: From seeds and from weeds and crop hosts.

Secondary spread: Through insect vector, aphids (*Aphids craccivora, A. evonymi, A. gossypii* and *Myzus persicae*), cucumber beetles, sap and grafting.

Favourable conditions: Presence of many collateral hosts and abundance of insect vectors (aphids and cucumber beetles).

Management

- ☆ Follow crop rotation.
- ☆ Use certified/disease free seeds.
- ☆ Eradicate weed hosts and diseased plants.
- ☆ Control the vector population using imidacloprid 0.5ml/lit or other systemic insecticides.

10. Drumstick/Moringa (*Moringa oleifera* Lam.)

1. Damping Off

Symptoms

- ☆ Pre emergence damping off: Seedlings disintegrate before they come out of soil surface leading to poor seed germination.
- ☆ Post emergence damping off: takes place after seedlings have emerged out of soil but before the stems are lignified.
- ☆ Water soaked lesions formed at the collar portion.
- ☆ Infected areas turned brown and rot
- ☆ Plant shrivels and collapse result of softening of the tissue.
- ☆ In *Rhizoctonia* infected stems become hard thin and infected seedlings are toppled.

Chromista: *Pythium aphanidermatum* (Edson) Fitzp., *Pythium debaryanum* R. Hesse.

Fungus: *Rhizoctonia solani* Kuhn.

Primary spread: From oospore in soil (*Pythium*).

From sclerotia in soil (*Rhizoctonia*).

Secondary spread: Through zoospore through irrigation water (*Pythium*).

Through mycelia in soil and Sclerotia in soil (*Rhizoctonia*).

Favourable conditions:

Pythium - heavy rainfall, excess irrigation, poorly drained soil and closer spacing, high soil moisture with temperature around 25-30°C.

Rhizoctonia - high soil moisture with temperature around 30-35°C.

Management

- ✰ Raise nursery in light soil with proper drainage.
- ✰ Burning farm trashes on the surface of the bed.
- ✰ Sowing seed on raised bed of 15 cm height.
- ✰ Use low seed rate.
- ✰ Apply *Trichoderma* and *Pseudomonas* as seed, nursery treatment and soil application.

2. Twig Canker

Symptoms

- ✰ Initial symptoms are clearing of the vein lets and chlorosis of the leaves.
- ✰ The younger leaves may die in succession and the entire may wilt and die in a course of few days.
- ✰ Soon and the petiole and the leaves droop and wilt.
- ✰ At later stage browning of vascular system occurs.
- ✰ Plants become stunted and die.

Fungus

Fusarium pallidoroseum (Cooke) Sacc.

Spread: From soil and implements.

Favourable conditions: High soil moisture and soil temperature.

Management

- ✰ Avoid density planting.
- ✰ Provide proper irrigation and sanitation.
- ✰ Apply *Trichoderma* and *Pseudomonas* as nursery treatment and soil application.

Other Diseases of Drumstick

3. **Root rot – *Diplodia* sp.**

4. **Fruit rot - *Cochliobolus hawaiiensis***

5. **Postharvest molds - *Aspergillus niger* van Tieghem, *Aspergillus flavus* Link, *Alternaria alternata* (Fr.) Keissl., *Fusarium oxysporum* Schlecht., *Macrophomina phaseolina* (Tassi) Goid, and *Rhizopus stolonifer* (Ehrenb.) Vuill.**

11. Pea (*Pisum sativum* L.)

1. *Fusarium* Wilt

Symptoms

- Symptoms appear as true wilt and near wilt which are described below:
- True wilt: The disease is mostly observed at or after flowering. Lower leaves look pale yellow and slowly plant loose their turgidity and the whole plant topples down. Pod formation is severely affected. The stem tissues in collar region are discoloured. Yellow to orange brown discoloration occurs in the vascular system.
- Near wilt: Rapid development of yellowing and drooping were observed. Brick red discoloration occurs in the vascular system. Cortical decay is observed in lower part of the stem.

Fungus

Fusarium oxysporum f. sp. *pisi* (Hall) Race 1 (true wilt) and Race 2 (near wilt).

The mycelium is septate white to purple color. Microconidia are oval to cylindrical on less branched conidiophores. Macroonidia are thin walled, 3-5 septate fusiod and pointed at both ends and borne on branched conidiophores or on the surface of sporodochia. Chylamydospores are both terminal and intercalary.

Mode of survival: In seeds (seed-borne) and infected plant debris in soil (soil-borne).

Primary spread: From infected seeds.

Secondary spread: Through irrigation water.

Favourable conditions: Temperature of 25-28°C and dry soil.

Management

- Adopt crop rotation.
- Grow resistant varieties.
- Treat the seeds with *Trichoderma viride* 4 g/kg or carbendazim 2 g/kg.
- Collect and destroy infected plants.

2. Powdery Mildew

Symptoms

- White powdery growth consisting of mycelia and conidia occur on upper surface of leaves and stem.
- Severely infected leaves become chlorotic and distorted before falling.
- Usually it occurs at pod formation stage.

- ☆ Pods and tendrils are also covered with powdery growth.
- ☆ The pods turn black, become small and malformed and seeds are wrinkled.

Fungus

Erysiphe polygoni DC.

The mycelium is hyaline and septate. Conidiophores arise vertically from the mycelium, bearing conidia singly or in chains. Cleistothecia are produced as dark colored, myceliod appendages, round, minute bodies scattered in the mycelial web, contains 2-8 asci. Each ascus contain 2-8 ascospores, which are elliptical, hyaline and unicellular.

Mode of survival: As cleistothecia in the plant debris in soil.

Primary spread: From plant debris.

Secondary spread: Through air-borne conidia.

Favourable conditions: Dry humid weather.

Management

- ☆ Grow resistant varieties.
- ☆ Sow early-maturing cultivars.
- ☆ Remove and destroy infected plant materials.
- ☆ Spray with dinocap 0.1 per cent or or sulphur dust 25 kg/ha.

3. Rust

Symptoms

- ☆ Large number of minute, raised, reddish brown pustules (uredia) appears on lower surface of leaves and stem.
- ☆ Several uredopustules merge and burst exposing a mass of brown spores on leaves, give a brownish appearance and can be seen even from a distance.
- ☆ Later uredopustules turn dark brown or black due to the production of telia.
- ☆ Severely infected leaves shrivel and dry up.
- ☆ Disease is also seen on green pods.

Fungi

Uromyces viciae-fabae (Schroet) and *U. pisi* (Pers.) de Bary.

U. viciae-fabae: It is macrocyclic autoecious pathogen producing all its spore stage on the same host.

U. pisi: It is heteroecious pathogen. *Euphorbia* is the intermediate host. Spermagonia and aecia develop on *Euphorbia.*

Mode of survival: In infected plant debris as telia and weed hosts (*Lathyrus* spp. and *Vicia* spp.).

Primary spread: From weed hosts and infected plant parts.

Secondary spread: Through air-borne uredospores.

Favourable conditions: Prevalence of low temperature (17-22°C) coupled with high humidity.

Management

- ☆ Use disease tolerant varieties.
- ☆ Adopt long crop rotation.
- ☆ Remove and burn plant debris and weed hosts.
- ☆ Treat the seeds with captan or thiram 4 g/kg.
- ☆ Spray mancozeb 0.2 per cent or bitertanol 0.1 per cent or triademefon 0.1 per cent or hexaconazole 0.1 per cent.

4. *Ascochyta* Blight or Foot Rot

Symptoms

- ☆ It appears on seedlings in cool and wet weather.
- ☆ Small purple spots on the leaves, increases in size, turn brown to black and become zonate.
- ☆ Infected leaves dry and hang down.
- ☆ Stem tissues blacken at soil line and later extends up to 15 cm.
- ☆ Young plants die or survive with weakened state.
- ☆ Later brown, irregular spots with dark margin are produced on immature green pods.
- ☆ Several spots cause shrinkage of pods and stained and deformed seeds.
- ☆ Foot rot occurs soon after germination below the soil-line dark brown spot appear on the stem, which gradually spread on both direction.

Fungus

Blight: *Ascochyta pinodes* (Berk and Blox)

Leaf and Pod spot: *Ascochyta pisi* Lib.

Foot rot: *Ascochyta pinodella* Jones. [Syn: *Phoma medicaginis* var. *pinodella*]

Mode of survival: In crop debris in soil (soil-borne) and in seeds.

Primary spread: From infected seeds.

Secondary spread: Through air-borne pycnidiospores.

Favourable conditions: Temperature of 16°C, cool weather with abundance of water on the leaves and frequent rains.

Management

- ☆ Adopt crop rotation.
- ☆ Use disease free seeds.
- ☆ Collect and burn infected plant debris.
- ☆ Treat the seeds with captan or thiram 4g/kg
- ☆ Spray mancozeb 2g/lit.

12. Potato (*Solanum tuberosum* L.)

1. Early Blight

Symptoms

- ☆ The disease mainly infects leaves and tubers.
- ☆ Initially the symptoms occur on the lower and older leaves in the form of small (1-2 mm) circular to oval brown spots.
- ☆ These lesions coalesce to form large and angular with concentric rings at later stage.
- ☆ Mature lesions on foliage look dry and papery.
- ☆ Symptoms are observed on petioles and stem as brown sunken lesions and the internal tissues turn brown and corky.
- ☆ On the tubers brown, circular to irregular and depressed lesions are formed and the tissues below the spots turn dry, brown and corky. Infected plants produce few and undersized tubers.
- ☆ These lesions on tubers enlarge during storage and affected tubers later become shriveled.

Fungus

Alternaria solani Sorauer

It is a polycyclic necrotrophic pathogen, produces dark colored, multinucleate muriform and beaked conidia arise singly in simple conidiophore.

Mode of survival: In tubers and in infected plant debris in soil.

Primary spread: Through air-borne conidia.

Favourable conditions: Temperature of 20-30°C, dry warm weather alternating with intermittent rains, reduced plant vigour and other crop hosts and weeds.

Management

- ☆ Grow moderately resistant varieties.
- ☆ Follow crop rotation with non-solanaceous crops.
- ☆ Plant disease free/certified seed tubers.
- ☆ Remove and destroy infected crop debris.

- Follow proper fertilization to maintain the vigour.
- Spray mancozeb 0.2 per cent or chlorothalonil 0.1 per cent or zineb 0.2 per cent or Bordeaux mixture 1 per cent.

2. Late Blight

Symptoms

- The disease affects all plant parts, *viz.*, leaves, stems and tubers.
- It appears on leaves as small, pale-green, water-soaked lesions which enlarge to cover the entire surface.
- A white cottony growth forms around the dead areas on the lower side of leaves.
- In dry weather, water soaked areas turn necrotic brown.
- These lesions turn purple brown and finally black in 2 to 3 days.
- Disease spreads to petioles, rachis and stem.
- On the stem, black lesions are observed at nodes. Stem breaks at this point and the plant topples over.
- Tubers develop reddish brown, shallow to deep, dry rot lesions and the rot continues even after harvest.
- Secondary fungi and bacteria causing soft rots subsequently invade infected tubers. Such rotting tubers emit putrid smell.

Fungus

Phytophthora infestans (Mont.) de Bary.

Mycelium is hyaline and coenocytic, endophytic, intercellular with haustoria. Sporangia are hyaline, pear-shaped or lemon-shaped with papilla at the tip. Each sporangium releases 3 to 8 zoospores. Zoospores are reniform and biflagellate. Oospores produced late in the season and they are the thick walled resting spores help in the survival of the pathogen.

Mode of survival: As oospores in infected haulms and tubers in soil and as mycelia in infected tubers in storage.

Primary spread: From infected tubers.

Secondary spread: Through air-borne sporangia.

Favourable conditions

Four important conditions for severe late blight (epidemic) incidence are:

- Night temperature below dew point for a minimum period of 4 h.
- Minimum temperature of 10°C or slightly above.
- Cloudy weather on the next day.
- Rainfall of atleast 0.1 mm during next 24 h.

Temperature of 12 to 15°C and relative humidity above 90 per cent with heavy dews or rain alternating with a warm temperature of 20°C favours the disease development.

In hills, occurrence of the disease generally coincides with the onset of monsoon rains.

Management

- Grow resistant/tolerant varieties- Kufri Swarna, Kufri Muthu, Kufri Neela and Kufri Neelima in southern hills.
- Select seed tubers from disease free crop.
- Eliminate self-sown potato plants.
- Remove and destroy weeds to avoid build-up of humidity.
- Burn the haulms after removal to reduce the build-up of primary inoculum in soil.
- Provide proper drainage in field.
- Prophylactic sprays with mancozeb or zineb 0.2 per cent or metalaxyl 0.2 per cent or metalaxyl+mancozeb 0.2 per cent or Bordeaux mixture 1.0 per cent.
- Avoid contact of tubers with the foliage during harvest for which the haulms are cut 2-3 weeks before digging of tuber or use herbicides/flame to kill the foliage.
- Dig only the sound healthy tubers in dry weather.
- Avoid injuries while harvesting and store only disease free tubers.

3. Black Scurf and Stem Canker

Symptoms

- The disease commonly affects the tubers, sprouts, stems and stolons.
- The most common symptom is black scurf. It is nothing but dark brown to black irregular lumps which stick on the surface of tubers.
- These lumps firmly adhere on the tuber surface and cannot be washed easily.
- Other symptoms on the tubers are skin cracks, crater like depressions, pitting, stem-end necrosis and shape deformity.
- The disease also causes sprout injury in storage and in the field after planting.
- The affected sprouts are discoloured and fail to emerge from soil leading to gappiness.
- The emerging sprouts when infected later, develop cankers causing girdling of stem bases (stem canker) and upward rolling of leaves in the plant.

- Small green or reddish aerial tubers are also formed in the axils.
- The roots rot resulting in poor root system.
- Infected stolons produce only deformed tubers.

Fungus

Rhizoctonia solani Kuhn.

The vegetative mycelium is brown colored at mature. The mycelium consists of septate hyphae partitioned by dough-nut shaped pore. It does not produce asexual spores and rarely produce basidiospores.

Mode of survival: As sclerotia in infected debris in soil.

Primary spread: From infected tubers.

Secondary spread: Through irrigation water.

Favourable conditions: Moderate temperature of 23°C, heavy and wet soil, poor fertilization and delayed harvest.

Management

- Follow crop rotation for 2 years.
- Grow disease tolerant varieties.
- Plant disease-free seed tubers.
- Treat seed tubers with thiabendazole or carbendazim + boric acid.

4. Wart

Symptoms

- It is restricted to Darjeeling district of West Bengal, Bhutan and Nepal (endemic disease).
- It is characterized by 'cauliflower-like' warty growths on tubers, stolons and stem bases but not on roots.
- Greenish-yellow crusts are formed on the stems and leaves at or near the soil line during wet weather.
- Only few tubers in a plant develop warts.
- Diseased tubers show formation of one or more warts.
- Severe infection causes complete transformation of tuber into warty mass and they turn brown to black with age.
- Old warts disintegrate in soil even before harvest of the crop.
- The fungal zoospores help in the spread of viral diseases (vector).

Fungus

Synchytrium endobioticum (Schilb) Perc.

The chytrids produces sporangia and resting spores. Resting spores germinate and a large number of zoospores are liberated. It develops a firm cell wall made up of a thick orange-coloured outer layer, the exospore and a thin, hyaline inner layer, the endospore.

Mode of survival: As resting spores in tubers (survive in soil for 25 to 30 years even in the absence of host crop).

Primary spread: From infected tubers.

Secondary spread: Through irrigation water, implements and animals.

Favourable conditions: Periodic flooding followed by lack of proper drainage and aeration, temperature 16-18°C and high soil moisture.

Management

- Ban the movement of potato tubers from infected to non-infected areas through quarantine regulations.
- Grow immune/resistant varieties such as Kufri Jyoti, Kufri Muthu, Kufri Anand, Kufri Sutlej, Kufri Jawahar and Kufri Surya.
- Follow a long crop rotation with non-host crops.
- Use only disease free tubers/certified seed tubers for planting.

5. Powdery Scab

Symptoms

- Powdery scab is a fungal blemish disease of potato.
- The powder comprises of spore balls (cystosori) that are released into the soil.
- There are no above ground symptoms.
- Roots develop distinctive light coloured lobed galls.
- It appears as small raised pimples beneath the skin. Later, they expand and the skin breaks open to expose a dark brown powdery mass of cystosori.
- The lesions are small, dark, round with shallow depressions surrounded by raised torn edges of ruptured skin.
- Cystosori under the skin give the scab a dark margin.
- With the advancement of disease, scabs appear as grouped individuals and they do not coalesce into giant scabs.
- Eye infections result in prominent outgrowths (cankers) covered in brown pitted depressions but without ragged skin edges.
- The disease reduces market value of the tubers.
- During storage the disease does not spread but lesions are more prominent on tubers.
- The motile zoospores released from spore balls act as a vector of Potato Mop Top Virus in the field.

Fungus

Spongospora subterranea (Wallroth) Lagerheim.

Mode of survival: As spore ball in soil for 3-5 years.

Primary spread: From infected tubers and soil.

Secondary spread: Through irrigation water.

Favourable conditions: Wet soil and temperature below 18°C, cool and humid weather and heavy soil.

Management

- ✰ Follow crop rotation.
- ✰ Grow resistant varieties like Kufri Jyoti, Kufri Anand and Kufri Jawahar.
- ✰ Provide proper drainage.
- ✰ Use disease free tubers for planting.
- ✰ Apply sulphur dust 900 kg/ha in soil.
- ✰ Do not throw infected rotten tubers into the manure pit.

6. Brown Rot/Bacterial Wilt/Ring Disease/Bangle Disease

Symptoms

- ✰ Wilting of the youngest leaflets during the hottest part of the day and recover at night.
- ✰ The disease is characterised by stunting of plants, general wilting and yellowing of foliage followed by death of plants especially during tuber formation.
- ✰ If the stem is cut a mass of white bacterial slime exude from the vascular tissues.
- ✰ When tubers are cut, browning of the vascular tissues in the form of ring is prominent (brown rot/ring disease) with grayish white sticky ooze.
- ✰ The tubers rot at harvest, transit and during storage.

Bacterium

Ralstonia solanacearum (Yubucchi *et al.*)

[Syn. *Pseudomonas solanacearum* (Smith), *Burkholderia solanacearum* (Smith)].

The bacterium is rod-shaped, gram negative, lophotrichus bacteria. (more than one polar flagellum), multiplication occurs by bacterial fission.

Mode of survival: In the tubers in soil (soil-borne) and several crop hosts.

Primary spread: From infected tubers and from other crop hosts.

Secondary spread: Through irrigation water, air and cutting knives.

Favourable conditions: Soil temperature of 25-35°C, soil moisture above 50 per cent, soil pH of 6.2-6.6 and injuries in the tuber.

Management

- Plant disease free seed tubers from disease free areas.
- Follow crop rotation with cereals (maize/wheat) or soybean.
- Apply sulphur 450 kg/ha during off-season followed by application of lime 1,350 kg/ha before planting (to reduce the acidic reaction) in order to reduce the disease incidence.
- In endemic areas treat seed tubers with streptocycline 0.02 per cent for 30 min.
- Remove solanaceous weeds from the field.
- Adopt sanitation measures like destruction of rotten tubers and disinfection of cutting knife in methyl alcohol or mercuric chloride 0.2 per cent.
- Don't allow irrigation water from infested field to healthy field.
- Spray formaldehyde in walls and floors of storage yards.

7. Soft Rot

Symptoms

- It is a storage disease.
- Water soaked and soft lesions appear around lenticel or at the point of stolon attachment in the tuber and they become dark and sunken.
- Under high humidity, the lesions enlarge and spread to larger area.
- In advanced stages, tubers decay due to invasion by other organisms.
- The decaying tissues become slimy with foul smell and brown liquid ooze.
- The tuber skin remains intact and the rotted tubers are swollen due to gas formation.

Bacterium

Pectobacterium carotovorum sub sp. *carotovorum* (Jones)

[*Erwinia carotovora* sub sp. *carotovora* (Jones) Bergey *et al.*].

The bacterium is Gram-negative, rod-shaped and motile with 1-6 peritrichous flagella.

Mode of survival: In infected tubers in the field and store houses.

Primary spread: From contact of diseased tubers with healthy tubers in storage.

Secondary spread: Through maggot flies (*Hylemyia* sp. and *Phorlia* sp.) spread bacteria

Favourable conditions: Temperature of 21-29°C, relative humidity of 94-100 per cent, high moisture in soil or during storage, harvesting of immature tubers and wetting of tubers, injuries to tuber and poor ventilation during storage.

Management

- Avoid injuries to tubers and injured tubers are to be sorted out and discarded before storage.
- Sanitation at harvest and storage.
- Drying and disinfecting seed tubers before storage.
- Store the potato tubers in well-ventilated store houses.
- Disinfect cutting knife in methyl alcohol.
- Spray formaldehyde in walls and floors of storage yard.
- Treat seed tubers with streptocycline 200 ppm for 30 min.

8. Common Scab

Symptoms

- Presence of mere abrasion of skin on potato tuber is called shallow scab or russet scab.
- Appearance of 3-4 mm deep, dark brown to black, circular, pitted, corky lesions around lenticels is called deep scab or pitted scab.
- One or both the types of scab can be seen in a single tuber.
- In case of severe attack, dark brown lesions may develop on roots and stolons.
- The disease disfigures the tubers and reduces the market value.

Bacterium

Streptomyces scabies (ex Thaxter) Lambert and Loria.

Aerial mycelium is branched threads. Sporogenous hyphae are spiral in form. Conidia are produced by the formation of septa at intervals along the hyphae, which contract to form narrow isthmuses between the cells.

Mode of survival: In infected rotten tubers in soil.

Primary spread: From infected tubers and manure of cattle fed with infected tubers.

Secondary spread: Through irrigation water

Favourable conditions: Soil pH from 5.5 to 7.5, monoculturing, long duration potato varieties suffer heavily and abundant host plants.

Management

- Avoid rotation with carrot, beetroot, turnip, radish, *etc.*

- Follow six year crop rotation with alfalfa, wheat, oats, barley, sorghum, bajra, soybean, pea, *etc*.
- Maintain soil pH below 5.5 since the bacterium is active at alkaline range.
- Avoid alkaline fertilizers (calcium ammonium nitrate) but apply sulfur or acid-forming fertilizers such as ammonium sulphate which lower the soil pH.
- Grow resistant varieties.
- Select scab-free tubers and treat seed tubers with mercuric chloride 0.1 per cent for 90 min.
- Irrigate to field capacity during initial tuber formation.

9. Leaf Roll

Symptoms

- The virus produces two types of symptoms *viz.*, primary or secondary symptoms in plants depending upon the age of infection.
- The primary symptoms confined to top young leaves, which usually stand upright, roll and turn slightly pale.
- The secondary symptoms develop when plants are grown from infected seed tubers and such symptoms are prominent in older leaves. These infected plants are pale, stunted and upright. The lower leaves roll turn yellow, brittle and leathery.

Virus

Potato leaf roll virus (PLRV).

Mode of survival: In tubers.

Primary spread: From infected seed tubers.

Secondary spread: Through aphid vector, *Myzus persicae.*

Favourable conditions: Susceptible varieties and abundant insect vector.

Management

- Plant certified disease-free seed tubers.
- Multiply seed potatoes in vector free areas
- Remove and burn diseased plants.
- Spray insecticide to control aphid vectors.

10. Spindle Tuber

Symptoms

- Plants are erect but stunted.
- Infected tubers have pointed ends giving them a spindle shape.

- ☆ Infected tubers are smaller than healthy ones.
- ☆ Above ground symptoms are reduced leaf size, thin stems with upright in position and increased internodal length.

Viroid

Potato spindle tuber viroid (PSTVd)

Mode of survival: As volunteer plants and infected tubers.

Primary spread: From Infected tubers.

Secondary spread: Through mechanical spread by cutting knives and other tools used during cultural operations or planting.

Management

- ☆ Use disease free seed tubers
- ☆ Maintain crop sanitation like disinfection of tools and machineries.

13. Radish [*Raphanus raphanistrum* subsp. *sativus* (L.) Domin]

1. *Alternaria* Blight

Symptoms

- ☆ All foliar parts are affected.
- ☆ Symptoms appear as small circular dark, black spots on the leaves.
- ☆ Irregular, oval to elongated lesions appear on siliqua and inflorescence.
- ☆ Infected seeds cause seed rot and post-emergence seedling death.

Fungus

Alternaria raphani Groves and Skolko.

Conidiophores are simple or branched, septate and olivaceous brown in color. Conidia are formed in chains of 2-3, ellipsoidal, generally with a short beak, dark golden brown or olivaceous brown in color.

Mode of survival: In seeds (seed-borne) and infected plant debris.

Primary spread: From infected seeds and plant debris.

Secondary spread: Through air-borne conidia

Favourable conditions: Infection progresses rapidly at 22-26°C.

Management

- ☆ Sow disease free seeds.
- ☆ Though hot water treatment of the seed kills the fungus, use of diseases-free seeds is recommended.

- ☆ Treat the seeds with captan 4g/kg.
- ☆ Remove and destroy severely infected leaves.
- ☆ Spray mancozeb 0.2 per cent.
- ☆ Regular spraying with difolatan 0.3 per cent or dithane 0.2 per cent or ridomil 0.1 per cent controls the disease effectively.

2. White Rust

Refer white rust in cabbage and cauliflower

3. Root Rot of Radish

Symptoms

- ☆ Symptoms appear as rotting of pith tissues resulting in cavity formation and wilting of plants.

Bacterium

Erwinia rhapontici (Millard) Burkholder

It is facultative, anaerobic gram negative, non spore forming, motile bacteria possess peritrichus flagella, produces pink pigements.

Spread: Through implements and irrigation water.

Management

- ☆ Dipping of the seeds in a solution of Agrimycin-100 (100 ppm) at the time of sowing is effective in checking the disease.

4. Radish Mosaic Virus (RMV)

Symptoms

- ☆ Symptoms first appear as small, circular to irregular, chlorotic lesion in between and adjacent to the veins.
- ☆ Little or no leaf distortion is noticed and stunting or abnormal formation rarely occurs.

Virus

Radish mosaic virus (RMV)

Spread: It is transmitted through aphids.

Management

- ☆ The disease can be effectively checked by controlling aphids with 2-3 foliar sprays of dimecron (0.05 per cent) at 10 days interval.

5. Radish Phyllody

Symptoms

- ✰ The diseased plant assumes a dull grey to light violet colouration.
- ✰ The symptoms of the disease appear at the time of flowering when all the floral parts become green violet and leafy.
- ✰ The sepals and petals become green thick knob headed leaves.

Bacterium

Candidatus Phytoplasma (Radish phyllody phytoplasma)

Spread: The disease is transmitted by jassid *Orosius albicinctus*.

Management

- ✰ Sprays of Phosphamidon 0.05 per cent or Oxydemetan Methyl 0.02 per cent is done to eradicate the jassids - the vector of the virus.
- ✰ Soil application of Thimet 10-G 1.5 kg a.i./ha is also recommended.

14. Sweet Potato [*Ipomoea batatas* (L.) Lam.]

1. *Cercospora* Leaf Spot

Symptoms

- ✰ Light green spots initially appear on both surfaces of the leaf.
- ✰ They gradually increase in size and turn brown with yellow green border.
- ✰ Later the centre becomes pale brown to whitish grey and the border is brown to purple black. The veins limit the lesions.
- ✰ The lesions increase in size, coalesce and blight the leaf.
- ✰ Severely blighted leaves turn yellow leading to defoliation.
- ✰ More spots are found on older leaves.
- ✰ Fruiting structures of the fungus are seen at the centre of the spots, especially after rainfall.

Fungus

Cercospora ipomoeae G. Winter; *Pseudocercospora timorensis* (Cooke) Deighton.

Conidiophores are septate, dark brown, geniculate. Conidia are acicular to obclavate, multiseptate, hyaline, straight curved.

Mode of survival: As weed hosts and infected plant debris.

Primary spread: From infected crop debris and weeds.

Secondary spread: Through air-borne conidia and by splashing rains.

Favourable conditions: Hot humid climate with relative humidity of 90-95 per cent and intermittent rain.

Management

- ☆ Remove and destroy severely diseased and fallen leaves.
- ☆ Spray zineb 0.2 per cent.

2. Black Rot

Symptoms

- ☆ This is mainly a storage disease.
- ☆ On underground roots, the first symptom appears as small circular dark brown, slightly sunken spots.
- ☆ These spots enlarge up to 5 cm in dia during transit, storage and marketing.
- ☆ Characteristic small, black fungal structures (perithecia) with long neck protrude in the spotted area and they appear to the naked eye as dark bristles.
- ☆ Spots enlarge and appear greenish black to black when wet and greyish black when dry and coalesce to cover the entire root.
- ☆ If secondary fungi or bacteria invade the tissue, the flesh beneath the spot turns black and go deep and this blackened area extend to the centre of the root.
- ☆ Tissue near the discoloured area has bitter taste but fruity odour and the entire root rots. (The bitter taste is due to ipomeamarone (phytoalexin) produced by the storage roots in response to *C. fimbriata* infection while the fruity odour is from the volatile compound, amyl acetate produced after infection)
- ☆ The disease is serious on young sprouts and adult plants also.
- ☆ Infected sprouts develop black, sunken, necrotic lesions or cankers on the stem under the soil.
- ☆ In the field, plants arc stunted and chlorotic due to the cankers present in the underground stem. Chlorotic leaves drop.

Fungus

Certatocystis fimbriata Ellis and Halst.

Mode of survival: As chlamydospores and perithecia in plant debris, in soil and tubers.

Primary spread: From infected roots and shoots.

Secondary spread: Through sweet potato weevil and rodents in store houses.

Favourable conditions: Temperature of 25°C, high soil moisture and injuries on storage roots during storage, transit and marketing.

Management

- ☆ Propagate plants from disease free planting material.
- ☆ Cultivate sweet potato once in 2-3 years.
- ☆ Dip tubers in thiabendazole 0.1 per cent solution.
- ☆ Do not manure the field with cattle fed with diseased tubers.
- ☆ Store clean sound and disease-free tubers in store house.
- ☆ Cure roots immediately after harvest at 85-95°F and 85-90 per cent relative humidity for 5-10 days.
- ☆ Do not wash and package the roots showing symptoms of black rot.
- ☆ Decontaminate the equipment that comes into contact with an infected crop.
- ☆ Fumigate the crates/containers and store room.
- ☆ Control the insects and rodents in store houses.

3. Surface Rot/Stem Rot/*Fusarium* Wilt

Symptoms

- ☆ Fleshy roots are infected usually in storage through wound made during harvest.
- ☆ Roots damaged by splitting or insect or nematode feeding at the time of harvest also show symptoms.
- ☆ Circular, light to dark brown, dry and firm lesions are seen on the storage roots.
- ☆ The infection may start at the end attached to the mother root or at wounds and slowly extends to the cortical tissues but do not usually get beyond the vascular ring.
- ☆ The surface of the root shrinks and the flesh underneath dries out.
- ☆ A white mycelial growth is observed on it and then the root becomes hard and mummified.

Fungus

Fusarium oxysporum f. sp. *batatas* (Wollenw.) W.C. Snyder and H.N. Hansen

Favourable conditions: Temperature of 32-37°C and low soil moisture.

Mode of survival: As chlamydospores in soil for many years.

Primary spread: From infected tubers.

Secondary spread: Through irrigation water.

Management

- ☆ Grow resistant cultivars.
- ☆ Adopt crop rotation once in 3-4 years.

- ☆ Raise the soil pH to 6.5 or 7.0 and use nitrate nitrogen rather than ammoniacal nitrogen to retard disease development.
- ☆ Use certified seed-roots for planting. Obtain cuttings from sprouts 5 cm above the soil line. Shoots pulled from mother roots should not be used.
- ☆ Dip seed material in 0.1 per cent carbendazim for 20 min.
- ☆ Harvest the roots when soil is neither too wet nor too dry to avoid wounding during harvest.
- ☆ Follow proper curing and handling of storage roots after harvest to avoid surface rot during storage.

4. Soft Rot/*Rhizopus* Rot

Symptoms

- ☆ Infection and decay commonly occur at one or both ends of the roots.
- ☆ Under humid conditions the affected sweet potatoes become soft and watery and the entire root rots within few days.
- ☆ If the humidity is high, the sweet potatoes show greyish black fungal growth. This feature distinguishes *Rhizopus* soft rot from other storage rots.
- ☆ The colour of the root is not changed, but an odour produced by it attracts fruit flies.
- ☆ Chilling and heat damage also predispose sweet potatoes to infection.
- ☆ Soft rot is very destructive when sweet potatoes are washed, packed or shipped to market during cold weather.

Fungus

Rhizopus stolonifer (Enrenb. ex. Fr.) Lind.

Mode of survival: As zygospores in crop refuse.

Primary spread: From infected tuber and in many crop hosts and through contaminated equipments.

Secondary spread: Through air-borne spores or infested soil.

Favourable conditions: Temperature of 23-28°C, relative humidity of 75-85 per cent during storage or transport, longer roots, chilling and heat damage when sweet potatoes are washed, packed, or shipped to market during cold weather. Long stored sweet potato tuber decay occurs at 15-23°C.

Management

- ☆ Avoid planting sweet potatoes in poorly drained, heavy soils and avoid very high rainfall areas.
- ☆ Intercrop with less susceptible crops or with cover crops to reduce fungal inoculum levels in the fields.

- ☆ Manage weevil population to reduce holes in the sweet potatoes and to avoid pathogen entry.
- ☆ Clean up crop debris in fields after harvest.
- ☆ Dry the tubers for 1-2 h before storage.
- ☆ Wash and clean storage house with copper sulphate 2.5 per cent soultion.
- ☆ Follow crop rotation.
- ☆ Carefully handle sweet potatoes before and after harvest and during transport or storage to prevent unnecessary wounding.
- ☆ Cure properly the roots immediately after harvest.
- ☆ Do not allow sweet potatoes to be exposed to sunlight for extended periods (to prevent heat damage) or to be chilled in the field.
- ☆ Avoid lengthy storage period before sale in store houses or in markets.
- ☆ Avoid marketing diseased sweet potatoes in plastic bags without aeration.

15. Taro/Dasheen/Cocoyam [*Colocasia esculenta* (L.) Schott.]

1. Corm Rot

Symptoms

- ☆ Infected young plants are killed without producing leaves.
- ☆ Some of the plants may become severely stunted.
- ☆ The whole plant collapse. The rotting of the corm usually starts at the base and progresses upwards.
- ☆ The rotted tissue is firm and may be whitish-yellow to grey and blue or dark purple.
- ☆ The diseased tissue crumbles like cheese and has an offensive smell.
- ☆ There is often secondary infection with *Erwinia chrysanthemi* which causes a strong-smelling white soft rot.

Chromista

Phytophthora cinnamomi Rands, *Phytophthora nictotianae* var. *parasitica* (Dastur) Waterhouse and *Pythium* spp.

Mode of survival: Iin soil and plant debris.

Primary spread: From infected plant debris.

Secondary spread: Through soil-borne oospores and mycelium.

Favourable conditions: High moisture, warm temperature and poor soil conditions.

Management

- ☆ Disinfestation of planting materials with hot water.
- ☆ Enrich soil with FYM and neem cake followed by apply calcium, magnesium or phosphorous.
- ☆ Add lime before plating to raise the pH to 6.0-6.8.
- ☆ Crop rotation with non-host crops.
- ☆ Rogue the diseased plants, taking them far from the diseased area and destrying them by burying and burning.

2. Root Rot/Corm Soft Rot

Symptoms

- ☆ On young plants root decays is followed by rot of the planting sett.
- ☆ On established plants are outer leaves wilt and dies prematurely.
- ☆ The leaf blade are crinkled slightly rolled or curled inwards, their colored in unhealthy grayish blue green and the margin are pale yellow.
- ☆ Plants remain stunted new leaf production is slow and show the corm are small.

Chromista

Pythium aphanidermatum (Edson) Fitzp., *P. carolinianum* V.D. Matthews, *P. graminicola* Subraman, *P. irregulare* Buisman, *P. myriotylum* Drechsler, *P. splendens* Hans Braun and *P. vexans* de Bary.

The mycelium is hyaline, aseptate. Sporangia are the aseual spores.

Primary spread: From sporangia in the infected plant parts and debris.

Favourable conditions: Temperature of 25°C, wetland and where the rainfall is high.

Management

- ☆ Diseased plants are easily removed from the soil by hand.
- ☆ Practice crop rotation.
- ☆ Use of bio-agent for corm treatment at 5g/kg before planting.
- ☆ The severity of soft rot may be reduced in soil by incorporating Captan 112 kg/ha into the acid soils before planting.
- ☆ Corms should be dipped in metalaxyl + mancozeb solution at 2 g/kg for ½ to 1 h before planting.

3. Taro Leaf Blight

Symptoms

- ☆ The disease begins as purple-brown water-soaked lesions on the leaf.

- A clear yellow liquid oozes from the lesions.
- These lesions then enlarge, join together and eventually destroy the entire lamina.

Chromista

Phytophthora colocasiae Rac.

Mode of survival: In planting material for three or more weeks.

Primary spread: From plant to plant by wind and splashing rain.

Secondary spread: Through soil-borne oospores.

Favourable conditions: High temperature and high humidity.

Management

- Collect and burn or bury the vines after harvest.
- Practice crop rotation.
- Spray metalaxyl + mancozeb.

4. Leaf Spot

Symptoms

- The spots on the leaves are oval or irregular in shape.
- The young spots are buff to reddish brown.
- Older spots are dark brown with a chlorotic region surrounding the lesion. The centers of the infected area frequently rot out to produce a shot-hole type lesion.

Fungi

Phyllosticta colocasiophila Weedon

The fungus is produces single celled, hyaline conidia within pycnidia.

Mode of survival: In infected leaves.

Spread: From spores of infected leaves.

Favourable conditions: Cloudy and rainy weather for a prolonged time.

Management

- Remove all dead plant material and allow for adequate air circulation between and around plants.
- Col1ecting and burning the diseased leaves seems to be of some value.
- Keep the growing environment clean.
- Applications of fungicides such as captan, ferbam and mancozeb or thiophanate methyl at 3 g/lit of water will help control infection levels and can prevent new infections in healthy plants.

5. Bacterial Soft Rot

Symptoms

- ☆ Bacterial soft rot is a strong smelling watery soft rot ranging in color from white to dark blue.
- ☆ Wounds and bruises caused by the feeding of insects and other animals and those inflicted at harvest are the most common infection courts for this disease.

Fungi

Erwinia carotovora sub sp. *carotovora* (Jones) or *E. chrysanthemi* Samson

The bacterium is a Gram-negative, rod-shaped that lives alone or aggregates into pair's chains, non spore forming and peritrichously flagellated. As a mesophilic bacterium, *Erwinia carotovora* thrives the most in the temperature range 27-30°C.

Mode of survival: In infected plant debris in the soil.

Spread: Through contaminated soil, rhizomes and maggot flies.

Favourable conditions: High relative humidity 94-100 per cent and temperature 21-29°C.

Management

- ☆ Sowing healthy seeds.
- ☆ Keep the crops weed free.
- ☆ Keep the growing environment clean.

16. Tomato (*Solanum lycopersicum* L.)

1. Damping Off

Symptoms

- ☆ The disease may attack the seedlings before the emergence of seedlings from the soil (pre-emergence damping off) or after the emergence of seedlings (post-emergence damping off).
- ☆ In pre-emergence damping off, young seedlings are killed before emergence which leads to poor seed germination and gappiness in the nursery.
- ☆ In post-emergence damping off, water soaked, soft, brown lesions appear on the stem at or below the ground level.
- ☆ The stem is constricted at the base and subsequently collapse (toppling over of seedlings).

Chromista

Pythium aphanidermatum (Edson) Fitzp., *Pythium butleri* Subram., *Phytophthora nicotianae* var. *parasitica.*

Fungi

Rhizoctonia solani Kuhn

Mode of survival: As oospores or sclerotia in soil or plant debris.

Primary spread: From soil-borne oospores or sclerotia.

Secondary spread: Through irrigation water.

Favourable conditions: Excess soil moisture due to water stagnation, high humidity, overcrowding of seedlings and high seed rate.

Management

- Provide proper drainage and use raised nursery bed.
- Sterilize the nursery soil with formalin 2.5 per cent spray and cover with polythene sheet for 7-10 days or fumigate the nursery bed with methyl bromide.
- Treat the seeds with thiram or captan 4g/kg or *Trichoderma viride* 4g/kg or *Pseudomonas fluorescens* at 10g/kg.
- Go for soil application of *Pseudomonas fluorescens* or *Trichoderma viride* 10g/sq.m along with FYM.
- Soil drench with Bordeaux mixture 1.0 per cent or copper oxychloride 0.25 per cent at 4 lit/sq.m nursery area.

2. Early Blight

Symptoms

- On leaves (mostly on lower matured leaves), circular to irregular, dark brown to black spots with concentric rings appear.
- The spots coalesce and cause drying and defoliation of leaves.
- Dark spots at the base of the stem near the ground level gradually girdle the stem.
- Spots appear in the stem at the juncture of the side branches which are easily broken by air.
- Dark brown sunken spots appear on fruit. Immature fruits are shed.

Fungus

Alternaria solani Sorauer

It is a polycyclic necrotrophic pathogen, produces dark colored, multinucleate muriform and beaked conidia arise singly in simple conidiophore.

Mode of survival: In plant debris and seeds

Primary spread: From infected plant debris and seeds.

Secondary spread: Through air-borne conidia.

Favourable conditions: June-July sowing and weak and older plants.

Management

- Follow crop rotation with non-solanaceous crops.
- Use disease free seeds. Collect healthy seeds from disease-free fruits.
- Treat the seeds with thiram or captan 4g/kg.
- Provide optimum spacing to avoid overcrowding.
- Avoid heavy N and dense planting to minimize relative humidity.
- Keep the field clean.
- Removee infected plants and debris.
- Spray zineb or chlorothalonil or mancozeb or difolatan 0.2 per cent.

3. *Fusarium* Wilt

Symptoms

- Affected plants exhibit yellowing and drooping of lower leaves and it continues to younger leaves also.
- Plants wilt and die in due course.
- Internal stem portion exhibits vascular browning.
- Though the disease appears in all stages, young plants are more vulnerable for infection.

Fungus

Fusarium oxysporum f. sp. *lycopersici* (Sacc.) Synder and Hans

Mycelium is septate. The fungus produces microconidia, macroconidia and chlamydospores. Microconidia are hyaline, oval, single celled or one septate. Macroconidia are sickle-shaped, 3-5 celled and hyaline. Chlamydospores are thick walled, round resting spores, produced terminally or intercalarily.

Mode of survival: In crop debris and soil as chlamydospores.

Primary spread: From soil and diseased seedlings.

Secondary spread: Through irrigation water.

Favourable conditions: Root-knot nematode predisposes the plant to disease incidence.

Management

- Follow deep summer ploughing.
- Follow crop rotation with non-solanaceous crops (cereals).
- Remove infected plants and plant debris.
- Treat the seeds with carbendazim at 2g/kg or *Trichoderma viride* 4g/kg or *Pseudomonas fluorescens* at 10g/kg.

- Nursery application of *Pseudomonas fluorescens* 20g/m^2 and seedling root dip with *Pseudomonas fluorescens* 10g/lit of water and soil application of *Pseudomonas fluorescens* 2.5 kg/ha + 50 kg of FYM.
- Spot drenching with carbendazim 0.1 per cent.

4. *Septoria* Leaf Spot/Defoliation Disease

Symptoms

- Numerous, circular spots with grey centre and dark brown margins are seen on the leaves, stem and calyx.
- Centre of the spots shows minute black fructifications.
- Severe infection causes defoliation during rainy season.

Fungus

Septoria lycopersici Speg.

The mycelium is hyaline, thin walled and sparingly septate. Pycnidia are subglobose, composed of 2-3 layers of brown cells. Pycnidiospores are filiform, slightly curved, hyaline and septate with pointed or rounded ends.

Mode of survival: In infected plant parts.

Primary spread: From infected plant parts.

Secondary spread: Through air-borne pycnidiospores.

Favourable conditions: Rainy season.

Management

- Follow crop rotation (avoid potato and brinjal).
- Treat the seeds with thiram 4 g/kg.
- Spray zineb or mancozeb 0.2 per cent or ccopper oxychloride 0.25 per cent or Bordeaux mixture 1.0 per cent.

5. Late Blight

Symptoms

- Infection occurs on all the above ground plant parts.
- Water-soaked lesions with faded green patches appear on the leaves.
- Infections spread fast to entire leaf and petiole. Dead areas appear in leaf tip and margins.
- Whitish mildew like growth appears on leaves under humid conditions. The leaves are blighted and dry.
- Dark olivaceous greasy spots with cracking and appearance of white fungal growth on the fruits are seen resulting in soft rot.

- ☆ Internal tissues become spongy with bad odour.
- ☆ In severe cases the whole plant dies.

Fungus

Phytophthora infestans (Mont.) de Bary.

Mycelium is hyaline and coenocytic, endophytic, intercellular with haustoria. Sporangia are hyaline, pear-shaped or lemon-shaped with papilla at the tip. Each sporangium releases 3 to 8 zoospores. Zoospores are reniform and biflagellate. Oospores produced late in the season and they are the thick walled resting spores help in the survival of the pathogen.

Mode of survival: In infected plant debris and soil as oospores.

Primary spread: From oospores in soil.

Secondary spread: Through air-blown sporangia and rain splashes.

Favourable conditions: High humidity and low temperature, cloudy weather and rainfall with splashing rains.

Management

- ☆ Follow crop rotation with cereals.
- ☆ Provide good drainage.
- ☆ Follow summer ploughing and apply green manures/biocontrol agents.
- ☆ Avoid dense planting and maintain proper aeration in the collar region.
- ☆ Follow staking of plants or mulching with straw to avoid the contact of fruits in soil.
- ☆ Remove and destroy infected plants and drench the soil with copper oxychloride 0.25 per cent.
- ☆ Spray mancozeb or difolatan 0.2 per cent or metalaxyl + mancozeb 0.2 per cent.

6. Bacterial Wilt

Symptoms

- ☆ Stunting, yellowing and wilting of foliage leading to collapse of the entire plant.
- ☆ Lower leaves droop before wilting.
- ☆ Cross section of the stem near the base show vascular browning.
- ☆ When the affected stem is cut bacterial ooze can be seen.
- ☆ Adventitious roots are formed from the stem.
- ☆ The affected plants collapse and die.

Bacterium

Ralstonia solanacearum (Yubucchi *et al.*)

[Syn. *Pseudomonas solanacearum* Smith]

It is Gram-negative and rod-shaped, occurring in pairs with 1-4 polar flagella.

Mode of survival: As survives for several years in soil.

Primary spread: From soil and seed infestation.

Secondary spread: Through irrigation water.

Favourable conditions: Humid weather and high temperature.

Management

- ✰ Follow crop rotation with cereal crops or cruciferous vegetables avoiding solanaceous crops.
- ✰ Incorporate bleaching powder in soil 15 kg/ha before transplanting.
- ✰ Seedling root dip in Streptocycline 25 ppm (2.5g/100 lit of water) for 30 min.
- ✰ Soil drench with Bordeaux mixture 1 per cent.

7. Bacterial Leaf Spot

Symptoms

- ✰ Small translucent water-soaked spots on the leaves enlarge to black greasy or dark-brown spots surrounded by yellow halo.
- ✰ Several spots cause chlorosis in the leaves and defoliation occurs.
- ✰ Black cankerous spots occur on stem and petiole.
- ✰ Water-soaked lesions on unripe green fruits become corky resembling small scabs with irregular margins.

Bacterium

Xanthomonas vesicatoria (Pammel 1895) Dowson

[Syn. *Xanthomonas campestris* pv. *vesicatoria*].

It is Gram-negative, rod-shaped motile with single polar flagellum.

Mode of survival: In plant debris and seed (seed-borne).

Primary spread: From infected seeds.

Secondary spread: Through air-blown rain splashes.

Favourable conditions: Temperature of 25-30°C, relative humidity of more than 90 per cent and humid weather.

Management

- ✰ Follow crop rotation.

- Follow summer ploughing.
- Collect seeds from disease-free plants.
- Dip the seeds in streptocycline 100ppm or treat seeds in hot water at 50°C for 25 min.
- Follow soil solarization in nursery bed to avoid seedling infection.
- Spray copper oxychloride 0.25 per cent.

8. Bacterial Canker

Symptoms

- Leaves shows wilting symptoms.
- Stem shows streaks and canker.
- Small brown scabby lesions surrounded by white halo appear on the fruits which resembles the birds eye.

Bacterium

Clavibacter michiganensis sub sp. *michiganensis* (Smith).

It is Gram-positive, aerobic, coryneform, rod-shaped motile with single polar flagellum.

Mode of survival: In plant debris and seed (seed-borne).

Primary spread: From infected seeds.

Secondary spread: Through air-blown rain splashes.

Favourable conditions: More severe during wet weather.

Management

- Follow crop rotation.
- Follow summer ploughing.
- Collect seeds from disease-free plants.
- Dip the seeds in streptocycline 100ppm or treat seeds in hot water at 50°C for 25 min.
- Follow soil solarization in nursery bed to avoid seedling infection.
- Spray copper oxychloride 0.25 per cent.

9. Mosaic

Symptoms

- Typical mosaic pattern with dark green and light green areas appear on leaves.
- Leaflets are distorted, puckered and are small.
- Leaflets exhibit, fern-leaf symptoms.

- Necrotic sunken lesions are seen on fruits.
- Infection on matured fruits shows internal necrosis.
- The affected plants are stunted and pale green.

Virus

Tomato mosaic virus (ToMV).

Mode of survival: As solanaceous hosts and on the seeds (externally seed borne).

Primary spread: From infected seeds and from other host plants.

Secondary spread: Through contacts (sap transmissible), hands of workers, plant debris, implements and on the surface of seeds.

Favourable conditions: Injuries to plant.

Management

- Follow crop rotation with non-solanaceous crops.
- Select seeds from disease free plants.
- Rogue out infected seedlings in the nursery.
- Raise protected nursery (in net house or green house).
- Raise barrier crops like sorghum or maize or pearlmillet 5-6 rows around the field before planting tomato.
- Remove weed hosts, volunteer plants and infected plants.

10. Leaf Curl

Symptoms

- Typical downward curling, crinkling and chlorosis of leaves are seen.
- Infected plants are stunted with shortened internodes giving bushy appearance.
- No flowering and fruiting at later stage of infection.
- Small leaf-like growth called enations on the midrib on the lower surface of leaf.

Virus

Tobacco yellow leaf curl virus (TYLCV).

Mode of survival: In susceptible host plants.

Primary spread: From susceptible crops and weed hosts.

Secondary spread: Through insect vector, wwhitefly (*Bemisia tabaci*).

Favourable conditions: Abundance of insect vector.

Management

- Remove of weed hosts and infected plants

- ☆ Raise protected nursery (in net house or green house) to prevent whitefly infection.
- ☆ Adopt mulching with straw to repel whiteflies.
- ☆ Treat the nursery bed with carbofuran 1.0 kg ai/ha and another dose at one week after transplanting at 1.5 kg ai/ha followed by foliar spray with dimethoate or methyl demeton at 2ml/lit on 15, 25, 45 DAT to control vector.
- ☆ Dip seedling roots in Imidacloprid solution at 4-5 ml/lit for one h before transplanting.
- ☆ Install yellow sticky traps to manage insect vector.
- ☆ Raise barrier crops of sorghum or maize (5-6 rows) around the field before planting tomato.

10. Tomato Spotted Wilt/Bronzy Wilt

Symptoms

- ☆ Necrotic irregular spots are seen on the leaf surface.
- ☆ Leaves are reduced in size and exhibit thickened, bronzy veins.
- ☆ The place where the branch arises from the main stem becomes weak and necrotic leading to bending of branches from the main stem.
- ☆ On fruits, numerous pale yellow or yellowish concentric circular rings are seen.
- ☆ Younger plants wilt and die but older plants survive with spotted fruits that may not ripe properly.
- ☆ Poor seed recovery and reduced lycopene content.

Virus

Tomato spotted wilt virus (TSWV).

Mode of survival: In susceptible crop hosts and weeds.

Primary spread: From susceptible crops and weeds.

Secondary spread: Through insect vector, thrips (*Frankliniella schultzei, Scirtothrips dorsalis,* and *Thrips tabaci*).

Favourable conditions: Abundance of thrips population.

Management

- ☆ Remove weed hosts and keep the field clean.
- ☆ Raise barrier crops like sorghum, maize or bajra 5-6 rows around the fieldd before planting tomato.
- ☆ Select virus free seedlings for planting.
- ☆ Use aluminium surfaced plastic mulch to repel vector.

- ✰ Rogue out infected plants upto 45 DAT.
- ✰ Spray AVPs (leaf extracts of sorghum or coconut 10 per cent) on 10, 17 and 24 DAP.
- ✰ Apply carbofuran at 33 kg/ha in nursery at sowing followed by second application on 10 DAP in the main field and spray phosalone 1.5 ml/lit on 25, 40 and 55 DAP.

17. Yam (*Dioscorea* spp.)

1. Anthracnose

Symptoms

Symptoms vary between species of yam. They are given below.

a. *Dioscorea alata*

- ✰ Initial infection is small, brown spots, with a yellow edge; appear on the lower leaves and stems.
- ✰ The spots enlarge, spread and give the stem a blackened appearance.
- ✰ Brown to black acervuli is observed on the leaves and stem in concentric rings.
- ✰ Stem is attacked leads to rapid death of the plant.

b. *Dioscorea bulbifera*

- ✰ The symptoms are similar to those on *D. alata*.
- ✰ However a greater number of leaves are attacked at the same time and necrosis is more rapid.
- ✰ On the stem infected zones are seen around the points of attachment of the petioles. The diseased stems dry up and the plants die.

c. *Dioscorea esculenta*

- ✰ In leaf, brown spots appear with yellow halo, the leaf droops and then falls off.
- ✰ The affected stem rapidly loses its chlorophyll, fades, dries and shrivels and then dies.

Dioscorea rotunda

Brown spots develop on the leaves. Stems are not attacked.

Fungus

Colletotrichum gloeosporioides f. sp. *alatae*.

Mycelium is septate, branched and light pink to brown. Acervuli are simple, globose and dark brown. Conidiophores are erect and hyaline. Setae are long, septate and dark in colour. Conidia are one celled, smooth, oval to oblong or cylindrical with one or two oil drops. They germinate by producing appressoria.

Mode of survival: As on crop debris as acervuli.

Spread: Spores are mainly dispersed by rain splash when there are heavy rains.

Favourable conditions: Heavy and well distributed rains.

Management

- ☆ Among different species of *Dioscorea, D. alata* is highly susceptible.
- ☆ Crop debris should be removed and burnt as it is a major source of inoculum.
- ☆ Srpay with mancozeb 0.2 per cent at weekly interval.
- ☆ Zineb or ferbam sprays at 10 days interval also increased the yield.
- ☆ Copper fungicides produce phytotoxic symptoms and they should not be used.

Fungus

Rhizoctonia solani Kuhn.

Mode of survival: As sclerotia in soil.

Primary spread: From soil-borne sclerotia.

Secondary spread: Through irrigation water.

Favourable conditions: Water-saturated soils.

Management

- ☆ Use well-drained soil mixes.
- ☆ Never store peat moss, sphagnum moss, chips, or potting media mixes directly on soil surfaces where they can be colonized by the fungus.
- ☆ Plants on raised benches to limit root contact with soil.

2. Leaf Spots

Symptoms

Symptoms vary between species of *Mycospharella*. They are given below.

Fungi

Cercospora carboanceae Miles, *C. golaghatii* Saikia and Sarbhoy, *C. dioscoreae* Ell. and Mart. and *Pseudocercospora contraria* (H. and P. Syd.) Deighton. *C. carbonacea* Miles

- ☆ *Cersopora* is most common and produces brown angular, spots appear on the leaves.

- They darken with age and coalesce as the disease progresses.
- Severely affected leaves dry and fall prematurely.

Cercospora golaghati Saikia and A. Sarbhoy produces brown to reddish brown discolouration on the upper surface of the leaves.

Pseudocercospora contraria (Syd. and P. Syd.) Deighton produces amphigenous leaf spots which are subcircular, light brown with hypophyllous fruit bodies.

Guignardia dioscoreae A.K. Pande appears as minute tan coloured spots on the leaves. These spots are dark tan in colour with a yellowish halo surrounded them.

Guignardia dioscoreae A.K. Pande previously known as *Phyllostricta dioscoreae* Brunaud and *Guignardia dioscoreae -bulbiferae*.

Primary spread: From spores ooze from the fruitbodies during wet weather and splashed by rain.

Secondary spread: Through air-borne conidia.

Management

- Collect and burn or bury the vines after harvest.
- Practice crop rotation.
- Spray copper products, chlorothalonil or mancozeb.

6. Viral Diseases

- *Yam mosaic virus* (YMV),
- *Dioscorea alata potyvirus* (DAV),
- *Dioscorea alata badnavirus* (DABV),
- *Dioscorea alata ring mottle virus* (DARMV),
- *Dioscorea trifida virus* (DTV),
- *Dioscorea dumeratum virus* (DDV),
- *Chinese yam necrotic mosaic virus* (CYNMV).

Management

- Dipping of seed pieces in carbofuran before planting.
- Use of virus-free planting materials and meristem culture has been recommended.

Postharvest Diseases of Vegetables

1. Alternaria Rot: *Alternaria tenuis*

Symptoms

- Brown to black lesion with definite margin.
- Lesion slightly sunken in advanced stage.
- Mould growth on the lesion.

2. Phoma Rot: *Phoma* sp.

Symptoms

- ☆ Slightly sunken brown spots at the edge of stem scar.
- ☆ Spots extend and become circular.
- ☆ In the center of the spot brown to black pycnidia form.

3. Phytophthora Rot: *Phytophthora infestans*

Symptoms

- ☆ Affected area less firm, water soaked.
- ☆ White fungal growth in advanced stage.

4. Bacterial Soft Rot: *Erwinia carotovora*

Symptoms

- ☆ Depressed water soaked spots.
- ☆ Spots enlarge become soft and watery.
- ☆ Emit foul odour.

Favourable conditions: High moisture, high temperature, harvest with wounds, latent infection and chilling injury.

Management

- ☆ Good agronomical practice.
- ☆ Good sanitation.
- ☆ Careful handling of vegetables.
- ☆ Harvest at correct stage.
- ☆ Pre-harvest spraying with 0.25 per cent Mancozeb before 10 days of harvest.

Chapter 3

Flower Crops

1. Anthurium (*Anthurium andraeanum* Linden ex André)

1. Root Rot

Symptoms

- ☆ Young, tender stems are girdled, become water soaked, and are unable to support the weight of the plant.
- ☆ It attacks the roots and lower stems of plants, but under wet conditions it spreads to upper leaf canopy.

Fungus

Rhizoctonia solani Kuhn.

Mode of survival: As sclerotia in soil.

Primary spread: From soil-borne sclerotia

Secondary spread: Through irrigation water.

Favourable conditions: Water-saturated soils.

Management

- ☆ Use well-drained soil mixes.
- ☆ Never store peat moss, sphagnum moss, chips, or potting media mixes directly on soil surfaces where they can be colonized by the fungus.
- ☆ Plants on raised benches to limit root contact with soil.

2. Black Rot and Leaf Blight

Symptoms

- ☆ Black spots with water soaked margins appear on leaves, petioles and spadices.
- ☆ On spathe, purple water soaked spots appear first and later turn black.
- ☆ Spots enlarge rapidly and cover the entire area.

Chromista

Phytophthora nicotianae var. *parasitica* (Dastur) Waterhouse and

P. citrophthora (R.E. Smith and E.H. Smith) Leonian.

Mode of survival: As oospores in soil.

Primary spread: From soil-borne oospores.

Secondary spread: Through zoospores.

Management

- Spray Fosetyl-Al 0.1 per cent or chlorothalonil 0.2 per cent.

3. Black Nose Disease/Anthracnose

Symptoms

- The pathogen is highly specific, only attacking the spadix (the nose) portion of the flower.
- Small, brown to black spots appear on the floral spadix.
- Later, they rapidly enlarge, become watery and turn brown to black and necrotic.
- In advanced cases, masses of orange coloured spores form on necrotic areas.
- The spadix may eventually fall off.

Fungus

Colletotrichum zingiberis (Sundar.) Butler and Bisby

[*Glomerella cingulata* (Stonem.) Spauld. and Schrenk.].

Mode of survival: In infected plant parts.

Primary spread: From diseased plant parts.

Secondary spread: Through splashing water, air movement and workers.

Favourable conditions: Humid and warm conditions; infection by bacterial blight pathogen (*Xanthomonas*).

Management

- Follow strict sanitation program.
- Foliar spray of mancozeb 0.2 per cent or carbendazim 0.1 per cent.

4. Bacterial Blight

Symptoms

- It is a contagious disease.
- Irregular water-soaked spots appear on the underside of leaves and spathes and faint chlorosis when viewed from the upper side.
- Later, the spots become necrotic, brown or black at the centre and bright yellow at the edges.
- The symptoms spread quickly and cover large sections of the leaf or spathe.
- If the infected leaves are not removed, the bacterium moves down the petiole, stem and will manifest itself in the systemic stage.

- ✰ The petioles of these leaves and systemically infected flowers break off readily from the stem.
- ✰ Leaves of systemically infected plants exhibit a bronzed appearance.
- ✰ Floral quality is reduced and flowers become unmarketable.
- ✰ The bacteria enter the vascular system systemically and cause vascular discolouration.
- ✰ In most cases, systemic infection leads to plant death.

Bacterium

Xanthomonas axonopodis pv. *dieffenbachiae* (McCulloch and Pirone) Vauterin.

Mode of survival: As In infected soil and planting materials as latent infection.

Primary spread: From infected planting materials and soil.

Secondary spread: Through rain splashes, splash irrigation, misting and contaminated tools.

Favourable conditions: Wet foliage, high rainfall coupled with high temperature and splash irrigation.

Management

- ✰ Use only disease-free planting materials.
- ✰ Follow good field sanitation.
- ✰ Install foot paths with disinfectant at entrances to green houses.
- ✰ Adjust plant density on beds to facilitate good aeration.
- ✰ Remove all leaves with foliar infection and burn.
- ✰ Avoid overhead or sprinkler or splash irrigation.
- ✰ Disinfect the harvesting implements, containers, tools and clothing materials.
- ✰ Provide good drainage.
- ✰ Keep the foliage dry.
- ✰ Spray affected beds thoroughly with streptomycin sulphate or oxytetracycline (Copper fungicides are phytotoxic).

5. Bacterial Wilt

Symptoms

- ✰ Yellowing (chlorosis) of outer leaves from the margin extends later to the entire leaf. Later it spreads to inner leaves also.
- ✰ Necrosis of leaf laminae, petioles, spathes and peduncles and eventually results in crown rot.
- ✰ As in any wilt, it shows systemic yellowing, wilting and death of plants even though adequate soil moisture is available.

- ☆ The disease spreads rapidly throughout the vascular system, turning veins and stems a brown, bronze colour.
- ☆ Root and collar region shows rotting.
- ☆ If the cuts are made in the stems of highly infected plants, bacterial ooze comes out.
- ☆ Black necrotic lesions with greasy margin appear in local infection of leaves leads to distortion.

Bacterium

Ralstonia solanacearum (Yubucchi *et al.*)

[Syn. *Pseudomonas solanacearum* (Smith), *Burkholderia solanacearum* (Smith)].

Mode of survival: As infected planting materials.

Primary spread: From diseased planting materials.

Secondary spread: Through irrigation water, cutting tools, contaminated soil, water or worker contact.

Favourable conditions: Hot weather conditions.

Management

- ☆ Use disease-free propagation materials.
- ☆ Follow proper and strict phytosanitary measures.
- ☆ Disinfect the cutting tools.

6. Bacterial Leaf Spot

Symptoms

- ☆ The disease appears as localized and systemic symptoms.
- ☆ In the localized symptoms, oily, angular lesions with chlorotic halo appear on lower leaf surface and spathes.
- ☆ These lesions become necrotic and cause deformity of leaves.
- ☆ Under high humid conditions white exudates ooze out from margins of the lesions.
- ☆ Infection spreads to veins leading to rotting and abscission of leaves and cut-flowers.

Bacterium

Acidovorax anthurii Gardan *et al.*

Bacterial characters: Gram negative, aerobic mesophyllic bacterium.

Mode of survival: As infected planting materials.

Primary spread: From contact of infected planting materials.

Secondary spread: Through rain splashes, cutting tools, contaminated soil, water or worker contact.

Favourable conditions: Hot weather conditions.

Management

- ☆ Use disease-free propagation materials.
- ☆ Follow proper field sanitation and regular disinfection of tools and clothes with 70 per cent alcohol or 10 per cent sodium hypochlorite.
- ☆ Remove and burn infected leaves, cut-flowers and plants.

2. Aster (*Aster amellus* L.)

1. *Fusarium* Wilt

Symptoms

- ☆ Young seedlings wilt and die rapidly.
- ☆ Two types symptoms are seen in older plants.
- ☆ In the first type, older plants are stunted with yellowing one side of the plant and vascular browning of stem is seen.
- ☆ In the second one, leaves become yellow, wilt and die and show vascular browning of affected stem.
- ☆ The basal stem portion becomes dark brown.

Fungus

Fusarium oxysporum f. sp. *callistephi* (Beach) W.C. Snyder and H.N. Hansen

Mode of survival: As chlamydospores and mycelia in soil and plant refuse.

Primary spread: From soil-borne inoculum.

Secondary spread: Through irrigation water, air blown rain splashes.

Favourable conditions: Full bloom stage of the crop.

Management

- ☆ Use fungicide treated potting soil to eliminate pathogen.
- ☆ Use soilless potting mixture.
- ☆ Grow resistant varieties.

2. Foot Rot

Symptoms

- ☆ Infected plants show rapid wilting of leaves with brownish-black discolouration seen on the lower portion of the stem.
- ☆ The stem shrivels and collapses at or near the soil line.

- The roots become soft with water-soaked decay symptoms.
- Adventitious roots are formed close to the diseased portion of the roots.
- Diseased plants can be pulled out easily from the soil, leaving the outer barks of the roots in the soil.

Fungus

Phytophthora cryptogea Pethybr. and Laff.

Mode of survival: As soil-borne.

Primary spread: From soil-borne spores.

Secondary spread: Through irrigation water and rain.

Favourable conditions: Poor drainage in the field.

Management

- Plant in well-drained soil.

3. Root and Stem Rot

Symptoms

- The pathogen affects roots, stems and lower leaves.
- Well-defined water soaked areas are seen on lower leaves that are in contact with the soil.
- Affected leaves wilt and turn dark brown.
- Water-soaked appearance is also seen on stem and roots. It becomes sunken and turns dark reddish-brown.

Fungus

Rhizoctonia solani Kuhn

Mode of survival: As soil-borne sclerotia.

Primary spread: From sclerotia in soil and plant materials.

Secondary spread: Through irrigation water.

Favourable conditions: Poor drainage/water stagnation.

Management

- Plant only in sterilized beds.
- Provide proper drainage.
- Drench the soil around the infected plants with carbendazim.

4. Leaf Spot

Symptoms

- Spots form on leaves.

- ☆ Leaves become yellow, wilt, and dry.
- ☆ Usually lower leaves are affected first and symptoms progress upward.
- ☆ Floral parts and stem are also affected.

Fungi

Alternaria sp., *Ascochyta* sp., *Cercospora* sp., *Septoria* sp.

Mode of survival: As diseased plant materials.

Primary spread: From diseased seeds and seedlings.

Secondary spread: Through air-borne spores.

Favourable conditions: Poorly drained soil and overhead irrigation.

Management

- ☆ Avoid sprinkler irrigation.
- ☆ Provide proper drainage.
- ☆ Spray mancozeb, iprodione, myclobutanil, copper hydroxide, thiophanate methyl + chlorothalonil.

5. Powdery Mildew

Symptoms

- ☆ White powdery growth develops on upper leaf surface.
- ☆ Affected young leaves become twisted and misshappened.
- ☆ Older leaves are little distorted.
- ☆ Young growing portions show complete white powdery growth with dwarfing and curling of leaves, stem and buds.
- ☆ Leaves wither and die.
- ☆ Flower buds fail to open.

Fungus

Golovinomyces cichoracearum DC (Syn. *Erysiphe cichoracearum* DC)

Mode of survival: In diseased plants as cleistothecia.

Primary spread: From ascospores in infected plant materials and diseased seedlings.

Secondary spread: Through air-borne conidia.

Favourable conditions: Warm and dry conditions.

Management

- ☆ Spray wettable sulphur 0.25 per cent or mancozeb 0.2 per cent or triadimefon 0.2 per cent.

6. Rust

Symptoms

- Reddish-orange masses of spores are seen on the underside of leaves.
- Severely affected leaves turn yellow and die.
- On the alternate host, pines, blisters of spores form on needles in the spring.

Fungus

Coleosporium solidaginis (Schwein.) Thüm.

7. *Botrytis* Blight

Symptoms

- Brown specks form on petals.
- The entire flower browns and dies.

Fungus

Botrytis cinerea Pers. Ex Fr.

Management

- Avoid sprinkler irrigation.
- Spray mancozeb, maneb or copper hydroxide or chlorothalonil.

8. Aster Yellows

Symptoms

- Leaves become yellow and the branching occurs abnormally.
- Yellow, leaf-like structures instead of flower parts.
- Reduced root system with stunting of plants.

Bacterium

Candidatus Phytoplasma

Mode of survival: As weed hosts.

Primary spread: From weed hosts around the cropping area.

Secondary spread: Through insect vector, leaf hopper (*Macrosteles quadrilineatus*)

Management

- Destroy infected plants and weeds in the vicinity.
- Spray insecticides to control leafhopper.

3. Carnation (*Dianthus caryophyllus* L.)

1. *Fusarium* Wilt

Symptoms

- In young plants, the leaves turn grey with wilting of leaves and young stems followed by collapse of the whole plant.
- When older plants are infected, older leaves show chlorosis followed by purple-red discolouration and the plants die and become straw coloured.
- The vascular tissues of stems turn dark brown.

Fungus

Fusarium oxysporum f. sp. *dianthi* W.C. Snyder and H.N. Hansen.

Mode of survival: As chlamydospores in soil.

Primary spread: From infected crop refuse in soil.

Secondary spread: Through irrigation water.

Favourable conditions: Humid and warm weather.

Management

- Soil fumigation with methyl bromide.
- Adopt soil solarization using clear transparent polyethylene film (0.1 mm thick) for 30 days.
- Plant disease-free materials.
- Adopt shallow planting which helps the plant to escape from the disease.
- Remove and destroy the diseased plants along with roots.
- Drench the soil with carbendazim 0.1 per cent or Bordeaux mixture 1 per cent.

2. *Phytophthora* Blight

Symptoms

- Infections on the basal portion of the stem cause a typical wilting and collapse of the stem or plant which resembles bromide toxicity symptoms.
- The fungus invades the stem at any point, causing stem collapse and eventual death of the branch.
- If the infection is on the upper leaves and stem, blighting occurs.

Chromista

Phytophthora parasitica Dastur

Management

- Soil drench with Bordeaux mixture 1.0 per cent

3. Basal Rot

Symptoms

- ☆ The stem rots at the soil level which extends to the leaves.
- ☆ Cottony growth of mycelium seen on infected plant materials.
- ☆ Dark brown spherical sclerotia are formed along with the mycelial mass.

Fungus

Sclerotium rolfsii Sacc. [Sexual stage: *Athelia rolfsii* (Curzi)].

It is a soil living saprophytic pathogen, produces ribbon like hyphae with clamp connections. Basidia are club-shaped, bearing four smooth, ellipsoid basidiospores. Small, brownish sclerotia (hyphal propagules) are also formed.

Management

- ☆ Remove old leaves which touch the soil.
- ☆ Regulate soil moisture.
- ☆ Avoid excess N application.
- ☆ Soil drench with carbendazim 0.1 per cent.

4. Root Rot/Stem Rot

Symptoms

- ☆ The fungus attacks the plants at the soil level. Stem shows lesions with a brown border and these affected stems are dry and shredded.
- ☆ Sudden wilting and rotting of roots of infected plants.
- ☆ The diseased plants die within a week.

Fungus

Rhizoctonia solani Kuhn.

Management

- ☆ Provide proper drainage facilities.
- ☆ Follow shallow planting of cuttings.
- ☆ Treat the cuttings with captan 0.2 per cent or carbendazim 0.1 per cent for 30 min before planting.
- ☆ Plant the cuttings in pasteurized, raised beds.
- ☆ Drench the soil with carbendazim 0.1 per cent or thiophanate-methyl 0.1 per cent.
- ☆ Apply *Trichoderma* in the soil.

5. *Alternaria* Leaf Spot

Symptoms

- ☆ The leaf bases and nodes blight and are girdled.
- ☆ The spots on leaves are ashy white or purplish and covered with dark brown fungal growth in the centre.
- ☆ Leaves shrink, twist and the leaf tips become necrotic.
- ☆ Branches show die back and cankers at the girdled area with black mass of conidia.
- ☆ Several lesions expand and coalesce to form large, irregular necrotic areas killing the entire leaf.

Fungus

Alternaria dianthi F. Stevens and J.G. Hall.

Mode of survival: In infected plants.

Primary spread: From infected cuttings.

Secondary spread: Through air-borne conidia and rain

Management

- ☆ Provide proper air circulation.
- ☆ Use disease-free planting materials.
- ☆ Disinfect the planting materials with formalin 5.0 per cent.
- ☆ Spray mancozeb or zineb or captan or chlorothalonil or copper hydroxide or iprodione 0.2 per cent or propiconazole 0.1 per cent.

6. *Alternaria* Flower Blight

Symptoms

- ☆ Tan to dark brown lesions appear on sepals and petals.
- ☆ These lesions are covered with dark brown powdery mass of spores.

Fungus

Alternaria dianthicola Neerg.

Management

As in *Alternaria* leaf spot.

7. Ring Spot

Symptoms

- ☆ Round to oval, sunken, greyish-white spots with dark coloured margin appear on leaves and stem.

- Infected leaves and shoots wither.
- Dark fruiting bodies of the fungus develop on the spots as concentric rings.
- High reduction in flower production.

Fungus

Didymellina dianthi C.C. Burt (Syn. *Heterosporium echinulatum* Berk.)

Mode of survival: In infected plants.

Primary spread: From infected cuttings.

Secondary spread: Through air-borne spores.

Management

- Spray Bordeaux mixture 1 per cent or copper oxychloride 0.25 per cent.

8. Ring Spot/Leaf Spot

Symptoms

- Circular, greyish spots with distinct reddish-purple margins develop on the leaves, stems and on sepals.
- The spots increase in size, coalesce and such leaves shrivel.

Fungus

Mycosphaerella dianthi (C.C. Burt) Jorst.

Management

- Spray mancozeb 0.2 per cent or zineb 0.2 per cent.

9. Rust

Symptoms

- Small, slightly raised blisters appear on leaves, stems, or flower buds which rupture and release powdery chocolate-brown spores.
- A yellow margin surrounds each pustule and in severe cases the entire leaves turn yellow, curl and die.
- Stems are girdled.
- Several pustules on the shoot, results in decreased flower production and quality.

Fungi

Uromyces caryophyllinus (Schrank) G. Winter, *U. dianthii* (Pers.) Niessl.

Mode of survival: In diseased plants.

Primary spread: From infected cuttings.

Secondary spread: Through air-borne uredospores and splashing water.

Favourable conditions: Cool nights alternating with warm humid days.

Management

- ☆ Collect and burn diseased plant parts.
- ☆ Spray wettable sulphur or chlorothalonil 0.2 per cent or plantvax or propiconazole 0.2 per cent or thiophanate methyl 0.2 per cent or triadimefon at 0.1 per cent.

10. *Septoria* Leaf Spot

Symptoms

- ☆ Spots on leaves or stem are scattered, more or less circular, light brown with a purplish brown border.
- ☆ These spots coalesce and cover larger leaf area.
- ☆ Minute black fruiting bodies of the fungus are seen on the spots.

Fungus

Septoria dianthi (Alb. and Schwein.) Desm.

Management

- ☆ Collect and burn diseased leaves.
- ☆ Spray carbendazim 0.1 per cent or zineb 0.2 per cent or thiram 0.2 per cent or copper oxychloride 0.25 per cent.

11. *Botrytis* Flower Blight

Symptoms

- ☆ The disease appears as tan to light brown soft rots on the petals of flower buds and opened flowers.
- ☆ Affected petals are covered with the greyish growth of the fungus with powdery mass of spores.

Fungus

Botrytis cinerea Pers. Ex Fr.

Management

- ☆ Space plants to ensure good air circulation.
- ☆ Maintain low humidity.
- ☆ Avoid watering late in the day.
- ☆ Remove crop debris.
- ☆ Spray carbendazim or triademefon 0.1 per cent or mancozeb 0.2 per cent.

- ☆ Apply chlorothalonil, copper hydroxide, iprodione, mancozeb or fludioxonil.

12. Bacterial Wilt

Symptoms

- ☆ Elongated discoloured stripes appear on the stem and the stripes show splitting of tissues.
- ☆ Cracks develop in internode tissue. Slimy ooze comes from these cracks when humidity is high.
- ☆ Vascular tissues turn brown.
- ☆ The roots rot partially and the cortical tissues become sticky and discoloured.
- ☆ The leaves are twisted, turn gray-green, then yellow and die.
- ☆ The bacteria invade xylem vessels and spread to infect the young shoots and leaves.

Bacterium

Burkholderia caryophylli (Burkholder) Yabuuchi

[Syn. *Pseudomonas caryophylli* (Burkholder) Yabuuchi]

Mode of survival: In diseased plants.

Primary spread: From infected cuttings.

Secondary spread: Through rain splashes and overhead watering.

Management

- ☆ Use disease free cuttings for planting
- ☆ Raise seedlings in pasteurized beds
- ☆ Collect and burn diseased plant debris
- ☆ Avoid overhead watering and splash watering
- ☆ Use clean and disinfested tools.

13. Bacterial Slow Wilt or Stunt

Symptoms

- ☆ Wilting, twisting, curling and stunting of lower leaves and side shoots followed by death of lower leaves.
- ☆ Wilted plants may recover only to wilt again.
- ☆ Stunted plants have narrow, yellow-green leaves.

Bacterium

Erwinia chrysanthemi Samson

Management

- ☆ Grow seedlings in raised pasteurized beds.
- ☆ Use pathogen free cuttings.
- ☆ Destroy infected plants.

4. China Aster [*Callistephus chinensis* (L.) Nees.]

1. Damping Off

Symptoms

- ☆ The fungus causes damping off disease in nursery and root rot disease in grown up plants.
- ☆ In the nursery, pre-emergence death occurs before the emergence of seedlings from the soil. After the emergence, patches of dead seedlings are seen.
- ☆ In grown up plants, brown or black lesions appear at collar region near the soil line.
- ☆ Yellowing of lower leaves followed by wilting and drying of leaves are seen.
- ☆ Diseased plants show shredding of bark and death of plants.

Fungus

Rhizoctonia solani Kuhn.

Mode of survival: As sclerotia in infected plant tissues in soil.

Primary spread: From soil borne inoculum.

Secondary spread: Through irrigation water.

Management

- ☆ Remove and destroy infected plants in the field.
- ☆ Apply *Pseudomonas fluorescens* at 2 g/kg of soil + 0.1 per cent thiophanate methyl.

2. Wilt

Symptoms

- ☆ The disease attacks the plant at all stages. But it is more severe from flowering phase.
- ☆ The pathogen causes damping-off of young seedlings at soil temperatures of 70-80°F.
- ☆ Diseased plants wilt, wither and die.
- ☆ Older plants are stunted.

- ☆ Lower leaves wilt; stem blackens and shows mass of pink spores at ground level.
- ☆ Plants show yellowing and wilt, often on one side.
- ☆ Brown or black discolouration of vascular system develops. The black colour of the basal stem extends into the pith.

Fungus

Fusarium oxysporum f. sp. *callistephi* (Beach) W.C. Snyder and H.N. Hansen.

Mode of survival: As seed-borne and in soil for many years.

Primary spread: From seed and soil-borne inoculum.

Secondary spread: Through irrigation water.

Favourable conditions: High soil and air temperature, less soil moisture, continuous cultivation of china aster.

Management

- ☆ Use sterilized soil for raising nursery bed.
- ☆ Obtain seeds from healthy plants.
- ☆ Avoid planting in wilt affected soils.
- ☆ Follow crop rotation.
- ☆ Treat the seeds in 0.1 per cent mercuric chloride solution or carbendazim 2 g/kg.
- ☆ Collect and destroy infected plants.
- ☆ Spot drench with carbendazim 0.1 per cent.

3. Collar Rot/Root Rot

Symptoms

- ☆ Stems and roots of affected plants appear water soaked and black.
- ☆ Infected plants turn pale and later wilt.

Fungus

Phytophthora cryptogea Pethybr. and Laff.

Management

- ☆ Avoid monocropping.
- ☆ Collect and burn diseased plants.
- ☆ Drench the soil with copper oxychloride 0.25 per cent.

4. Root Rot

Symptoms

- ☆ Plants wilt and suddenly collapse.

- ☆ Black discolouration of leaves, stems and roots occur.
- ☆ The roots of diseased plants decay.
- ☆ Damping-off of seedlings also occur.

Chromista

Pythium sp. and *Phytophthora* sp.

Mode of survival: As oospores in soil.

Primary spread: From soil-borne inoculum.

Secondary spread: Through irrigation water.

Favourable conditions: Heavy soil, water logged conditions.

Management

- ☆ Avoid planting in poorly drained soils.
- ☆ Plant on raised beds.
- ☆ Do not over irrigate the field.
- ☆ Drench the seedlings with copper oxychloride0.25 per cent or Fosetyl-Al or metalaxyl 0.1 per cent.

5. *Sclerotinia* Rot/Cottony Rot

Symptoms

- ☆ Infection girdles the stems.
- ☆ Cottony white fungal growth and large sized black sclerotial bodies develop on and inside the stem.
- ☆ Stems look like bleached-white colour.

Fungus

Sclerotinia sclerotiorum (Lib.) de Bary

Mode of survival: As sclerotia in soil.

Primary spread: From soil-borne sclerotia.

Secondary spread: Through air-borne spores.

Favourable conditions: Wet weather.

Management

- ☆ Remove and destroy infected plants.
- ☆ Avoid overhead irrigation.
- ☆ Treat the planting area with carbendazim 0.1 per cent.
- ☆ Spray thiophanate-methyl 0.1 per cent before rainy periods.

6. Stem Canker and Flower Blight

Symptoms

- ☆ Usually the plants at maturity stage are seriously affected.
- ☆ Depressed dark brown lesions appear on the stem which girdle it resulting in death of plants.
- ☆ The disease attacks ray florets and produce dark brown irregular spots. Later these spots turn grey.
- ☆ Brown to grey spots appear on the peduncles resulting in drying.
- ☆ Numerous pycnidia are observed on the diseased and dead tissues.

Fungus

Phomopsis callistephi Tehon and E. Y. Daniels

Favourable conditions: Humid weather

Management

- ☆ Use disease-free seeds for sowing.
- ☆ Collect and destroy diseased plant debris.
- ☆ Avoid injuries to plants during cultural operations.

7. Leaf Spots

Symptoms

***Stemphylium* leaf spot**: Brown, circular, concentrically zonate spots with dark margins appear on leaves, bracts, petals and stems. The disease adversely affects the quality of flowers.

***Ascochyta* and *Septoria* leaf spots**: At first lower leaves are affected. The spots are yellowish at first and later become dark brown and black. These spots increase in size, coalesce and cause blighting of leaves.

***Alternaria* leaf spots**: Circular to irregular, pale brown necrotic spots with dirty yellow margins and concentric zonations appear on leaves. Water soaked irregular brown lesions are also seen on petals. The fungus also infects chrysanthemum, *Dahlia* sp., *Gerbera* sp., *Nasturtim* sp., sunflower and *Zinnia* sp. The fungus, *Alternaria tenuissima* is seed-borne.

Fungi

Stemphylium callistephi K.F. Baker and L.H. Davis, *Ascochyta asteris* (Bres.) Gloyer., *Septoria callistephi* Gloyer, *Alternaria alternata* (Fr.) Keissl. (syn. *A. tenuis*) and *Alternaria tenuissima* (Nees) Wiltshire.

Management

- ☆ Avoid low lying areas where air movement is poor.

- ☆ Do not use overhead irrigation.
- ☆ Protect foliage with fungicidal spray of mancozeb or maneb or zineb 0.2 per cent.

8. Rust

Symptoms

- ☆ It is a serious disease of china aster.
- ☆ Bright yellow orange pustules appear on the under surface of the leaves of young plants.

Fungus

Coleosporium solidaginis (Schwein.) Thüm. and *C. asterum* (Dietel) Syd. and P. Syd.

Mode of survival: As infected plants.

Primary spread: From diseased plants and from alternate host (three-needle pines).

Secondary spread: Through air-borne spores.

Favourable conditions: Free moisture from rain, dew or fog.

Management

- ☆ Grow seedlings away from main crop.
- ☆ Avoid overhead irrigation.
- ☆ Spray wettable sulphur 0.2 per cent.

9. Grey Mould

Symptoms

- ☆ The disease appears 3-4 days after sowing or 10-12 days after prickling out the seedlings.
- ☆ Brown, water soaked decay appear on flowers.
- ☆ Wooly grey fungus spores form on rotted tissues.
- ☆ It causes blighting of flower buds and blossoms.
- ☆ The affected flower buds fail to open and rot.
- ☆ Leaves and shoots discolour, wilt, decay and drop.
- ☆ Twigs show die back symptoms.

Fungus

Botrytis cinerea Pers.

Mode of survival: As sclerotia and mycelia in dead plant materials.

Primary spread: From infected plant tissues.

Secondary spread: Through air-borne or rain splashed conidia.

Favourable conditions: Cloud and high humid weather.

Management

- ☆ Use well-drained soil for sowing.
- ☆ Remove and dispose fallen leaves and debris around the plants.
- ☆ Prune out diseased and dead tissues.
- ☆ Avoid overhead watering.
- ☆ Thin the plant canopy to improve air circulation and avoid overcrowding of plants.
- ☆ Provide proper cultural care.
- ☆ Spray the plants with zineb or maneb or mancozeb 0.2 per cent or iprodione 0.1 per cent.

10. Mosaic

Symptoms

- ☆ Extreme distortion of blooms and dwarfing is seen.

Virus

Chrysanthemum mosaic virus. It is a seed-borne virus.

Management

- ☆ Use disease free seeds.
- ☆ Remove and destroy infected plants.

11. Aster Yellows

Symptoms

Diseased plants become dwarf and blossoms are off-colour.

The ray florets are usually yellowish green.

Flower parts develop into leaf-like structures and are yellow or green and dwarfed.

Clusters or tufts of spindly, yellow upright shoots develop around the base of infected plants, often on one side of the plant.

Corms or tubers mature early and are undersized with stunted roots.

Bacterium

Candidatus Phytoplasma

Mode of survival: As infected plants.

Primary spread: From diseased plants and weeds.

Secondary spread: Through leaf hopper vector, *Macrosteles fascifrons.*

Management

- ☆ Locate seed bed away from weedy areas.
- ☆ Plant pathogen-free plants.
- ☆ Rouge out diseased plants.
- ☆ Follow good sanitary measures in the field.
- ☆ Spray the crop with systemic insecticides to control leaf hopper.
- ☆ Root dip or spray with oxytetracycline or tetracycline at 100 ppm to delay the development of disease symptom.

5. Chrysanthemum (*Chrysanthemum indicum* L.)

1. Blotch/Leaf Spot

Symptoms

- ☆ Infection first appears on lower leaves.
- ☆ Blackish-brown, circular to irregular spots on leaves with yellow halo around each spot.
- ☆ Spots coalesce and form large patches. In severe infection, the leaves become small, curl and wither prematurely.
- ☆ Dead leaves hang on the stem.
- ☆ Flower bud infection causes rotting.

Fungus

Septoria chrysanthemella Sacc., *Septoria chrysanthemi* Allesch

Pycnidia are brown to dark brown, globose to pyriform. Conidia are hyaline, smooth, ampulliform. Spermatia are hyaline, smooth, ellipsoid.

Mode of survival: As infected plant debris.

Primary spread: From infected plant debris.

Secondary spread: Through air-borne conidia and by rain splashes.

Favourable conditions: Cool-wet periods of rainy season.

Management

- ☆ Destroy diseased plant debris.
- ☆ Avoid excessive irrigation.
- ☆ Spray carbendazim 0.05 per cent or captafol 0.2 per cent six times at 15 days intervals from the end of July with sticking agent.

2. Leaf Spot

Symptoms

- ✰ Dark brown spots with yellow margin appear on the leaves causing premature withering of leaves.
- ✰ Flower bud infection causes rotting.

Fungus

Cylindrosporium chrysanthemi Ellis and Dearn

Produce white masses of conidia grouped in concentric circles (acervuli).

Mode of survival: As Infected plant debris and weeds

Primary spread: From infected plant debris

Secondary spread: Through air-borne conidia and by rain splashes

Favourable conditions: Cool-wet periods of rainy season

Management

- ✰ Destroy diseased plant debris.
- ✰ Avoid excessive irrigation.
- ✰ Spray carbendazim 0.05 per cent or captafol 0.2 per cent six times at 15 days intervals from the end of July with sticking agent.

3. Wilt

Symptoms

- ✰ The affected plant shows yellowing and browning of leaves.
- ✰ Leaves die from the base of the plant upward.
- ✰ Plants are stunted which often fail to produce flowers.
- ✰ The basal part of the stem and/or roots rots.

Fungus

Fusarium oxysporum f. sp. *chrysanthemi* Littrell, G.M. Armstr. and J.K. Armstr.

It has white to pink colored septate mycelium bearing micro-conidia, macro-conidia and chlamydospores. Chlamydospores are globose, aseptate, intercallary and individual or in chains of two.

Mode of survival: As chlamydospores and mycelia in the infected plant debris.

Primary spread: From infected plant debris.

Secondary spread: Through irrigation water and contaminated tools

Management

- ✰ Follow rotation with non-host crops.

- Rogue out diseased plants with roots.
- Dip rooted cuttings in a solution of thiram 0.15 per cent before planting.
- Plant only disease-free planting material in pasteurized soil.
- Drench soil with carbendazim 0.1 per cent or thiophanate methyl 0.2 per cent.
- Use nitrate rather than ammonium forms of fertilizer.

4. Brown Rust

Symptoms

- Brown powdery blister-like swellings or pustules seen on the undersides of leaves.
- Blisters mature and burst open to release masses of brown, powdery spores.
- Infected plants become weak and fail to bloom properly and dry up.

Fungus

Puccinia chrysanthemi Roze

It is an autocious, brachy-form rust with uredinial and telial states.

Mode of survival: In infected plant debris and weed hosts.

Primary spread: From infected plant debris and weed hosts.

Secondary spread: Through air-borne uredospores.

Management

- Collect and remove infected leaves/plants.
- Spray dinocap 0.025 per cent or wettable sulphur 0.2 per cent.

5. White Rust

Symptoms

- It is an autoecious rust.
- Pale-green to yellow spots up to 4 mm develop on the upper leaf surface with brown and necrotic centre with ageing.
- On corresponding lower surface, raised, buff or pinkish, waxy pustules (telia) are found.
- Later the spots on the upper surface become sunken and brown, the pustules on the lower surface become prominent and turn whitish when basidiospores are produced.
- Telia can also be found on the upper leaf surface.
- Pustules are more common on young leaves.

- Severely attacked leaves wilt, hang down the stem and gradually dry up completely. Sori may also appear on bracts and stems during severe infection.
- Flower infection shows necrotic flecking with occasional pustules.

Fungus

Puccinia horiana Henn.

The bicellular teliospores germinate *in situ* to produce unicellular basidiospores which are dispersed in air currents. No other spores are known.

Mode of survival: In *Chrysanthemum* species.

Primary spread: From infected cuttings and plants with latent infection (including cut flowers) of glasshouse chrysanthemums and contaminated soil.

Secondary spread: Through air-borne basidiospores, water splashes.

Favourable conditions: High humidity and a film of water on leaf surface.

Management

- Follow strict quarantine measures.
- Select and plant disease free planting materials.
- Avoid foliage wetting.
- Remove infected plants and debris from greenhouse between crops.
- Sterilize hands before field operation.
- Spray propiconazole or hexaconazole or tebuconazole + trifloxystrobin or oxycarboxin or triadimefon.

6. Stem Rot

Symptoms

- Symptoms like leaf chlorosis, necrosis, decay and discoloration of internal tissues of stem are seen.
- At flower opening period, small dark streaks are seen at the base of the stem.
- Roots decay in advanced stages.
- Stem and roots may rot on one side leaving other side healthy.

Fungus

Fusarium solani (Mart.) Sacc. [*Nectria haematococca* Berk. and Broome]

Mode of survival: As chlamydspores and mycelia in infected plant debris.

Primary spread: From infected plant debris.

Secondary spread: Through irrigation water.

Favourable conditions: Rainy weather.

Management

- ✰ Follow rotation with non-host crops.
- ✰ Plant cuttings from healthy plants.
- ✰ Rogue out diseased plants.
- ✰ Avoid excessive irrigation.
- ✰ Spot drench with carbendazim 0.1 per cent or Bordeaux mixture 1 per cent.

7. Powdery Mildew

Symptoms

- ✰ White powdery coating on the upper leaf surface.
- ✰ Leaves turn yellow and dry out.
- ✰ Infected plants remains stunted and fail to flower.

Fungi

Golovinomyces chrysanthemi (Rabenh.) (Syn. *Oidium chrysanthemi* Rabenh) and

Golovinomyces cichoracearum DC (Syn. *Erysiphe cichoracearum* DC)

Mode of survival: As cleistothecia in plant debris.

Primary spread: From infected plant debris.

Secondary spread: Air-borne conidia.

Favourable conditions: Dry weather with high humidity and over-crowding.

Management

- ✰ Good ventilation and proper spacing.
- ✰ Remove and destroy diseased leaves.
- ✰ Spray wettable sulphur 0.3 per cent or captan 0.2 per cent or bayleton or propiconazole or triademefon or thiophanate methyl.

8. *Rhizoctonia* Stem Rot

Symptoms

- ✰ Young infected plants wilt during the day and recover at night.
- ✰ Reddish-brown dead areas develop at the soil line and girdle the plant.

Fungus

Rhizoctonia solani Kuhn [Sexual stage*: Thanatephorus cucumeris* (Frank) Donk]

Rhizoctonia fungi do not produce conidia (asexual spores) and only rarely produce basidiospores.

Mode of survival: As sclerotia in soil

Primary spread: From soil borne sclerotia and infected plant materials.

Secondary spread: Through irrigation water.

Favourable conditions: Warm, moist condition.

Management

- Plant in pasteurized soil which is free of pathogen.
- Apply thiophanate methyl + granular etridiazole or thiram as soil drench.

9. *Botrytis* Blight/Grey Mould

Symptoms

- The disease appears at any stage from rooting bed to flowers in transit.
- Stem infection leads to death of upper part of the plant.
- Leaf infection starts as brown water soaked spots which may enlarge, coalesce causes blighting of leaves.
- Infected tissues become covered with grey, fuzzy mould.
- Light brown spots form on lower petals and the browning spreads to other petals.
- Affected flower buds do not open.

Fungus

Botrytis cinerea Pers. Ex Fr.

Botrytis species produce sclerotia, which are melanized resting bodies. Which produce prolific grey conidiophores and (macro) conidia.

Mode of survival: As sclerotia and mycelia in dead plant materials.

Primary spread: From infected plant tissues.

Secondary spread: Through air-borne or rain splashed conidia.

Favourable conditions: Cool temperature, high relative humidity and rainy weather.

Management

- Adjust the spacing to reduce the incidence.
- Reduce humidity below 98 per cent in green house.
- Apply chlorothalonil or mancozeb or captan 0.2 per cent or thiram 0.3 per cent as foliar spray.

10. Ray (Flower) Blight/Black Rot

Symptoms

Cuttings and flowers are highly susceptible.

Cuttings

- Cuttings are usually attacked at terminal bud and it spreads downwards to whole plant. Unopened buds, bracts and stem tissues become dark.
- Leaves show irregular, brownish black blotches of 2-3 cm dia which rapidly coalesce causing leaf rots.
- Symptoms on stem are seen at the point of attachment of the leaves or at cutting base.

Adult Plants

- Stem lesions are found at the base or node and cause girdling.
- The terminal shoot droops, the leaves become small, chlorotic and brittle and the plant is stunted.

Flowers

- Red or brown spots develop on one-side of the blossom only.
- Infection spreads rapidly causing rotting of entire flower head. The infected florets stick together.
- The fungus spreads through peduncle and causes blackening and weakening of tissues, drooping and wilting of head.

Fungus

Ascochyta chrysanthemi F. Stevens (Syn. *Mycosphaerella* ligulicola K.F. Baker, Dimock and L.H. Davis)

Mode of survival: In infected plants, soil and weed hosts.

Primary spread: From infected plants, flowers, soil adjoining roots and weed hosts.

Secondary spread: Through air-blown rain drops.

Favourable conditions: High humid weather.

Management

- Dip the cuttings in carbendazim (0.05 per cent).
- Spray ziram 0.2 per cent or carbendazim 0.1 per cent or mancozeb (0.2 per cent).

11. *Alternaria* Leaf Spot and Blossom Blight

Symptoms

Leaf Spot

- Small, dark brown spots develop with yellow halos and they coalesce to form large blotches resulting in withering and defoliation of leaves.

Blossom Blight

- This phase causes more damage than the leaf spot phase.
- On young buds, small brown to black spots appear. These buds shrivel, turn brown and drop prematurely.
- The bud blast symptom develops rapidly on moist warm weather.
- On open flower, small, brown to black spots appear on ray florets.
- These flowers are deformed and shriveled with reduced number of ray florets.

Fungi

Alternaria alternata (Fr.) Keissl. and *A. chrysanthemi E.G.* Simmons and Crosier

It has elongated dark brown color conidiophores. In favourable conditions conidia are produced (growing as buds from the conidiophores) from the conidiophores asexually.

Mode of survival: In infected plants and collateral hosts like *Centuria cyanus* and *Bougainvillea spectabilis.*

Primary spread: From infected plants and collateral hosts.

Secondary spread: Through air-borne conidia.

Favourable conditions: Moist warm weather.

Management

- Use disease free cuttings.
- Spray thiram or captan or mancozeb 0.2 per cent at 10 days interval.

12. Bacterial Blight

Symptoms

- Water-soaked lesions on the stem causes splitting.
- The upper part of stem rots.
- The leaves show margin scorching.
- Infected plants wilt and collapse.

Bacterium

Erwinia chrysanthemi pv. *chrysanthemi* (Burkholder *et al.*) Dye

Mode of survival: In infected plants, soil and weed hosts.

Primary spread: From infected plant parts and weed hosts.

Secondary spread: Through air-blown rain drops.

Favourable conditions: High temperature and high humidity.

Management

- ☆ Disinfect propagation beds between crops.
- ☆ Plant disease free cuttings.
- ☆ Dip the cuttings in streptomycin solution for 4 hour before planting.
- ☆ Remove and burn diseased plants.

13. Aspermy

Symptoms

- ☆ Mild mosaic mottling with marked distortion of the flowers.
- ☆ Flowers are small in size, distorted with irregular curlings and waviness of the ray florets.
- ☆ The central disc florets remain green and stunted.
- ☆ Line pattern with diffuse chlorotic mottling appear on new leaves.

Virus

Chrysanthemum aspermy virus (ChAV)

Mode of survival: In infected plants.

Primary spread: From infected cuttings.

Secondary spread: Through insect vectors, aphids (*Aphis gossypii, A. craccivora*).

Management

- ☆ Give heat therapy to plants at 37.2°C for 4 weeks and produce virus free seedlings by meristem culture.

14. Stunt

Symptoms

- ☆ Plants dwarfed with smaller and crinkled leaves.
- ☆ Flowers are fewer, smaller, appear bleached and inferior in quality.
- ☆ Flowers from diseased plants open 7-10 days earlier than flowers from healthy plant.
- ☆ Axillary buds often grow prematurely and produce excess number of branches and stolons.
- ☆ Stems become very weak and break easily.
- ☆ Cuttings from diseased plants root poorly.

Viriod

Chrysanthemum stunt viroid (ChSVd)

Mode of survival: In infected plants.

Primary spread: From infected seedlings and cuttings.

Secondary spread: Through sap, dodder and grafting.

Management

- ☆ Purchase and use viroid-indexed plants that are free of the pathogen.
- ☆ Destroy infected plants and disinfest tools used to handle them.
- ☆ Do not handle healthy chrysanthemums after handling infected plants.
- ☆ Follow strict phytosanitary measures.
- ☆ Use meristem tip culture combined with heat therapy.

6. Crossandra [*Crossandra infundibuliformis* (L.) Nees]

1. Wilt

Symptoms

- ☆ Disease occurs at random in the field one month after transplanting.
- ☆ Leaves turn pale green or yellow and later into pink colour.
- ☆ Drooping of leaves, wilting and death of plants.
- ☆ The stem shrivels.
- ☆ Dark lesions on roots extend up to collar region resulting in sloughing off cortical tissues.

Fungus

Fusarium solani (Mart.) Sacc. [*Nectria haematococca* Berk. and Broome]

Mode of survival: As chlamydospores and mycelia in soil.

Primary spread: From soil-borne inoculum.

Secondary spread: Through irrigation water.

Favourable conditions: Root lesion nematode (*Pratylenchus delattrei*) in the soil enhances wilt incidence.

Management

- ☆ Remove and destroy the infected plants.
- ☆ Application of phorate at 1 g/plant 10 DAT to control nematode and reduce disease incidence.
- ☆ Soil drenching with carbendazim 0.1 per cent or copper oxychloride 0.25 per cent at 25 days interval from 30 days after planting.

2. Root Rot

Symptoms

- ☆ Sudden wilting of plants in patches.

- ☆ Leaves droop, dry and the plant dies.
- ☆ The roots rot, barks shred and decay.
- ☆ Barks show large number of small, black sclerotia.

Fungus

Macrophomina phaseolina (Tassi) Goid.

Mode of survival: As sclerotia in soil and plant debris.

Primary spread: From soil-borne sclerotia.

Secondary spread: Through irrigation water.

Favourable conditions: Rain after long spell of high temperature.

Management

- ☆ Remove diseased plant and soil drench with carbendazim 0.1 per cent.

3. Stem Rot

Symptoms

- ☆ Pre-emergence damping off causing seed rot and death of seedlings before emergence.
- ☆ Brown to black lesions develops on stem just above soil line results in girdling of the stem.
- ☆ The lesions spread to the upper part of the stem result in collapse and death of seedlings.
- ☆ The roots of diseased plants rot.
- ☆ Shredding of root bark with numerous minute dark brown sclerotia found on the tissues.

Fungus

Rhizoctonia solani Kuhn.

Mode of survival: As sclerotia in soil and plant debris.

Primary spread: From soil-borne sclerotia.

Secondary spread: Through irrigation water.

Favourable conditions: Rain after long spell of high temperature.

Management

- ☆ Remove and destroy infected plants.
- ☆ Spot drench with carbendazim 0.1 per cent or captan 0.25 per cent.

4. Root and Crown Rot

Symptoms

- ☆ The leaves of infected plants are discoloured to violet.
- ☆ The basal stem region (foot) and the roots rot and the infected plants die.

Chromista

Phytophthora nicotianae Breda de Haan (Syn: *Phytophthora parasitica* Dastur)

Mode of survival: As oospores in soil and plant debris.

Primary spread: From soil-borne oospores.

Secondary spread: Through irrigation water.

Favourable conditions: Rain after long spell of high temperature.

Management

- ☆ Remove and destroy infected plants.
- ☆ Spot drench with Fosetyl-Al 0.2 per cent.

5. Leaf Blight

Symptoms

- ☆ Brown, sunken necrotic spots surrounded by reddish raised margins.
- ☆ The lesions are more prominent on lower leaves and confined to the margins.
- ☆ Infected leaves roll up, shrivel and drop off, leaving a barren stem.

Fungus

Colletotrichum crossandrae Patel. Kamat and C. B. Pande

Mode of survival: In infected plant debris.

Primary spread: From infected plant parts.

Secondary spread: Through air-borne conidia.

Management

- ☆ Spray carbendazim 0.1 per cent or mancozeb 0.2 per cent.

6. *Alternaria* Leaf Spot

Symptoms

- ☆ Small, circular or irregular yellow spots seen on the upper leaf surface.
- ☆ They enlarge, turn brown with concentric rings.
- ☆ Infected leaves become yellow and drop off prematurely.

Fungus

Alternaria amaranthi var. *crossandrae*

Mode of survival: In infected plant debris.

Primary spread: From infected plant parts.

Secondary spread: Through air-borne conidia.

Management

- ☆ Spray plants with Mancozeb (0.2 per cent) or Bavistin (0.1 per cent) at 15 days interval during the growing season.

7. Dahlia (*Dahlia pinnata* Cav.)

1. Wilt

Symptoms

***Fusarium* Wilt**

- ☆ Yellowing and drooping of leaves and wilting of plant.
- ☆ The vascular bundles show reddish brown or black discolouration.

***Verticillium* Wilt**

- ☆ Single branch or the entire plant wilts and dies.
- ☆ Vascular tissues in affected stems become black, dark brown or greenish-brown.
- ☆ Stored tubers rot if infected.

Fungi

Fusarium sp. and *Verticillium albo-atrum* Reinke and Berthold.

Mode of survival: As chlamydospores/microsclerotia in soil.

Primary spread: From soil-borne inoculum and infected tubers.

Secondary spread: Through irrigation water.

Management

- ☆ Plant healthy tuber in pathogen free soil.
- ☆ Treat the tuber with carbendazim 0.1 per cent for 15 minutes before planting.
- ☆ Rogue out and destroy infected plants.
- ☆ Drench soil in root region with carbendazim 0.1 per cent.

2. Powdery Mildew

Symptoms

- ☆ Occur in all stages but serious at flowering stage.
- ☆ Dry, white powdery fungal growth occurs on upper leaf surface of lower leaves, later turns grey.
- ☆ Leaves distorted and fall prematurely.

Fungus

Golovinomyces cichoracearum DC (Syn. *Erysiphe cichoracearum* DC)

Mode of survival: As cleistothecia in infected leaves.

Primary spread: From ascospores present in infected plant debris.

Secondary spread: Through air-borne conidia.

Management

- Spray wettable sulphur 0.2 per cent or carbendazim 0.1 per cent or triademefon or thiophanate methyl + mancozeb.

3 Leaf Smut

Symptoms

- The disease occurs at flowering stage.
- Yellow greenish, circular or angular spots occur mostly on lower leaves.
- The spots become brown surrounded by yellow halo.
- In severe infection, shot-holes are seen on leaves and affected leaves shrivel, dry and fall off.
- The petioles and stem are also affected.

Fungus

Entyloma calendulae f.sp. *dahliae* Syd. and P. Syd.

Mode of survival: As chlamydospores in soil and leaf debris.

Primary spread: From infected soil and plant debris.

Secondary spread: Through air-borne sporidia.

Favourable conditions: Deficiency of calcium in the soil, high humidity, late planting, damp conditions and shady areas.

Management

- Collect and destroy infected crop debris.
- Adopt crop rotation.
- Grow tolerant varieties.
- Spray Bordeaux mixture 1 per cent or zineb or copper oxychloride 0.2 per cent.

4. *Botrytis* Blight/Bud and Flower Blight

Symptoms

- In wet weather, masses of grey spores form on brown spots on buds, leaves or stems.

- ✰ They initially cause brown spots which enlarge in humid atmosphere and turn tan to grey fuzzy mold.

Fungus

Botrytis cinerea Pers. Ex Fr.

Mode of survival: As sclerotia and mycelia in dead plant materials.

Primary spread: From infected plant tissues.

Secondary spread: Through air-borne or rain splashed conidia.

Favourable conditions: Cool temperature, high relative humidity and rainy weather.

Management

- ✰ Maintain low relative humidity.
- ✰ Remove affected parts.
- ✰ Spray iprodione or fludioxonil or mancozeb or captan 0.2 per cent.

5. Leaf Spot

Symptoms

- ✰ Dark brown, angular to irregular spots of 4-5 mm dia appear on the leaves.
- ✰ They enlarge, coalesce and large patches.
- ✰ Diseased foliage gives charred appearance.

Fungus

Cercospora grandissima Rangel (Syn. *Cercospora dahliae, C. dahliicola*)

Mode of survival: In infected plant parts and plant debris.

Primary spread: From infected soil and plant debris.

Secondary spread: Through air-borne conidia.

Management

- ✰ Collect and destroy infected crop debris.
- ✰ Adopt crop rotation.
- ✰ Grow tolerant varieties.
- ✰ Spray Bordeaux mixture 1 per cent or zineb or copper oxychloride 0.2 per cent.

6. Bacterial Stem Rot

Symptoms

- ✰ The affected stem becomes black and soft.
- ✰ The layers of stem pith are moist and blackish with rot, which extends into the bark.

- ☆ The rot emits foul odour.
- ☆ Tubers and roots are also attacked.

Bacterium

Erwinia carotovora sub sp. *carotovora* (Jones) Bergey *et al.*

Management

- ☆ Remove and destroy the infected plants.
- ☆ Improve drainage facilities.
- ☆ Avoid overcrowding of seedlings.
- ☆ Adopt crop rotation.

7. Crown Gall

Symptoms

- ☆ Large abnormal growths are seen at the base and or on tubers of the infected plant.
- ☆ The plants are stunted and the shoots become spindly.

Bacterium

Rhizobium radiobacter (Beijerinck and van Delden)

[Syn. *Agrobacterium tumefaciens* (Smith and Townsend) Conn].

It is a Gram-negative, rod shaped (bacilliform), aerobic, motile bacterium having 1-6 peritrichous flagella. A large extrachromosomal piece of DNA, commonly referred to as a tumor-inducing (Ti) plasmid, is carried by *A. tumefaciens*.

Mode of survival: In soil and infected roots.

Primary spread: From infected soil and plant parts.

Secondary spread: Through air-borne conidia.

Management

- ☆ Collect and destroy infected plants, including tubers.
- ☆ Adopt crop rotation with cereals or legumes (non-woody crops).
- ☆ Avoid wounding.
- ☆ Fumigate soil.

8. Mosaic

Symptoms

- ☆ Leaves show mosaic mottling and produce vein clearing symptoms.
- ☆ The leaves become small.

Virus

Dahlia mosaic virus and *cucumber mosaic virus*

Mode of survival: In infected plants.

Primary spread: From infected plants.

Secondary spread: Through aphids (*Myzus persicae, Aphis gossypii* and *A. fabae*).

Management

- ☆ Plant disease free tubers.
- ☆ Rogue out and destroy infected plants.
- ☆ Spray dimethoate or methyl demeton 2 ml/lit.
- ☆ Use meristem tip culture to eliminate virus.

9. Aster Yellows

Symptoms

- ☆ Leaf-like tissue forms where flower parts should be located. Flower-like parts remain green or light green.

Bacterium

Candidatus Phytoplasma

Management

- ☆ Destroy infected plants.
- ☆ Spray insecticides to control leafhoppers.

8. Gerbera (*Gerbera aurantiaca* Sch. Bip.)

1. Anthracnose

Symptoms

- ☆ Circular, scattered reddish brown spots appear on leaves.
- ☆ They enlarge and coalesce with each other during moist weather involving large areas.
- ☆ Later, affected leaves wither, roll and dry.
- ☆ Black dot-like acervuli are seen on the leaf spots.

Fungus

Colletotrichum gloeosporioides Pat and Penz.

[Sexual stage: *Glomerella cingulata* (Stoneman)].

Acervuli (asexual fruiting body) with black setae, conidiophores are hyaline, septate, branched. Conidia are hyaline, aseptate, oval to elliptical or straight cylindrical and 11.0 to 17.0 × 3.0 to 6.0 µm.

Mode of survival: In infected plant debris.

Primary spread: From affected plants.

Secondary spread: Through air-borne conidia.

Favourable conditions: Crowding of plants, excess watering.

Management

- ☆ Collect and burn diseased leaves.
- ☆ Avoid excessive watering and crowding of plants.
- ☆ Spray carbendazim 0.1 per cent.

2. Leaf Spot

Symptoms

- ☆ Spots are light buff in the beginning which later turns dark saffron with smooth margin.
- ☆ They are scattered on the leaves.
- ☆ Numerous fruiting bodies (pycnidia) are seen on the spots

Fungus

Phyllosticta gerbericola Bat. (Sexual stage: *Pleospora gerberae* Dippenaar)

Mode of survival: As pycnidia in plant debris and in other infecting hosts *viz.*, china aster, chrysanthemum and marigold.

Primary spread: From pycnidia from infected hosts.

Secondary spread: Through air-borne pycnidiospores.

Management

- ☆ Foliar spray of carbendazim or zineb 0.2 per cent.

3. Powdery Mildew

Symptoms

- ☆ White powdery coating is seen on the upper surface of leaves and on tender stems.
- ☆ Later, white powdery coating turns red, distort, discolour and wither.
- ☆ Powdery coating on the stems and flowers.

Fungus

Golovinomyces cichoracearum DC (Syn. *Oidium erysiphoides* Fr.)

Mode of survival: In diseased plants, volunteer plants and weeds.

Primary spread: From infected plant parts and weeds.

Secondary spread: Through air-borne conidia.

Management

- ☆ Use resistant cultivars.
- ☆ Avoid overcrowding of plants and provide good air movement.
- ☆ Spray wettable sulphur 0.2 per cent or thiophanate methyl 0.1 per cent or pyrozophos 0.1 per cent or azoxystrobin 0.1 per cent.

4. Downy Mildew

Symptoms

- ☆ Chlorotic spots appear on the upper leaf surface.
- ☆ The infected tissues turn yellow and then brown with age with blighting of leaves.
- ☆ Correspondingly on the lower leaf surface, dense whitish sporulation occurs.

Chromista

Bremia lactucae Regel and *Plasmopara halstedii* (Farlow) Berlese and de Toni

Mode of survival: As oospores in infected plant debris.

Primary spread: From infected crop debris.

Secondary spread: Through zoospores.

Management

- ☆ Spray metalaxyl 0.1 per cent 3-6 times at 15 days interval.
- ☆ Drench the soil with metalaxyl 0.1 per cent or captan 0.2 per cent twice at 3-4 weeks interval.

5. Blossom Blight and Stalk Rot

Symptoms

- ☆ Light brown irregular water soaked spots appear on flower stalks and petals.
- ☆ The spots size increase rapidly, coalesce with one another and form distinctly depressed lesions.
- ☆ Under humid conditions, infection becomes severe involving entire flower head resulting in blossom blight and stalk rot.

Chromista

Phytophthora palmivora (E. J. Butler)

Mode of survival: As oospores in infected plant debris.

Primary spread: From soil-borne oospores.

Secondary spread: Through zoospores, rain or overhead irrigation.

Favourable conditions: Drizzling rain and cool moist weather.

Management

- Use disease-free soil.
- Rogue out and destroy infected plant parts and flowers.
- Avoid excess watering.
- Spray difolatan 0.2 per cent or copper oxychloride 0.25 per cent.

6. Root Rot

Symptoms

- Stem at soil level shows brown lesions which spread both upwards and downwards.
- Plants get stunted or wilt and die.
- Roots show wilting

Fungi

Rhizoctonia solani Kuhn and Chromista: *Pythium irregulare* Buisman

Mode of survival: As sclerotia or oospores in soil.

Primary spread: From soil-borne inoculum.

Secondary spread: Through irrigation water.

Favourable conditions: High humidity and water logged conditions.

Management

- Plant in sterilized soil.
- Avoid dense sowing.
- Apply fertilizers judiciously.
- Regulate irrigation.
- Adopt strict sanitary measures.
- Remove and destroy infected plants.

7. *Botrytis* Blight

Symptoms

- Blight occurs on all plant parts except the root system.
- Petioles exhibit long brown spots.
- Leaves become yellow and dry.
- Petals show tan coloured spots.
- Stem lesions at soil level leads to death of plants.
- Infected tissues are covered with grey, fuzzy growth containing masses of fungal spores.

Fungus

Botrytis cinerea Pers. Ex Fr.

Mode of survival: As sclerotia and mycelia in dead plant materials.

Primary spread: From infected plant tissues.

Secondary spread: Through air-borne or rain splashed conidia.

Favourable conditions: Rain, cool weather, deep planting, poor drainage and poor ventilation.

Management

- ☆ Space plants to ensure good air circulation.
- ☆ Maintain low humidity.
- ☆ Avoid over watering.
- ☆ Remove and destroy infected plant parts.
- ☆ Spray captan 0.2 per cent at weekly intervals.

9. Gladiolus (*Gladiolus communis* L.)

1. Wilt and Corm Rot/*Fusarium* Yellows

Symptoms

- ☆ Corms decay with brownish to black dry rot, leaves turn yellow and brown.
- ☆ Roots in affected plants are killed.
- ☆ The plants show stunted growth, late flowering, greener and smaller buds and poor opening.
- ☆ The number and quality of flower spikes is also reduced.
- ☆ Latent infection in corms leads to active rotting of corms/cormels during storage and in field.

Fungus

Fusarium oxysporum f. sp. *gladioli* (Massey) W.C. Snyder and H.N. Hansen

Mode of survival: In the corm/cormels and soil.

Primary spread: From infected corm and soil.

Secondary spread: Through irrigation water.

Favourable conditions: Poor soil aeration, flooding or hot weather.

Management

- ☆ Grow disease tolerant cultivars.
- ☆ Use of clean soil or fumigated soil.

- ☆ Treat corms/cormels in hot water at 50°C for 30 min before planting.
- ☆ Treat the corms/cormels with fungicide carbendazim 0.02 per cent solution for half an hour before planting followed by soil application of carbofuran at 3 g/sq.m at the time of planting which is very helpful in preventing the incidence of this dreaded disease.
- ☆ Use nitrate form of nitrogenous fertilizers.
- ☆ Harvest corm in dry weather.

2. Botrytis Soft Rot

Symptoms

- ☆ The disease occurs in the cold storage when the temperature stays below 13°-15° C.
- ☆ The disease is characterized by the formation of white-moulded, soft, spongy corms.
- ☆ At early stage, the infection is confined to surface only, but later it travels to the core as well. At high temperature, corms seal off infection, escaping further rotting.
- ☆ In the field, infection is characterized by circular brown leaf spots. The smallest spots are visible only on the upper surface of leaf.

Fungus

Botrytis gladolorum

Mode of survival: As sclerotia and mycelia in plant materials/corm.

Primary spread: From infected plant tissues.

Secondary spread: Through air-borne or rain splashed conidia.

Favourable conditions: Low temperature below 15° C, Lack of air circulation.

Management

- ☆ Spray mancozeb at 0.2 per cent twice a week and dust healthy corms with mancozeb during storage.

3. *Penicillium* Corm Rot

Symptoms

- ☆ This is commonly known as green mold.
- ☆ It occurs when the corms are stored in heaps after having been dug in the cold weather.
- ☆ Large reddish-brown sunken lesions with greenish mycelial growth is seen on the affected tissues in the corm.
- ☆ Affected corms are mummified.

Fungus

Penicillium gladioli L. McCulloch and Thom

Mode of survival: In the affected corms.

Primary spread: From infected corms.

Secondary spread: Through air borne spores.

Favourable conditions: Hot humid conditions.

Management

- ☆ Examine periodically the diseased corms in the storage and discard the infected ones.
- ☆ Store only the healthy corms and avoid dampness and injuries to corms in the store house.
- ☆ Dust the corms with mancozeb, which will prevent the disease during storage.

4. Rust

Symptoms

- ☆ Small yellow raised pustules seen on either side of the leaves which become orange coloured at later stage.
- ☆ The yellowish orange typically transverse (horizontal) uredosori seen across the width of the leaf burst and releases uredospores.
- ☆ The teleutosori are black in colour. Spermagonia and aecia are unkown.

Fungus

Uromyces transversalis (Thümen) Winter

Mode of survival: In infected Gladiolus plants and other members of the Iridaceae: Anomatheca, Crocosmia, Melasphaerula, Tritonia, Watsonia.

Primary spread: From infected plants.

Secondary spread: Through air-borne uredospores.

Management

- ☆ Grow resistant varieties.
- ☆ Spray bitertanol or triademefon at weekly intervals.

5. Leaf Spots and Corm Rot

Symptoms

Curvularia

- ☆ Leaf spots are tan coloured surrounded by dark brown rings and yellowish halos.

- Small leaves born on cormels are curled and twisted at the tips.
- Brown to black elongated blotches appear on the stem.
- Plants are girdled near the soil level and die.
- Brown to black, round, elongated or irregular lesions seen on the corm and vascular tissues turn brown.

Septoria

- It is also called as hard rot of corm.
- Brown or purple brown circular spots seen on leaves, later turn to dark brown with greyish centre.
- Minute pycnidia of black colour are seen on the spots.
- Infected corms show water soaked, circular, reddish brown to brownish black lesions on the lower parts and are concealed by the scales.
- Lesions enlarge and the centre become sunken and colour changes to dark brown or black.
- Diseased tissues are hard and corm becomes mummified.
- Diseased corm do not flower but die prematurely.
- The disease is favoured by low fertility and humid weather.

Stemphylium

- Circular, light green or yellow spots with prominent red centre appear on leaves and coalesce to form irregular patches.
- Severe infection destroys all the above ground parts during or near flowering.

Fungi

Curvularia trifolii f.sp. *gladioli* Parmelee and Luttr., *Septoria gladioli* Pass., *Stemphylium botryosum* Wallr.

Management

- Grow resistant varieties.
- Spray zineb 0.2 per cent or copper oxychloride 0.3 per cent or mancozeb 0.2 per cent.

6. Bacterial Scab

Symptoms

- Infection starts as pale yellow, circular, water-soaked spots on corm, deepening to black or nearly black.
- These spots later become sunken with raised margins that are scab-like and exude yellow to dark brown gummy ooze.

- ✰ Tiny reddish raised specks appear on leaves, mostly near the base immediately after planting. These specks enlarge to dark sunken spots which later cause either firm or soft rots in the neck or basal parts.
- ✰ Affected plant dies.

Bacterium

Burkholderia gladioli Yabuuchi *et al.* (Syn. *Pseudomonas marginata* McCulloch)

Mode of survival: In the corm.

Primary spread: From infected corm.

Secondary spread: Through irrigation water.

Favourable conditions: Injury to the corm by bulb mites, grubs or wireworms.

Management

- ✰ Examine corms before planting.
- ✰ Plant only quality corms that are free from disease or insect injury.
- ✰ Adopt crop rotation.
- ✰ Dry corms well before storage and over-winter in a cool (40-45°F) well-ventilated place.
- ✰ Take control measures for bulb mite.

7. Viral Diseases

Symptoms

Bean Yellow Mosaic Virus

- ✰ Pale yellow mottling appear on leaves.
- ✰ Thin dark streaks seen on flowers.
- ✰ Beans, peas, and sweet peas are also attacked by this virus.
- ✰ Aphids, *Aphis fabae* and *Myzus persicae* transmit the virus.

Cucumber Mosaic Virus

- ✰ Greyish mottling on leaves and white blotches are seen on the flowers.
- ✰ Virus attacks cucurbitaceous plants.
- ✰ It is transmitted by Aphids, *Myzus persicae.*

Tobacco Ring Spot Virus

- ✰ Yellow ring spots on leaves and white streaks on the flowers are seen.
- ✰ The virus is transmitted by nematodes.

Viruses

Bean yellow mosaic virus, Cucumber mosaic virus, Tobacco ring spot virus

Management

- ☆ Plant disease free plants.
- ☆ Rogue out and destroy diseased plants.
- ☆ Spray insecticides to control insect vectors.

8. Grassy Top

Symptoms

- ☆ Affected plant mature early and produce only small sized corms.
- ☆ Root development is arrested.
- ☆ In the ensuing year, corms produce many, thin, weak shoots giving the name "Grassy top".
- ☆ Flowers produced from diseased plants are green.

Bacterium

Aster Yellows Phytoplasma

Management

- ☆ Plant disease free plants.
- ☆ Rogue out and destroy diseased plants.

10. Jasmine (*Jasminum* sp. L.)

1. *Fusarium* Wilt

Symptoms

- ☆ Plants of all ages are attacked and wilt occurs in patches.
- ☆ Top leaves wilt and bottom leaves show yellowing.
- ☆ Yellowing spreads upward resulting in death of the plant
- ☆ Vascular discolouration in stem is common.
- ☆ Roots become black.

Fungus

Fusarium solani (Mart.) Sacc. [*Nectria haematococca* Berk. and Broome]

Mode of survival: As chlamydospores in soil and as mycelium in infected plant debris.

Primary spread: From infected plant debris and soil.

Secondary spread: Through irrigation water.

Management

- ☆ Remove and destroy the infected plant.
- ☆ Spot drench with Bordeaux mixture 1 per cent or carbendazim 0.1 per cent.

2. Sclerotium Wilt/Southern Blight/Collar Rot

Symptoms

- ☆ Diseased plants wilt suddenly.
- ☆ Whitish mycelial growth is seen adhering to the surface of the roots.
- ☆ Sclerotia found along with the whitish mycelia in the soil at the root region or in collar region of the affected plant.
- ☆ Affected plants wilt and die.

Fungus

Sclerotium rolfsii Sacc. [Sexual stage*: Athelia rolfsii* (Curzi)].

It is a soil living saprophytic pathogen, produces ribbon like hyphae with clamp connections. Basidia are club-shaped, bearing four smooth, ellipsoid basidiospores. Small, brownish sclerotia (hyphal propagules) are also formed.

Mode of survival: As sclerotia in soil and in plant debris.

Primary spread: From soil-borne sclerotia.

Secondary spread: Through irrigation water.

Management

- ☆ Remove and burn infected plants along with the roots, mycelial mat and sclerotia of the fungus.
- ☆ Apply well rotten FYM with *Trichoderma viride* at 25-50 g/plant.
- ☆ Spot drench with Bordeaux mixture 1 per cent or carbendazim 0.1 per cent.

3. *Alternaria* Leaf Blight

Symptoms

- ☆ Both young and old leaves are affected.
- ☆ Dark brown spots with concentric rings appear on leaves and the number of spots vary with variety and environment.
- ☆ Spots increase in size and the whole leaf is blighted.
- ☆ Blighted leaves dry and become brittle and exhibit burnt up appearance and fall off prematurely.
- ☆ Oval to elongated light brown spots develop on petioles, stem, calyx and tubular corollas.
- ☆ In severe cases of infection vegetative buds and young branches dry up.
- ☆ The disease is severe during winter months (Oct - Dec).

Fungus

Alternaria jasmini T.K. Kandasw and *A. alternata* (Fr.) Keissl.

Mode of survival: In infected plant debris.

Primary spread: From diseased plant parts.

Secondary spread: Through air-borne conidia.

Favourable conditions: Temperature 21-23°C and relative humidity 80 per cent.

Management

- Spray mancozeb or zineb 0.2 per cent or azoxystrobin 0.1 per cent.

4. *Cercospora* Leaf Spot

Symptoms

- The disease commences after the onset of monsoon in June and continues till February.
- All aerial parts of the plant are infected.
- Circular to irregular reddish brown spots of 2 to 8 mm dia appear on upper leaf surface. Later, these spots increase in size to form irregular patches.
- Leaf margins exhibit inward curling, become hard and brittle.
- Oval to elongated spots also appear on stem, petioles, calyces and corolla.
- Vegetative buds and young branches are also affected leading to heavy defoliation and death.
- Flower production is adversely affected to the tune of 50 per cent.

Fungus

Cercospora jasminicola var. *khandalensis* Chidd.

(Syn. *C. jasmini* Sawada and *C. odoratissimi* Sawada)

Mode of survival: In infected plant debris.

Primary spread: From diseased plant parts.

Secondary spread: Through air-borne conidia.

Management

- Prune the crop and spray with thiophanate-methyl 0.1 per cent or mancozeb 0.2 per cent or carbendazim 0.1 per cent.

5. Anthracnose

Symptoms

- Large, circular, distinct, brownish to greyish spots with brown to yellowish haloes appear on upper surface of the leaves.
- Infected tissues turn greyish, become papery and fall off.
- Twig infection causes dieback and drying of branches.

Fungus

Colletotrichum jasminicola Tilak

Mode of survival: As acervuli in infected plant debris.

Primary spread: From diseased plant parts.

Secondary spread: Through air-borne conidia.

Management

- ☆ Collect and destroy diseased leaves and twigs.
- ☆ Spray zineb 0.2 per cent or copper oxychloride 0.3 per cent or carbendazim 0.1 per cent at fortnightly intervals.

6. Rust

Symptoms

- ☆ The disease is observed during rainy months of June-July and persists until the flowering in March.
- ☆ It attacks all aerial parts and cause blisters.
- ☆ Oval shaped orange pustules appear on lower leaf surface in the beginning and they become dark due the production of teleutospores.
- ☆ Infected plant parts are hypertrophied.
- ☆ Severe infection leads to defoliation.
- ☆ Blisters or tumours are seen on stem and twigs.
- ☆ Stem barks of branches split and subsequent such branches die.
- ☆ Flower buds are swollen, deformed and do not open.

Fungus

Uromyces hobsonii Vize

It is autoecious rust. The fungus does not produce uredial stage.

Management

- ☆ Prune and burn infected plant parts.
- ☆ Dust sulphur at 20 to 25 kg/ha or spray mancozeb 0.2 per cent or triadimefon 0.05 per cent or Bordeaux mixture 1 per cent or copper oxychloride 0.25 per cent.

7. Leaf Spots

Symptoms

Myrothecium

- ☆ Circular to irregular, brown to dark brown spots surrounded by yellow halo appear on leaves.

Phyllosticta

- ☆ Circular, pale brownish spots with smooth margin appear on leaves.
- ☆ The spots enlarge, coalesce and cause yellowing and premature defoliation.

Cladosporium

- ☆ Light yellow areas appear on the leaf lamina and turn light brown, coalesce forming large patches resulting in blight symptom.
- ☆ Effuse, velvety, olive brown colonies appear on the corresponding lower surface of the leaves.

Fungi

Myrothecium roridum Tode, *Phyllosticta jasminina* V.G. Rao and *Cladosporium fulvum* m [syn. *Passalora fulva* (Braun *et al.*, 2003)].

Management

- ☆ Collect and destroy diseased plant parts.
- ☆ Avoid dense planting, excessive watering and shading.
- ☆ Spray copper oxychloride 0.25 per cent or mancozeb 0.2 per cent or carbendazim 0.1 per cent.

8. Bacterial Leaf Spot

Symptoms

- ☆ Yellowish green water soaked irregular spots appear on leaves.
- ☆ Severe infection leads to stunting of plants.

Bacterium

Xanthomonas campestris pv. *jasmini*

Management

- ☆ Collect and destroy severely diseased plant parts.

9. Phyllody

Symptoms

- ☆ Clustering of the leaves and bushy appearance of the plants.
- ☆ Leaves become linear and small and are closely arranged.
- ☆ In the place of normal flowers, green, leafy and malformed flowers are seen.
- ☆ Loss of flower yield is total.

Bacterium

Candidatus Phytoplasma

Mode of survival: In diseased bushes.

Primary spread: From infected planting materials.

Secondary spread: Through grafting and whitefly vector (*Dialeurodes kirkaldii*).

Favourable conditions: High temperature.

Management

- ☆ Rogue out diseases bushes.
- ☆ Control vector with methyl demeton or dimethoate 2 ml/lit.
- ☆ Use disease free planting materials.
- ☆ Spray tetracycline to reduce disease severity.

11. Lilium (*Lilium candidum* L.)

1. Basal Rot or Bulb Rot

Symptoms

- ☆ Unlike other Fusarium diseases, it is not a vascular wilt but it is a rot of the cortical tissues of the root.
- ☆ The disease causes severe root rot and destruction of the basal plate.
- ☆ Chocolate or bluish-grey rot that extends from the basal plate into the scales.
- ☆ The surface of rotting scales remains smooth until disintegration begins.
- ☆ Scales detach at the basal plate and bulbs fall to pieces after digging.
- ☆ Premature yellowing of the foliage, stunting and premature senescence.
- ☆ Flower buds wilt and fail to open.
- ☆ The disease also occurs often as a combination with *Cylindrocarpon.*

Fungi

Fusarium oxysporum f. sp. *lilii* Imle

Mode of survival: In soil indefinitely.

Primary spread: From infected bulbs and soil inoculum.

Secondary spread: Through irrigation water.

Favourable conditions: Excessive wet condition of the soil, acidic soils.

Management

- ☆ Disinfect the infected soil by using suitable fungicides.
- ☆ Remove and destroy infected bulbs along with its remnants and plant debris.
- ☆ Add lime in acidic soils to suppress the pathogen.

- ✰ Plant bulbs with rot-free scales.
- ✰ Plant lilies in well-drained soils or improve soil drainage.
- ✰ Avoid damaging the bulbs during planting, transplanting or weeding.
- ✰ Mulch around plants to reduce soil temperature and suppress the pathogen.
- ✰ Avoid over watering in hot summer months.
- ✰ Avoid excess nitrogenous fertilizers.
- ✰ Postharvest dip of bulbs in thiophanate-methyl for 15-30 min within 48 h after digging at 80°F – 85°F.
- ✰ Soil drenching with carbendazim 0.1 per cent or difenoconazole 0.05 per cent.

2. *Botrytis* Blight

Symptoms

- ✰ This disease is more common in hills.
- ✰ Early infection causes cessation of apical growth.
- ✰ Orange to reddish brown, oval spots with yellow halo or with water soaked margin appear on leaves.
- ✰ Under damp conditions, they enlarge, coalesce and may blight the whole leaf.
- ✰ Infection usually starts with lowest leaves.
- ✰ Brown spots appear on stem near soil level.
- ✰ Lesions on flowers are brown and in cool and moist weather flowers are converted into wet slimy masses covered with powdery spores.

Fungus

Botrytis elliptica (Berk.) Cooke (Syn. *Botrytis liliorum* Fujik)

Mode of survival: As sclerotia and mycelia in plant materials.

Primary spread: From infected soil or from infected plant materials.

Secondary spread: Through air-borne or rain splashed conidia.

Favourable conditions: Temperature of 21°C and moisture on plant parts.

Management

- ✰ Grow resistant varieties.
- ✰ Remove and destroy diseased flowers or leaves or plants.
- ✰ Provide good air circulation.
- ✰ Avoid dense planting in low lying and shaded areas, excessive watering and water stagnation in the beds.

- ☆ Maintain hygienic conditions and remove weeds to avoid infection.
- ☆ Spray Bordeaux mixture 1 per cent or carbendazim 0.1 per cent or chlorothalonil 0.2 per cent when the foliage is dry.

3. Anthracnose or Brown Scale

Symptoms

- ☆ The disease infects outer scales of the bulbs in the soil.
- ☆ Small, brown spots appear on linear scale.

Fungus

Colletotrichum lilii Boerema and Hamers

Mode of survival: In infected plants and bulbs.

Primary spread: From diseased plant parts.

Secondary spread: Through air-borne conidia.

Management

- ☆ Remove and destroy infected plants.
- ☆ Dip bulbs in carbendazim 0.1 per cent.
- ☆ Avoid excessive watering.

4. Foot Rot

Symptoms

- ☆ Stem at collar region is mostly attacked.
- ☆ Affected plants wilt and die.
- ☆ Sometimes leaves and flowers on the upper plant parts are also affected.

Fungus

Phytophthora cactorum (Leb. and Cohn) Schröeter and *P. nicotianae* Breda de Haan

Mode of survival: In infected plants and bulbs.

Primary spread: From diseased plant parts.

Secondary spread: Through Air-borne conidia.

Management

- ☆ Remove and destroy infected plants.
- ☆ Plant seedlings in clean pasteurized soil.
- ☆ Avoid excessive watering and provide good drainage.
- ☆ Spot drench the soil with metalaxyl or Fosetyl-Al 0.1 per cent.

5. Viral Diseases

Symptoms

Lily Mottle Virus (LMV)

- Vein clearing, leaf mottling, mosaic, chlorotic and yellow streaking, leaf curling and narrowing are the common symptoms.
- Colour breaking in flowers is also common.

Lily Symptomless Virus (LSV)

- Mild mottle, severe mottle, basal striping and twisting of leaves are the characteristic symptoms.
- Stunting of plants is common.
- Infected plants take longer time to flower.
- It is spread by *Aphis gossypii.*

Cucumber Mosaic Virus (CMV)

- It does not affect growth drastically.
- Light green mosaic symptom is seen early followed by brownish necrotic spots.
- Leaves become brittle and flowers are malformed.

Tulip Breaking Virus (TBV)

- Severe leaf mottling, distortion of flowers and stunting of plants.
- Transmitted by aphids, *Aphis fabae, Macrosiphum euphorbiae and Myzus persicae.*

Viruses

Lily mottle virus, Lily symptomless virus, Cucumber mosaic virus and *Tulip breaking virus.*

Management

- Plant virus disease free planting materials.
- Rogue out and destroy diseased plants.
- Spray insecticides to control insect vectors.

12. Marigold (*Tagetes erecta* L.)

1. Damping Off

Symptoms

- The disease is most prevalent at the seedling stage.

- Necrotic spots and rings develop on the young seedlings causing collapse of the seedlings.
- Considerable loss is sustained if seedlings are not properly looked after.

Chromista

Pythium sp.

Mode of survival: As oospores in soil.

Primary spread: From soil-borne oospores.

Secondary spread: Through zoospores.

Favourable conditions: High moisture and poor drainage in the nursery.

Management

- Sterilize the nursery soil with formalin at 2 per cent before sowing.
- Raise the seedlings in raised beds.
- Use optimum seed rate.
- Provide better drainage.
- Drench the nursery area with Bordeaux mixture 1 per cent or copper oxychloride 0.25 per cent or metalaxyl 0.1 per cent.

2. Wilt and Stem Rot

Symptoms

- The fungus affects the collar portions of the plants.
- In nursery, the infection results in damping off and is aggravated by soil moisture.
- In the field, the infected plants show wilting.
- French marigold and Dwarf varieties are less susceptible whereas the African types are highly susceptible to the disease.

Fungus

Phytophthora cryptogea Pethybr. and Laff.

Mode of survival: As oospores in soil.

Primary spread: From soil-borne oospores.

Secondary spread: Through irrigation water.

Favourable conditions: Wet soil.

Management

- Remove the infected plants.
- Drench the soil with Bordeaux mixture 1.0 per cent or copper oxychloride 0.25 per cent or metalaxyl 0.1 per cent or Fosetyl-Al 0.1 per cent.

3. Root Rot

Symptoms

- ☆ The disease occurs from seedling stage onwards.
- ☆ The above ground symptoms are non-specific and include a general wilting, decline and collapse of the foliage and the entire plant.
- ☆ The general droopiness or flaccid appearance is often accompanied by browning and rotting of the roots and the crown.
- ☆ Yellowing and death of the outer leaves follows, until finally the entire plant is dead.

Fungus

Rhizoctonia solani Kuhn.

Mode of survival: As sclerotia in infected plant debris and in soil.

Primary spread: From soil-borne inoculum.

Secondary spread: Through irrigation water.

Management

- ☆ Avoid over-watering especially in heavy soils and directly onto the plant.
- ☆ Rouge out the infected plants.
- ☆ Provide proper spacing to improve air circulation.

4. *Verticillium* Wilt

Symptoms

- ☆ Leaves turn yellow, first at the margins and between veins.
- ☆ Foliage turns tan or brown and dies, progressing upwards from the base to the tip of the plant or branch.
- ☆ Browning of older leaves while younger leaves remain green.

Fungi

Verticillium dahlia Kleb and *V. albo-atrum* Reinke and Berthold

Mode of survival: In infected plant parts and in other herbaceous plants including dahlia, gerbera, peony, snapdragon and *Catharanthus roseus.*

Primary spread: From infected plants and other host plants.

Secondary spread: Through irrigation water.

Favourable conditions: Cool weather.

Management

- ☆ Adopt proper sanitary measures.
- ☆ Plant only pathogen-free plants in disease-free soil.

- Avoid planting of susceptible cultivars.
- Practice soil solarisation to reduce soil inoculum.

5. Powdery Mildew

Symptoms

- White powdery coating on the upper or lower leaf surface.
- Powdery coating seen on petiole, tender stems and flowers also.
- In severe cases, leaves dry up and flower production is reduced.
- Infected buds do not open.

Fungus

Oidium sp. and *Leveillula taurica* (Lév.) G. Arnaud

Mode of survival: As cleistothecia in infected plants.

Primary spread: From infected plants.

Secondary spread: Through Air-borne conidia.

Management

- Spray carbendazim 0.1 per cent or wettable sulphur 0.2 per cent.

6. *Botrytis* Blight

Symptoms

- Flowers turn a papery brown and become covered with grey, fuzzy masses.
- Maturing flowers are particularly susceptible.
- Tan to brown spots with target-like appearance also develop on the leaves.
- These patches are often associated with flowers which have dropped onto the leaf surface.

Fungus

Botrytis cinerea Pers. Ex Fr.

Mode of survival: As Sclerotia

Primary spread: From infected debris.

Secondary spread: Through airborne or rain-splashed conidia.

Favourable conditions: Extended cloudy, humid and wet weather.

Management

- Follow good sanitary measures including trimming of plants and removing the spent or senescent flowers.
- Remove carefully the affected plant tissues and discard properly.
- Avoid overhead irrigation and wetting of flowers when watering.

- Provide adequate spacing between plants to promote good air circulation.
- Spray iprodione or chlorothalonil or thiophanate-methyl.

7. Flower Bud Rot

Symptoms

- The unopened flower buds are mostly affected and they show dark brown spots on the buds.
- The infected buds shrivel and do not open.
- Flower quality is affected if opened flowers are affected.
- Leaves develop spots of dead tissue on their edges.

Fungi

Alternaria dianthi and *A. alternata* (Fr.) Keissl.

Management

- Remove and discard the infected flowers.
- Spray the crop with mancozeb 0.2 per cent.

8. Leaf Spots

Symptoms

Septoria Leaf Spot

- Infection starts from lower leaves and progresses upwards.
- Numerous, minute, oval to irregular, smoky grey to black spots appear on leaves.
- Minute black fruiting bodies are formed on the spots.
- In severe infection, the plant vigour and flower production is reduced.

Alternaria Leaf Spot

- Dark brown, irregular spots appear on leaves, flowers and stem.
- When the disease is severe, drying of leaves and death of plants are common.

Cercospora Leaf Spot

- Circular to oblong spots of 0.1 to 0.5 cm dia surrounded by blackish brown margin are seen on the leaves.
- They increase in size, coalesce and cause blighting.

Fungi

Septoria tageticola Changsri and G.F. Weber, *Alternaria tageticola* Fli, *Cercospora tageticola* Ellis and Everh. and *C. thunbergii.*

Mode of survival: In infected plants.

Primary spread: From infected parts and plant debris.

Secondary spread: Through air-borne conidia.

Management

- Avoid overhead irrigation.
- Remove and destroy infected plant parts.
- Spray copper oxycholoride 0.25 per cent or mancozeb 0.2 per cent or carbendazim 0.1 per cent.

9. Bacterial Blight

Symptoms

- Light green water-soaked lesions appear on the leaves.
- Many spots cause drying and defoliation.

Bacterium

Xanthomonas tagetis Rangaswami and Sanne Gowda

Management

- Spray antibacterial antibiotic, streptomycin at 0.01 per cent.

10. Mosaic

Symptoms

- Mosaic mottling and leaf distortion are the common symptoms.
- The infected plants are stunted with poorly developed flowers.

Virus

Marigold mosaic virus

Mode of survival: In infected plants.

Primary spread: From diseased plants.

Secondary spread: Through aphid vectors, *Aphis craccivora, Aphis gossypii, Myzus persicae.*

Management

- Remove and destroy diseased plants.
- Spray methyl demeton 0.25 per cent or dimethoate 0.3 per cent to control insect vector and disease spread.

11. Aster Yellows

Symptoms

- Newly infected leaves exhibit yellowing or vein clearing symptoms.

- The internodes in the main stem are shortened and produce long axillary breaks which create a yellowish witches'broom appearance.
- If the plants are infected young, they remain stunted, flowers are small, abnormal and often exhibit phyllody symptoms.
- Symptoms may appear only on one stem.

Bacterium

Candidatus Phytoplasma

Mode of survival: In infected plants and many other hosts like Daisies, Chrysanthemums, Plantain, Echinacea and Rudbeckia.

Primary spread: From diseased plants and other host plants.

Secondary spread: Through insect vector, leafhopper.

Management

- Rogue out and destroy diseased plants.
- Spray systemic insecticides to control insect vectors.

13. Oleander (*Nerium oleander* L.)

1. Bacterial Gall

Symptoms

- Leaf galls are formed on both the leaf surfaces.
- Deep pits develop on opposite side of leaves to each gall leading to distortion of leaves.
- Young infected leaves get twisted.
- Young shoots develop longitudinal swellings which split.
- On the old stems, galls are circular and isolated.
- Severely infected plants do not flower.

Bacterium

Pseudomonas syringae sub sp. *savastanoi* Janse pv. *nerii* (Garden)

Mode of survival: In the galls of diseased plants.

Primary spread: From infected plants.

Secondary spread: Through water splash, insects, and pruning tools.

Management

- Avoid overhead irrigation.
- Apply fertilizers judiciously.
- Prune diseased stems 100 cm below galls.

- ☆ Remove or prune severely diseased plants at ground level.
- ☆ Disinfect pruning shears between each cut and between each shrub with 70 per cent methylated spirit as disinfectant.

14. Orchids

a. *Cattleya* spp.

1. Black Rot

Symptoms

- ☆ The disease appears as water soaked patches on leaves which quickly turn black from the tip.
- ☆ The rot spreads down the leaf and makes it completely black.
- ☆ In moist weather, it becomes dry and brittle.
- ☆ The infection begins at the rhizome but the rotten patch is seen at the tip of the leaf.

Chromista

Pythium splendens Hans Braun

The mycelium is hyaline, septate, aerial. The asexual reproduction takes place by means of zoosporangia and zoospores. The sexual reproduction takes place by means of oogonia and antheridia.

Mode of survival: As oospores in soil were still viable after 8 months to 12 years.

Primary spread: From infected plant debris.

Secondary spread: Through soil borne oospores and rarely chlamydospores.

Favourable conditions: High humidity and temperature.

Management

- ☆ Remove and destroy diseased leaves and plants.
- ☆ Avoid excessive watering and over crowding.
- ☆ Spray Bordeaux mixture 1 per cent or captan 0.2 per cent or zineb 0.2 per cent.

2. Wilt

Symptoms

- ☆ The symptoms include wilting of leaves, abscission of roots, decay of roots followed by the production of few, small short-lived flowers.

Fungus

Fusarium oxysporum f. sp. *cattleyae*

Management

- Use of disease free planting materials in disease free soil.

3. Rust

Symptoms

- The disease is characterized by yellow pustules covered with fine powdery masses of spores.

Fungus

Hemileia americana Massee

It does not produce all five states (microcyclic). Pycnial/spermagonial and aecial stages have not been observed.

Mode of survival: As survives primarily as mycelium in the living tissues.

Primary spread: From infected plant debris.

Secondary spread: Through dry uredospore can survive about 6 week and cause infection dispersed by wind and rain.

Favourable conditions: Temperature between 15-30°C.

Management

- Spray copper fungicides to control the disease.

4. Bacterial Soft Rot

Symptoms

- The disease starts at the upper end of the leaf as a small, water soaked dark green spots.
- Pseudobulbs of infected plants turn soft and pulpy and become yellow in colour.
- A foul smelling liquid oozes from the bulbs.

Bacterium

Pectobacterium carotovorum sub sp. *carotovorum* [*Erwinia carotovora* sub sp. *carotovora* (Jones) Bergey *et al.*].

Management

- Spray streptomycin to control the disease.

5. Bacterial Brown Spot

Symptoms

- The disease infects both seedlings and grown up plants.
- It appears as small, water soaked spots on leaves.

- These spots enlarge and soon the leaf becomes soft.
- The disease progresses into the growing point and kills the plant.

Bacterium

Acidovorax avenae sub sp. *cattleyae* Pavarino (*Pseudomonas cattleyae* Pavarino)

Primary spread: From rain splashes.

Secondary spread: Through contaminated propagation tools.

Favourable conditions: Warm, moist conditions and high nitrogen fertility.

Management

- Remove and destroy the affected plants.
- Spray streptomycin.

6. Viral Diseases

Symptoms

Blossom Brown Necrotic Streak

- It produces brown spots, streaks or whole flower-break.
- Leaves show yellow streaks.
- The disease is transmitted through cutting knives.

Mosaic Flower-Break/Odontoglossum Ring Spot

- It appears either as a mild colour breaks with flower variegation but no distortion or as a severe colour break with flower distortion and variegation.
- Leaves are mottled and twisted.
- Virus transmission by aphid, *Myzus persicae.*

Cattleya Leaf-necrosis

- Sunken, brown to black leaf patterns appear. Sometimes rings or elongated streaks are formed on the older leaves. If leaves are killed before maturity only few small and normal flowers are formed.

Management

- Use of disease free planting materials.
- Disinfect cutting knives.
- Remove and destroy diseased plant materials.
- Spray systemic insecticides to control insect vector..

b. *Cymbidium* spp.

1. Wilt

Symptoms

- ✰ It attacks *Paphiopedilum* plants also.
- ✰ The leaf base initially become yellow and later turns brown.
- ✰ Warm humid weather favours the spread of the disease.

Fungus

Sclerotium rolfsii Sacc. [Sexual stage: *Athelia rolfsii* (Curzi)].

It is a soil living saprophytic pathogen, produces ribbon like hyphae with clamp connections. Basidia are club-shaped, bearing four smooth, ellipsoid basidiospores. Small, brownish sclerotia (hyphal propagules) are also formed.

Management

- ✰ Remove and destroy the infected plants.
- ✰ Treat the pot soil with zineb.

2. Leaf Spot

Symptoms

- ✰ The disease appears as buff-coloured areas with dark mousy-grey margins.
- ✰ The spots coalesce and form elongated to irregular areas with smooth margins.

Fungus

Haplosporella cymbideii Yadav and Rao

c. *Dendrobium* spp.

1. Stem Rot

Symptoms

- ✰ Rapid collapse and rotting of the stem, roots and the leaf tissues.
- ✰ When the disease attacks the plants from the top, rapid collapse and rotting of top growth occurs.
- ✰ The disease then advances either up or down the stem.
- ✰ Presence of small yellow cream to brown spherical sclerotia which are embedded in a dense white fluffy mat of fungal growth is seen.

Chromista

Pythium ultimum Trow, *Pythium splendens* Hans Braun, *Phytophthora palmivora* (E. J. Butler) and *Phytophthora parasitica* Dastur.

Mode of survival: As dormant mycelium and oospores in soil.

Primary spread: From infected plant debris.

Secondary spread: Through soil borne oospores.

Favourable conditions: High humidity and temperature.

Management

- ☆ Remove and destroy the diseased plants.
- ☆ Sterilize the potting media for planting.
- ☆ Spray or drench the soil with metalaxyl 0.2 per cent.

2. Leaf Blight/Anthracnose

Symptoms

- ☆ The infected leaf tissue turns brown with distinct yellow or green halo around it.
- ☆ Later, the brown dead tissue forms concentric ring like appearance.

Fungus

Colletotrichum gloeosporioides Pat and Penz., *C. dendrobii* (Cooke and Massee) Hansf. and *C. Orchidacearum* Allesch.

Favourable conditions: Injuries to the leaf by mechanical or insect damage or sun burn.

Management

- ☆ Avoid exposure of plants to direct sunlight as sun burn.
- ☆ Maintain proper growth with adequate nutrition.
- ☆ Eradicate the diseased plant parts or plants.
- ☆ Spray captan 0.2 per cent or thiophanate-methyl 0.1 per cent at 15 days interval.

3. Leaf Spot

Symptoms

Cercospora Leaf Spot

- ☆ The disease starts as yellowish pale brown spots, turns to purple black.
- ☆ Later large areas of leaves turn pale green with slightly sunken purple coloured spots, present on both upper and lower leaf surfaces.

Phomopsis Leaf Spot

- ☆ The spots are salmon coloured with chestnut coloured smooth margins.
- ☆ The spots are sub-circular to angular usually appearing at the central region of the lamina. Infected areas are sunken with dark brown and punctiform pycnidia.

Fungus

Pseudocercospora dendrobii Goh and W.H. Hsieh and

Phomopsis dendrobii Yadav and Rao.

Mycelium internal and external, branched, septate, pale olivaceous, thin-walled, smooth. Conidia are olivaceous brown 3-8 septate.

Mode of survival: As dormant mycelium.

Primary spread: From infected plant debris.

Secondary spread: Through air-borne conidia.

Favourable conditions: Humid weather and mild temperature.

Management

- Follow good sanitary measures.
- Remove and destroy diseased leaves and plants.
- Spray carbendazim 0.1 per cent at periodical interval.

4. Petal Blight

Symptoms

- Flowers are most susceptible to this disease which causes very prominent small brown spots on the petals.
- These spots reduce the marketability of such flowers.

Fungus

Botrytis cinerea Pers. Ex Fr.

Mode of survival: As sclerotia in infected debris.

Primary spread: From infected plant debris.

Secondary spread: Through air borne and rain splashes conidia.

Favourable conditions: Rainy weather.

Management

- Avoid overhead watering during blooming period.
- Adopt wider plant spacing and proper ventilation.
- Remove and destroy infected plant parts.
- Spray carbendazim 0.1 per cent or mancozeb or thiophanate methyl 0.2 per cent at 7-10 interval.

5. Inflorescence Blight

Symptoms

- The disease creates blights in the individual floret and these florets do not open.

- ☆ Later the tip of the florets turns black and is covered with brown fungal mass and drops off before open.

Fungus

Fusarium sp.

Management

- ☆ Spray carbendazim 0.1 per cent at periodical interval.

6. Bacterial Soft Rot

Symptoms

- ☆ The disease causes rapid collapse of stem tissues resulting in leaf loss and the infected tissues become very soft.
- ☆ When the stems are cut open a pungent odour comes.

Bacteria

Erwinia carotovora sub sp. *carotovora* (Jones) Bergey *et al.,* and *Erwinia chrysanthemi* Burkholder *et al.*

Mode of survival: In infected plant parts.

Primary spread: From diseased plants.

Secondary spread: Through sap on hands or cutting tools, water splashes.

Favourable conditions: Continuous drizzling and moderate rainfall.

Management

- ☆ Use disease free planting materials.
- ☆ Use sterilized pot mixture using streptomycin or oxytetracycline.
- ☆ Remove and destroy the infected plant parts and maintain sanitation.

7. Viral Disease

Symptoms

- ☆ The symptoms include chlorotic mosaic patterns, which seen clearly on the youngest leaves in many orchid varieties.

Fungus

Cymbidium mosaic virus (CyMV)

Mode of survival: In infected orchids.

Primary spread: From diseased plants.

Secondary spread: Through mechanical transmission of sap on hands or tools.

Management

- Adopt strict crop sanitary measures.
- Remove and destroy the infected plants as it cannot be cured.

d. *Eulophia* spp.

1. Leaf Spot

Symptoms

- Circular to irregular, brown spots of 3-5 mm dia with greyish centres appear on the leaves.

Fungus

Cercospora eulophiae Patil

Management

- Good sanitation and removal of diseased leaves and plants.
- Spray of Bavistin at 0.1 per cent at periodical interval controls the disease.

e. *Phalaenopsis* spp.

1. Black Rot and Brown Rot

Symptoms

- Small seedlings die suddenly.
- The disease appears as brown watery patches.
- Severely infected plants do not show normal growth but dry up.

Chromista

Pythium ultimum Trow

Mode of survival: In infected plant debris in the soil.

Primary spread: From soil-borne inoculum.

Secondary spread: Through rain or water splashes.

Favourable conditions: Crowded plants, high humidity, low temperature

Management

- Spray Bordeaux mixture 1 per cent to control the disease.

2. Anthracnose

Symptoms

- The disease appears on the leaves as small, oblong, brown spots with dark brown raised margins.
- In severe cases, the leaves dry up and fall off.

Fungus

Colletotrichum orchidacearum Allesch

Mode of survival: The fungus also survives *Bulbophyllum, Coelogyne, Cymbidium* and *Dendrobium.*

Favourable conditions: Rainy season.

Management

- ✫ Spray copper oxychloride 0.3 per cent or difolatan 0.2 per cent or carbendazim 0.1 per cent.

3. Leaf Blight

Symptoms

- ✫ Small, oblong to globose, brown, zonate spots with lighter centres and deep brown margins appear on the leaves.

Fungus

Pestalotiopsis versicolor (Speg.) Steyaert.

4. Sobralia-Blight or Black Spot

Symptoms

- ✫ Dark discolouration starts from the leaf tips and advances towards the base.

Fungus

Glomerella cincta (Stoneman) Spauld. and H. Schrenk

[Conidial stage: *Colletotrichum cinctum* (Berk. and M.A. Curtis) Stoneman]

Management

- ✫ Spray carbendazim 0.1 per cent or zineb 0.2 per cent to control the disease.

f. *Paphiopedilum* spp.

1. Bacterial Brown Rot

Symptoms

- ✫ The bacterium attacks the plants through wounds and cracks.
- ✫ Yellowing and browning of leaves indicate the presence of the disease.
- ✫ The infection spreads into the crown and causes death of the plant.

Fungus

Erwinia cypripedii (Hori)

Mode of survival: In the infected plant debris.

Primary spread: From infected plant debris.

Secondary spread: Through air and rain splashes.

Favourable conditions: Temperature of 25-30°C.

Management

- ☆ Application of *E. herbicola* or Copper oxychloride 0.25 per cent.

g. *Pholiota imbricata*

1. Leaf Spot

Symptoms

- ☆ Irregular, ochrraceous to fulvous spots surrounded by blackish-brown haloes with wavy margins initiate from the leaf tips and progress downwards.

Fungus

Phyllosticta sp.

Management

- ☆ Spray carbendazim 0.1 per cent or zineb 0.2 per cent or difolatan 0.3 per cent to control the disease.

2. Petal Blight

Symptoms

- ☆ It attacks *Cattleya, Dendrobium, Oncidium* and *Phalaenopsis.*
- ☆ The disease occurs under cool, damp and still conditions.
- ☆ It produces small, dark brown spots on petals of older flowers.
- ☆ The fungus sporulates heavily on dead and diseased flowers.

Fungus

Botrytis cinerea Pers.

Management

- ☆ Remove and destroy diseased flowers.
- ☆ Spray carbendazim 0.1 per cent or mancozeb 0.2 per cent to control the disease.

h. *Rhynchostylis retusa*

1. Leaf Spot

Symptoms

- ☆ Conspicuous blackish-grey lesions appear which coalesce and form large dark patches.

Fungus

Macrophoma sp.

2. Stem Canker

Symptoms

- ✰ Symptoms appear as whitish-grey, irregular spots on the stem.

Fungus

Stagnospora sp.

3. European Anthracnose

Symptoms

- ✰ The fungus usually attacks the leaf tips.
- ✰ Necrotic tissues shrivel and dry up.
- ✰ The infected regions have smooth margins and are surrounded by yellow zones.

Fungus

Gloeosporium affine Sacc.

i. *Vanda caerulea*

1. Heart Rot

Symptoms

- ✰ The fungus attacks Cattleya and Phaloenopsis also.
- ✰ The affected leaves turn yellow and drop off.
- ✰ Pseudo bulbs have dark rotted areas inside.

Chromista

Phytophthora palmivora (E. J. Butler)

Management

- ✰ Remove the infected pseudo bulbs and repot the plants in fresh media after dipping fungicidal solution.

2. Leaf Spot

Symptoms

- ✰ Brownish to dark, irregular blotches which turn grey later appear on the leaves.
- ✰ Severely infected leaves dry up.

Fungus

Phomopsis orchidophila E.K. Cash and A.M.J. Watson

15. Pyrethrum [*Chrysanthemum cinerariifolium* (Trevir.) Vis.]

1. *Fusarium* Wilt

Symptoms

- ☆ Yellowing of older leaves is seen first and it extends to upper leaves also.
- ☆ Infected plants are stunted and produce few branches only.
- ☆ Infected plants droop and wilt.
- ☆ Roots become brown to black.

Fungus

Fusarium oxysporum Schltdl

Mode of survival: As chlamydospores in infected plant debris in soil.

Primary spread: Infected seeds and chlamydospores in soil.

Secondary spread: Through irrigation water and rain splashes.

Favourable conditions: Low soil moisture acidic pH (5.5-6.5) sandy loam soil, temperature of 23-32°C.

Management

- ☆ Follow deep summer ploughing.
- ☆ Grow resistant varieties.
- ☆ Follow rotation with non-hosts like cereals.
- ☆ Use healthy certified seeds.
- ☆ Neutralize soil pH with lime 200-300 g/plant.
- ☆ Treat seeds with carbendazim 2 g/kg.
- ☆ Uproot and destroy diseased plants.
- ☆ Drench the soil with carbendazim 0.1 per cent.

2. Leaf Spot/Leaf Sheath

Symptoms

- ☆ Leaves show circular, irregular blackish brown spots surrounded by yellow halo.
- ☆ Severe infection causes curling of leaves.
- ☆ Diseased leaves become yellow, wither, dry and hang on the stem for sometimes.
- ☆ Infected of flower buds leads to rotting.

Fungus

Septoria chrysanthemella Cavara

Spread: Through wind-borne pycnidiospores, rain splashes.

Favourable conditions: Cool cloudy weather, temperature of 18-20°C and relative humidity of 95-99 per cent, intermittent rainfall.

Management

- ☆ Collect and destroy infected leaves and buds.
- ☆ Avoid excess irrigation.
- ☆ Spray carbendazim 0.1 per cent or mancozeb 0.2 per cent.

3. *Sclerotinia* Collar Rot/Crown Rot

Symptoms

- ☆ Pale or dark brown bleached lesions appear on the collar region which rapidly cause rotting of stem and collapse of the plant.
- ☆ White cottony fungal mycelium is seen on the collar region with white sclerotia becoming black and hard later.
- ☆ Drooping of leaves occur followed by drying and death of entire plant.

Fungus

Sclerotinia minor Jagger

Mode of survival: As sclerotia, apothecia and ascospores.

Primary spread: From sclerotia in diseased plant debris in soil.

Secondary spread: Through mycelial strands.

Favourable conditions: Cool wet weather.

Management

- ☆ Collect and destroy infected plant materials.
- ☆ Follow crop rotation with non-hosts like corn, cotton, onion or sweet potato.
- ☆ Apply optimum dose of fertilizers.
- ☆ Drench the root region of plants with carbendazim 0.1 per cent.

4. Ray Blight

Symptoms

- ☆ Attacks all the above ground plant parts.
- ☆ Flower bunds and ray florets of flowers are affected.
- ☆ Necrosis of ray florets.
- ☆ Distortion of stem and necrosis of growing tips.

Fungus

Phoma ligulicola var. *inoxydablis*

Mode of survival: As mycelium.

Primary spread: From pycnidia in infected tissue.

Management

- ✰ Use healthy certified seeds.
- ✰ Follow rotation of crops including cereal crops.
- ✰ Collect and destroy diseased flower/plants.
- ✰ Spray carbendazim 0.1 per cent or azoxystrobin or chlorothalonil.

16. Rose (*Rosa* sp.)

1. Powdery Mildew

Symptoms

- ✰ Greyish white powdery growth appears on tender leaves, young shoots, flower buds and on open-flowers.
- ✰ Foliage may be distorted, and shoots stunted or swollen.
- ✰ Diseased flower buds may not open and dry later.

Fungus

Podosphaera pannosa (Wallr.:Fr.) de Bary.

Formerly, *Sphaerotheca pannosa* var. *rosae* Woron.

Conidia oval-shaped and produced in chains (5 to 10) on short, erect conidiophores. Cleistothecia are with myceloid appendages. Each cleistothecium (ascocarp) contains only one large ascus. Ascospores are spherical.

Mode of survival: As overwinters as mycelium in dormant buds and shoots. Rose plants are the only host for the pathogen.

Primary spread: From ascospores or conidia.

Secondary spread: Through air-borne conidia.

Favourable conditions: Immature leaves, dry weather with maximum day temperature of 20-25°C with cool humid nights.

Management

- ✰ Grow resistant varieties.
- ✰ Collect and burn diseased leaves.
- ✰ Avoid excess nitrogen application and crowding of plants.
- ✰ Spray wettable sulphur 0.3 per cent or carbendazim 0.1 per cent or dinocap 0.07 per cent or dust sulphur at 25 kg/ha at 10 days interval.
- ✰ Sulphur should not be used when the temperature reaches above 30°C.

2. Black Spot

Symptoms

- ☆ Coal black, more or less circular spots with fimbriate or fringed margins (diagnostic character) are seen mostly on the lower leaves.
- ☆ The spots vary from one or two to a dozen or more on a leaf, usually on the upper surface.
- ☆ With close examination, small black dots or pimples in the center of the spots represent the acervuli (fruiting bodies of the fungus), which glisten when wet.
- ☆ Yellow halo appears around each matured spot.
- ☆ Affected leaves turn yellow and fall off prematurely leaving the canes completely defoliated.
- ☆ Flower size and production are reduced.
- ☆ The affected plants are devitalized and those bushes may die during the following winter.

Fungus

Diplocarpon rosae F.A. Wolf (Asexual stage*: Marssonina rosea*). Conidia fusiform, hyaline and two celled. Ascospores are oblong, hyaline, unequally 2-celled, constricted at the septum.

Spread: Air-borne conidia or ascospores.

Mode of survival: As On infected fallen leaves, cane (stem) lesions and in wild roses.

Primary spread: From ascospores or conidia.

Secondary spread: Through rain splashes and overhead watering.

Favourable conditions: Winter frosts; low temperature of 21°C and high humidity.

Management

- ☆ Grow resistant/tolerant varieties. *Rosa bracteata* is tolerant.
- ☆ Avoid shade and excess irrigation.
- ☆ Remove and burn the affected and dried leaves.
- ☆ Prune diseased canes.
- ☆ Spray captan 0.2 per cent or carbendazim 0.1 per cent.

3. Die back

Symptoms

- ☆ Die back generally occurs after pruning.

- ☆ Dieback of canes or shoots occurs from tips, wounds created by removing buds or thorns and other mechanical injuries.
- ☆ Lesions start as small yellow or reddish discoloration around wounds that turn tan with a dark brown border.
- ☆ Cracks form in these areas due to drying and the wounds girdle the stems killing them.
- ☆ The spores producing fruiting bodies (pycnidia) are black and appear in the dead tissues.

Fungi

Leptosphaeria coniothyrium, Diplodia rosarum, Botryodiplodia theobromae and *Collectotrichum* sp.

Mode of survival: In dead twigs and the stalks of the withered blooms.

Primary spread: From ascospores.

Secondary spread: Through air-borne pycnidiospores.

Favourable conditions: Older plants, neglected and weak bushes, defoliated black spot affected plants, wet weather, poorly drained soil and poor nutrition.

Management

- ☆ Avoid damage by digger wasp.
- ☆ Prune affected shoots and apply Bordeaux paste or copper oxychloride.
- ☆ Spray copper oxychloride 0. 25 per cent or difolatan or chlorothalonil or mancozeb 0.2 per cent.
- ☆ Maintain vigour by proper fertilization.

4. Rust

Symptoms

- ☆ In early spring, minute, glistening, orange-red spots (spermagonia or pycnia) appear on the upper surfaces of the leaves.
- ☆ Within a few days, yellow spots (aecia) appear on the under leaf surfaces which rupture and give bright orange coloured powdery masses of aeciospores.
- ☆ Petioles, green stems, calyxes, sepals, and fruit are infected.
- ☆ The lesions on young stems are long and narrow. The stem lesions cause distortion, curving, and destruction of the buds.
- ☆ The summer stage of the disease (uredial stage) is the most conspicuous and serious.
- ☆ Very small, reddish orange pustules (uredia) containing powdery masses of bright yellow-orange spores (urediospores) form on the lower leaf surfaces and petioles.

- ☆ Heavily rusted leaves become twisted and wilt, wither, and fall off within 5 to 10 days.
- ☆ Infected young shoots may become distorted and turn reddish followed by a general decline in plant vigour.
- ☆ Masses of black spores (teliospores) begin to form during late summer or early fall gradually replacing the orange uredospores in the old pustules, turning them a dark chocolate-brown to black.
- ☆ The black pustules appear as crust-like on the stems.

Fungus

Phragmidium mucronatum (Pers.) Schltdl.

Uredospores are unicellular, orange yellow, ovoid, pedicellate and echinulate. Teleospores are dark coloured, cylindrical, 6-8 celled with a pointed papillum.

Mode of survival: As mycelium and teleutospores in dormant buds and shoots.

Primary spread: From basidiospores formed in overwintered black pustules (teleutosori) on dead leaves and stems.

Secondary spread: Through air-borne basidiospores and uredospores.

Favourable conditions: Temperature of 18-21°C

Management

- ☆ Collect and burn diseased fallen leaves.
- ☆ Use disease-free planting materials/resistant hybrids or cultivars.
- ☆ Destroy wild or uncared rose bushes.
- ☆ Spray wettable sulphur 0.3 per cent or dinocap 0.1 per cent or carbendazim 0.1 per cent or dust with sulphur at 25 kg/ha at 15 days interval.

5. Downy Mildew

Symptoms

- ☆ The disease occurs on all types of roses including cut flowers, garden roses, potted miniature roses and carpet type roses used as ground covers.
- ☆ Yellowish to brown blotches occur on upper leaf surface along the margins.
- ☆ Under moist, cool conditions leaves may turn yellow and drop.
- ☆ White, downy fungal growth occurs on the lower leaf surface.
- ☆ Reddish streaks occur on flower stems and canes and flowers are also infected.

Chromista

Peronospora sparsa Berkeley

Mode of survival: As oospores and dormant mycelium in plant debris.

Primary spread: From infected plant debris.

Secondary spread: Through soil-borne oospores.

Favourable conditions: Winter months with a temperature of 60-70°F and a relative humidity of 85 per cent.

Management

- Spray mancozeb 0.2 per cent or fosetyl - aluminium 0.1 per cent or metalaxyl + mancozeb 0.2 per cent.

6. *Botrytis* Blight

Symptoms

- One of the most common diseases affecting rose flowers from the garden to cut flower production to storage.
- The disease affects flowers which may not open and become covered with greyish brown fungal growth.
- Tiny water-soaked, translucent spots appear as flecks, blisters or sunken areas on petals.
- They rapidly turn tan brown with a dark margin, wither and fall-off (flower blight).
- Under Favourable conditions, petals become matted and stick together before buds open.
- Infection spreads from the blossom, leaf or petiole into the stem or enters through injuries during pruning.
- Lesions on rose canes tend to be black, sunken and elongated with a definite outline.
- Infections can weaken and girdle the stem which leads to wilting and death of the foliage above the lesion.
- These infections often result in cane blight.
- *Botrytis* leaf spot appears when infected petals or other debris falls on the leaf and the fungus can then invade the healthy tissue. The leaf spot will take the shape of the debris.
- Under ideal conditions the fungus sporulates to produce masses of grey-brown powdery spores.

Fungus

Botrytis cinerea Pers. Ex Fr.

Botrytis species produce sclerotia, which are melanized resting bodies. Which produce prolific grey conidiophores and (macro) conidia.

Mode of survival: As overwinters as sclerotia or mycelia or conidia in infected plant materials, plant debris and soil.

Primary spread: From conidia or infection hyphae arising from sclerotia.

Secondary spread: Through air-borne conidia and rain splashes.

Favourable conditions: Early spring season, wounds made during pruning or cutting of flowers, younger and succulent plants.

Management

- ☆ Follow strict sanitation by removing old blooms, canes showing die back and other infected plant materials.
- ☆ Collect fallen leaves and flowers and burn.
- ☆ Prune out infected canes, buds and flowers and maintain good sanitation.
- ☆ Follow proper spacing to improve air circulation.
- ☆ Avoid splashing of water during irrigation.
- ☆ Avoid excess nitrogenous fertilizers.
- ☆ Spray chlorothalonil or mancozeb 0.2 per cent.

7. Crown Gall

Symptoms

- ☆ Galls appear on stems and roots.
- ☆ Galls are pale green or white but turn brown and become corky with age.
- ☆ These gall are formed after one week to many months of infection.
- ☆ Galls are 0.6 cm to over 2.5 cm in diameter on stems and roots.
- ☆ Plants with crown galls are stunted and have weakened stems and foliage.

Bacterium

Rhizobium radiobacter (Beijerinck and van Delden)

[Syn. *Agrobacterium tumefaciens* (Smith and Townsend) Conn].

It is a Gram-negative, rod shaped (bacilliform), aerobic, motile bacterium having 1-6 peritrichous flagella. A large extrachromosomal piece of DNA, commonly referred to as a tumor-inducing (Ti) plasmid, is carried by *A. tumefaciens*.

Mode of survival: In soil

Primary spread: From soil-borne bacteria.

Secondary spread: Through irrigation water,

Favourable conditions: Weakened and dying plants, injuries made during grafting or cultural operations or by rodents or insects.

Management

- ☆ Use only crown gall free plants and avoid wounding as much as possible.
- ☆ Use sterilized tools for field operations.
- ☆ Rouge out infected plants and destroy.

8. Mosaic

Symptoms

- ☆ The symptoms are highly variable, depending on the cultivar, the virus and the environment and also from year to year.
- ☆ The more common symptoms are: chlorotic bands or ringspots, wavy lines, yellow vein banding, oak-leaf pattern and general mosaic and colour-breaking.
- ☆ Some infected plants never express symptoms.
- ☆ Infected plants are less vigorous, produce fewer flowers on shorter stems and have poor transplant survival rates.

Virus

Prunus necrotic ringspot virus (PNRSV) and *Apple mosaic virus* (ApMV).

They may be found separately or together in diseased plants.

Mode of survival: In infected bushes.

Primary spread: From vegetative propagation of infected buds, scion or root stocks.

Management

- ☆ Use only disease free planting materials.
- ☆ Do not use diseased plants for propagation.
- ☆ Remove and destroy infected plants.

9. Rosette

Symptoms

- ☆ Roses are the only plants known to be susceptible.
- ☆ Increased growth/rapid elongation of shoots with red discolouration.
- ☆ Witches' broom (prolific clustering of small shoots).
- ☆ Spiral pattern of cane growth.
- ☆ Shortening of internodes with distorted or dwarfed leaves.
- ☆ Abundance of thorns.
- ☆ Atypical flower coloration (mottling) and also with deformed buds and flowers.
- ☆ Infected plants are highly susceptible to powdery mildew.

Virus

Rose rosette virus

Mode of survival: In infected bushes.

Primary spread: From grafting.

Secondary spread: Through eriophyid rose leaf curl mite (*Phyllacoptes fructiphylus*).

Management

- ☆ Use only disease free planting materials.
- ☆ Do not use diseased plants for propagation.
- ☆ Provide sufficient space between plants to avoid crawling of mites to adjacent plants.
- ☆ Uproot the infected plants and destroy. Care must be taken to avoid scattering of mites to healthy plants.
- ☆ Spray any acaricide to control vector population.

17. Tuberose (*Polianthes tuberosa* L.]

1. Wilt and Stem Rot

Symptoms

- ☆ Disease appears in patches.
- ☆ Yellowing and drooping of leaves followed by drying of plants.
- ☆ Fungal mycelia mat is seen at the base of the infected plants near the soil line.
- ☆ Rotting extends and covers the entire leaf. The infected leaves get detached from the plant. As a result, the infected plant becomes weak and unproductive.
- ☆ The roots and tubers of diseased plants rot and plant wilts and die.
- ☆ Brown coloured mustard seed-like sclerotia are seen on the surface of the diseased stem tissues and at the soil level near the root region.

Fungus

Sclerotium rolfsii Sacc. [Sexual stage: *Athelia rolfsii* (Curzi)].

It is a soil living saprophytic pathogen, produces ribbon like hyphae with clamp connections. Basidia are club-shaped, bearing four smooth, ellipsoid basidiospores. Small, brownish sclerotia (hyphal propagules) are also formed.

Mode of survival: sclerotial bodies in infected plant parts and in soil.

Primary spread: From soil-borne sclerotia.

Secondary spread: Through irrigation water.

Favourable conditions: Rainy season, wet soil and high soil moisture.

Management

- ☆ Drench the soil with carbendazim 0.1 per cent or *Pseudomonas fluorescens* at 2 g/lit.

2 *Botrytis* Blight/Leaf and Flower Spot

Symptoms

- The disease appears during rainy season.
- The infection starts at lower leaves and stalks as orange to reddish brown, oval spots and these spots coalesce and cause blighting of leaves.
- Flower buds rot or open with distorted flowers.
- Infected flowers show dark brown spots and ultimately the entire inflorescence dries up.

Fungus

Botrytis elliptica (Berk.) Cooke

Mode of survival: As survives as sclerotia in fallen flowers and leaves.

Primary spread: From infected debris.

Secondary spread: Through rain, air current and gardeners.

Favourable conditions: Rainy season

Management

- Avoid dense planting.
- Collect and burn diseased plant parts.
- Spray Bordeaux mixture 1.0 per cent or carbendazim 0.1 per cent.

3. Blossom Blight

Symptoms

- The disease appears as light brown lesions on petals.
- These lesions soon darken and results in drying up of the affected portion.
- The blighted blosssoms drop off.
- The pathogen also infects the flower stalk.
- Under humid conditions, the flower tips turn brown.

Fungus

Fusarium equiseti (Corda) Sacc.

Management

- Collect and remove diseased fallen leaves and flowers.
- Spray carbendazim 0.1 per cent.

4. Leaf Spot

Symptoms

- The disease appears as brown spots with faint concentric rings on midrib.

- ✰ Circular to oval spots are seen on peduncle.
- ✰ Infected leaves and peduncles become necrotic and dry up.

Fungus

Alternaria polyanthi Mariappan

Mode of survival: In infected plants.

Primary spread: From diseased plant parts and debris.

Secondary spread: Through air-borne conidia.

Management

- ✰ Collect and remove diseased leaves.
- ✰ Spray mancozeb or chlorothalonil or iprodione 0.2 per cent at 10 days interval.

5. Flower Bud Rot

Symptoms

- ✰ It occurs mainly on young flower buds results in dry rot of buds.
- ✰ Brown, scorched and necrotic discolouration of peduncles is seen.
- ✰ Affected buds shrivel and dry.

Bacterium

Erwinia sp.

Mode of survival: In infected plants.

Primary spread: From infected plants.

Secondary spread: Through water splashes, irrigation water, contaminated tools and implements.

Management

- ✰ Eliminate infected plants.
- ✰ Avoid mist spray or overhead watering.
- ✰ Disinfect tools and implements.
- ✰ Spray streptomycin 0.01 per cent.

18. Zinnia (*Zinnia elegans* Jacq.)

1. Damping Off and Root Rot

Symptoms

- ✰ The disease is most prevalent at the seedling stage.

- Necrotic spots and rings develop on the young seedlings causing collapse of the seedlings.
- Considerable loss is sustained if seedlings are not properly looked after.

Chromista

Pythium spp. and *Rhizoctonia solani* Kuhn

Mode of survival: As oospores/sclerotia in soil.

Primary spread: From soil-borne oospores/sclerotia.

Secondary spread: Through zoospores/irrigation water.

Favourable conditions: High soil moisture and Cool climate.

Management

- Use disease-free seeds.
- Provide better drainage.
- Drench the nursery with Bordeaux mixture 1 per cent or Copper oxychloride 0.25 per cent.

2. Southern Blight

Symptoms

- White mycelial web is found at the base of the stem and it spread over the ground during wet weather.
- Numerous, brown, mustard seed-like sclerotia are found with the mycelium.
- Diseased plants wilt and die.

Fungus

Sclerotium rolfsii Sacc. [Sexual stage: *Athelia rolfsii* (Curzi)].

It is a soil living saprophytic pathogen, produces ribbon like hyphae with clamp connections. Basidia are club-shaped, bearing four smooth, ellipsoid basidiospores. Small, brownish sclerotia (hyphal propagules) are also formed.

Mode of survival: In infected plant tissues, in other plant species *viz.*, Calendula, Canna, Carnation, Cosmos, Chrysanthemum, Dahliae, Gladiolus, Jasmine, Lily, Marigold, Tulip and in soil as sclerotia.

Primary spread: From soil-borne inoculum.

Secondary spread: Through irrigation water.

Favourable conditions: Sandy loam soil.

Management

- Remove and destroy affected plants.

- ☆ Add more organic amendments to soil to suppress the pathogen.
- ☆ Adopt crop rotation with non-host crops.
- ☆ Drench the soil with captan (0.2 per cent) or carbendazim (0.1 per cent) at root region.

3. Powdery Mildew

Symptoms

- ☆ The disease first appears in August or September as circular, white powdery patches on the upper surface of lower leaves.
- ☆ As the disease advances all the leaves in the plants are covered with powdery coating.
- ☆ Affected leaves curl, twist and turn yellow and dry.
- ☆ Flower buds stem and petals are covered with powdery coating of the fungus.
- ☆ In severe infection, the quality of flowers and its market value is reduced.

Fungus

Golovinomyces cichoracearum DC (Syn. *Erysiphe cichoracearum* DC)

Mode of survival: In infected plant tissues.

Primary spread: From diseased plant parts.

Secondary spread: Through air-borne conidia.

Favourable conditions: Humid conditions.

Management

- ☆ Avoid shady conditions.
- ☆ Remove severely diseased plants.
- ☆ Spray wettable sulphur 0.2 per cent or dinocap 0.2 per cent.

4. Twig Die back/Blight

Symptoms

- ☆ Flowering branches show die back symptoms.
- ☆ Infected leaves show dark brown, irregular necrotic lesions in which pycnidia are embedded in the host tissues.
- ☆ Later, they become erumpent, globose to sub-globose and ostiolate.

Fungus

Phomopsis zinnia

Pycnidia are erumpent and globose, Beta spores are absent.

Management

- ☆ Collect and burn diseased plant debris.
- ☆ Use disease-free seeds for sowing.
- ☆ Spray zineb 0.2 per cent or copper oxychloride 0.3 per cent or carbendazim 0.1 per cent.

5. *Alternaria* Leaf Blight

Symptoms

- ☆ Small, reddish brown spots of 2-10 mm dia with greyish white centres appear on leaves which later enlarge to irregular, brown dry areas with concentric zonations.
- ☆ Severely affected leaves become brown and dry up. The centre of an older spot drop out leaving a ragged hole.
- ☆ Similar spots appear on petioles, stems, internodes and nodes.
- ☆ Dark brown, sunken spots on stem may girdle the stem with drying back of upper portions.
- ☆ Affected plants often wilt.
- ☆ Roots may turn grey, rot and slough off resulting in wilting and death of plants.
- ☆ Small brown spots on flowers enlarge and occupy whole petals causing the flower blight and it gives unsightly appearance.
- ☆ Damping off of seedlings is also common.

Fungus

Alternaria zinniae Ellis

Mode of survival: As Fungus survives on seed and in the soil associated with host debris.

Primary spread: From soil and seed-borne inoculum.

Secondary spread: Through air-borne conidia

Favourable conditions: Mid-summer and early fall season.

Management

- ☆ Collect and burn diseased plant debris.
- ☆ Follow atleast 2 year crop rotation.
- ☆ Use resistant varieties.
- ☆ Treat the seeds in hot water at 51.7°C for 30 min or treat with captan or thiram (4g/kg) which eliminates seed-borne infection.
- ☆ Do not use overhead irrigation.

- ☆ Give foliar spray with zineb or copper oxychloride or mancozeb or chlorothalonil 0.2 per cent or azoxystrobin 0.1 per cent.

6. *Cercospora* Leaf Spot

Symptoms

- ☆ Numerous, small, conspicuous, circular, white to greyish spots with reddish brown or purple margins appear on leaves.
- ☆ Infected tissues in the centre of the spot drop off leaving a shot-hole symptom.
- ☆ Dead tissues may crack and tear.
- ☆ In severe cases of infection, the leaves turn brownish and dry off.
- ☆ Fruiting bodies are seen on both sides but more on the upper surface.

Fungus

Cercospora zinnia Ellis and Martin.

Management

- ☆ Use healthy seeds for sowing.
- ☆ Collect and burn diseased plant materials.
- ☆ Spray carbendazim 0.1 per cent or zineb 0.2 per cent.

7. Leaf Spots

Symptoms

Glomerella Leaf Spot

- ☆ Numerous, irregular, slightly sunken spots are found on both surfaces of leaves.
- ☆ These spots are dark brown with violet shade and form irregular necrotic zones.
- ☆ They increase rapidly under warm and humid conditions.
- ☆ In older spots, numerous black acervuli are found.

Phyllosticta Leaf Spot

- ☆ Small, reddish brown spots appear on leaves which later turn greyish and coalesce to form irregular patches.
- ☆ Severe infection results in drying of leaves.
- ☆ It also attacks flowers and seeds.

Septoria Leaf Spot

- ☆ Infection starts from lower leaves and progresses upwards.
- ☆ Oval to irregular, smoky-grey to black spots are seen on the leaves.
- ☆ Minute, black fruiting bodies are seen on these spots.

Fungus

Glomerella cingulate (Stoneman), *Phyllosticta zinnia* and *Septoria tageticola* Changsri and G.F. Weber

Management

- ☆ Select healthy seeds for sowing.
- ☆ Collect and burn diseased plant materials.
- ☆ Foliar spray with carbendazim 0.1 per cent or zineb or mancozeb 0.2 per cent.

8. Bacterial Leaf Spot

Symptoms

- ☆ The disease first appears as small (1-2 mm dia), diffuse, translucent spots surrounded by broad yellowish halos usually seen in lower leaves and later spread to upper leaves.
- ☆ Under wet conditions, these lesions enlarge, become angular to irregularly circular and develop a reddish centre.
- ☆ The lesions merge to form irregular dead areas that may crack as they dry.
- ☆ Heavily spotted leaves wither and die.
- ☆ During humid weather, small brown spots form on the ray flowers.
- ☆ In severe conditions, the flower heads are seriously disfigured and decay completely.
- ☆ Since it is seed-borne, disease development on newly emerged or transplanted plants is often seen.

Bacterium

Xanthomonas campestris pv. *zinnniae* (Hopkins and Dowson) Dye

Mode of survival: In seed and diseased plant debris.

Primary spread: From seed-borne inoculum.

Secondary spread: Through rain splashes and irrigation water.

Favourable conditions: Frequent showers and overhead watering.

Management

- ☆ Use only disease-free, healthy transplants or disease-free seeds.
- ☆ Increase space between plants for good air circulation.
- ☆ Practice strict sanitation.
- ☆ Collect and burn diseased leaves and plant debris.
- ☆ Follow crop rotation for 3 years.

- ✰ Advise workers to wash their hands after handling diseased plants or soil.
- ✰ Avoid handling of wet foliage.
- ✰ Avoid splashing when watering.
- ✰ Treat seeds with bleaching powder 10 per cent for 15-30 min.

9. Mosaic

Symptoms

- ✰ Light and dark green patches appear on the foliage accompanied by vein clearing and green vein banding.
- ✰ The diseased plants are pale in appearance.
- ✰ Infected plants are stunted with fewer low grade flowers.

Virus

Dahlia mosaic virus (DMV)

Mode of survival: In infected plant tissues and in other host plants like bitter gourd, cucumber, *Petunia hybrida, Nicotiana glutinosa* and tobacco.

Primary spread: From infected plants and other host species.

Secondary spread: Through sap and by aphids, *Aphis gossypii* and *Myzus persicae.*

Management

- ✰ Collect and destroy diseased plants.
- ✰ Spray insecticides to control insect vectors.

10. Leaf Curl

Symptoms

Thickening of veins and veinlets on lower surface, curling of leaves, dwarfing of plants, distorted flowers with pale colour and partial sterility are the common symptoms of leaf curl disease. It is not a seed-borne viral disease.

Virus

Zinnia leaf curl virus

Mode of survival: In infected plants.

Primary spread: From infected plants.

Secondary spread: Through insect vector, whitefly (*Bemisia gossypiperda*).

Management

- ✰ Collect and destroy diseased plants.
- ✰ Spray insecticides at 15 days interval to control insect vectors.

Chapter 4

Spices and Condiments

1. Betelvine (*Piper betle* L.)

1. Foot Rot or Leaf Rot or Wilt

Symptoms

- ✰ Initial symptom is sudden wilting of vines. The affected vines show yellowing and drooping of the leaves from tip downwards.
- ✰ The affected plant dry up completely within 2 or 3 days. The succulent stem turns brown, brittle and dry as stick.
- ✰ The diseased internodes undergo 'wet rot' and the tissue become soft, slimy with a fishy odour.
- ✰ The leaves turn brown to dark brown or dirty black and defoliation occurs.
- ✰ The leaves within 2-3 feet height of the vine show the leaf rot symptom.

Chromista

Phytophthora parasitica var. *piperina* Dastur

The oomycete produces hyaline, non septate mycelium. The sporangia are thinwalled, hyaline ovate or learn shaped with papillae. Zoospores, which are liberated from the sporangia, are kidney-shaped and biflagellate. Oospores are dark brown, globose and thick walled.

Mode of survival: As soil-borne and survives as facultative saprophyte in the infected plant debris and in the soil as oospores and chlamydospores.

Primary spread: From irrigation water.

Secondary spread: Through sporangia and zoospores disseminated by splash irrigation and wind-borne rains.

Favourable conditions: high atmospheric humidity and low night temperature (23°C).

Management

- ✰ Grow tolerant cultivars Halisahar Sanchi, Pachaikodi and Karapaku.
- ✰ Select well matured (more than 1 year old) seed vines free from pest and diseases.
- ✰ A combined effect of flooding followed by hot weather and deep plowing has been found to reduce the incidence of the disease.
- ✰ Soak the seed vines for about 30 minutes in Streptocyline 500 ppm or Bordeaux mixture 0.5 per cent.
- ✰ Apply 150 kg N/ha/year through Neem cake (75 kg N) and Urea (75 kg N) and 100 kg P_2O_5 through Super phosphate and 30 kg Muriate of potash in 3 split doses first at 15 days after lifting the vines and second and third

dose at 40 - 45 days intervals. Apply on beds, shade dried neem leaf or *Calotropis* leaves at 2 t/ha and cover it with mud (2 t in 2 split doses).

- Drench Bordeaux mixture 0.25 per cent in basins formed around the vine at monthly intervals starting from October - January, three times soil drench and six times spray from June - July.
- Spray near the root zone with fosetyl-Al or deconil or peronox 0.35 per cent, fytolan 0.2 per cent, Dithane Z-78 0.2 per cent or Blitox-50 0.25 per cent three to four times at 8–10 days interval
- During winter season avoid frequent irrigation.
- Remove the affected vines away from the garden and burn them.
- Application of *Trichoderma viride* at 5 g/vine.

2. Leaf Spot/Anthracnose

Symptoms

- Presence of circular to irregular, light to dark brown lesions on the leaves that is surrounded by a yellow halo.
- The center of such spots are later turned straw yellow in color.
- The spots often coalesced to form bigger patches.
- On stems, branches and petiole small black, irregular specks are seen which occasionally ruptured the cortex underneath.

Fungi

Colletotrichum capsici (Syd.) Butler and Bisby; *Colletotrichum piperis* Petch

The fungus colony was grayish black and smooth. Conidia were oblong hyaline, non-septate with rounded ends, having oil globules in the center.

Favourable conditions: The disease appears after rains and only leaves are affected.

Management

- Grow seedlings in raised pasteurized beds.
- Use pathogen free cuttings.
- Spray bitertanol 0.05 per cent, followed by mancozeb 0.2 per cent, ziram 0.1 per cent (as Cuman-L) and Bordeaux mixture 0.5 per cent.

3. Sclerotium Foot Rot and Wilt

Symptoms

- The infection usually starts at the collar region.
- Whitish cottony mycelium is seen on the stem and roots.
- The stem portion shows rotting of tissues at the point of attack and the plants show dropping of leaves and withering finally dry up.

Fungi

Sclerotium rolfsii Dastur.

The fungus produces white to grey mycelium which have profuse branching. Sclerotia are spherical smooth and shiny. Brown coloured mustard like sclerotia are seen on the infected stem and soil near the vines.

Mode of survival: As fungus is soil-borne and grows saprophytically in the dead plant tissue in soil. The fungus also survives as sclerotia in the infected plant debris in the soil for more than one year.

Spread: The sclerotia spread through irrigation water.

Favourable conditions: May-July months with high temperature of 28-30°C.

Management

- ☆ Remove the affected vines along with the roots and burn.
- ☆ Soil solarization during summer followed by application of *Trichoderma* 10 kg/ha (set and soil application), neem cake 200 kg/ha, NPK at 100:50:50 kg/ha and deep irrigation.
- ☆ Spray fungicides, Metalaxyl + Mancozeb 0.1 per cent, Difenoconazole 0.05 per cent, Propiconazole 0.05 per cent, Hexaconazole 0.05 per cent and Metiram 0.1 per cent.

4. Powdery Mildew

Symptoms

- ☆ Appearance of yellow spots, which are slightly raised and irregular in outline.
- ☆ White powdery patches of mildews on upper surface of leaves.

Fungi

Oidium piperis Uppal, Kamat and Patel

The fungus is an ectophytes, which feeds by sending globular haustoria into epidermal cells of the leaves.

Favourable conditions: Spread of conidia during winter season.

Management

- ☆ Use resistant clones Kapoori, Tellaku, Vellairettala, Ambadi Badam and Kulgedu.
- ☆ Use pathogen free cuttings.
- ☆ Spray bitertanol 0.05 per cent, followed by mancozeb 0.2 per cent, ziram 0.1 per cent (as Cuman-L) and Bordeaux mixture 0.5 per cent.

5. Bacterial Leaf Spot/Stem Canker/Bacterial Blight

Symptoms

- Initially minute water soaked spots all over the leave blade delimited by veins.
- Several of these coalesce to form larger irregular brownish spots.
- The advance lesions are invariably accompanied by yellowish halos and result in ultimate defoliation.
- Often grayish black lesions are seen on stem and petioles.

Bacterium

Xanthomonas axonopodis pv. *betlicola* by Vauterin *et al.*

The color of colony is yellow and slimy, short rods, motile, temperature 26-30°C, gram negative, growth on nutrient broth-turbid yellow growth.

Favourable condition: Continuous rainfall.

Management

- Spray Bordeaux mixture 0.5 per cent or copper oxychloride 0.1 per cent.
- Spray Copper oxychloride 0.25 per cent with streptocycline 500 ppm.

6. Viral Diseases

a. *Piper Yellow Mottle Virus (PYMoV)*

Mild to severe chlorotic mottling, vein clearing, intervienal chlorosis, reduction in leaf size, leaf puckering and deformation.

Transmission: Mechanically and by mealy bug - *Planococcus citri.*

b. *Cucumber Mosaic Virus (CMV)*

Alternate dark green and light green patches was observed on leaf surface.

Transmission: non-persistently by aphids, such as *Aphis craccivora, Myzus persicae,* and *Acyrthosiphon pisum.*

c. *Canna Yellow Mottle Virus (CYMoV)*

Propagated through cuttings.

d. *Pepper Ringspot Virus (PeRSV)*

Transmitted by Aphids.

e. *Pepper Mild Mottle Virus (PMMoV)*

f. *Pepper Mottle Virus (PMoV)*

g. *Bean Yellow Mosaic Virus (BYMV)*

2. Chilli (*Capsicum annuum* L.)

1. Damping Off

Symptoms

- Seeds rot, seeds germinate but rot before emergence (pre-emergence damping off).
- Infection is observed in young seedlings.
- Collar region becomes water-soaked and the affected portion is soft and weak.
- Seedlings collapse and topple down at ground level and die (post-emergence damping off).
- Gappiness in the nursery is noticed due to death of seedlings in patches.

Chromista

Pythium aphanidermatum (Edson) Fitzp.

Mycelium is coenocytic, hyaline and branched. Sporangia are lobed and hyaline. The oospores are spherical and thick walled.

Mode of survival: As oospores present in the soil and in plant debris.

Primary spread: From infected plant debris in the soil.

Secondary spread: Through irrigation water and implements used.

Favourable conditions: Excessive soil moisture, poor drainage, high humidity and overcrowding of seedlings.

Management

- Sterilize nursery soil by burning trashes or by drenching the soil 10 cm deep with formaldehyde diluted 50 times with water.
- Adopt optimum seed rate and sow seeds on raised seed-bed.
- Treat seeds with captan or thiram or *Trichoderma viride* 4 g/kg or *Pseudomonas fluorescens* at 10g/kg.
- Prefer light-textured soils which provide better drainage and aeration.
- Avoid stagnation of water in the nursery especially during rainy season.
- Apply *Pseudomonas fluorescens* as soil application at 2.5 kg/ha mixed with 50 kg of FYM.
- Drench the soil with Bordeaux mixture 1 per cent or copper oxychloride 0.25 per cent at 4 lit/sq.m on 10 and 20 days after sowing.

2. Die back and Ripe Fruit Rot

Symptoms

Die back

- Symptoms are seen on stem and fruits.
- Branches dry and die from tip downward (die back), leaves fall off.
- Raised black dot-like acervuli are found scattered on dried stem.
- Infection of pedicel causes shedding of flowers and flower buds.

Fruit Rot

- Ripe fruits turning red are affected. Green fruits do not show visible symptoms.
- Water-soaked, elongated lesions on fruits turn dirty white.
- Fruits shrivel, dry and fall off.
- Acervuli seen on the surface of fruits are arranged in a concentric manner.
- Many lesions on the fruits change them into white and such fruits loose their pungency.
- Seeds show mass of fungal mycelia and they are poorly developed.
- Seed germination is also affected.

Fungus

Colletotrichum capsici (Syd.) Butler and Bisby.

The asexual fruiting body is acervulus. In the acervulus, dark brown, septate, long and rigid sterile structures called setae arise from the stroma. In between setae, a layer of hyaline, single celled conidiophores are produced. They bear single celled, hyaline and sickle-shaped conidia.

Favourable conditions: Temperature of 22-28°C and relative humidity of 96 per cent, humid weather with rainfall at frequent intervals during fruit ripening.

Mode of survival: As the dead twigs and in seeds in infected fruits (externally seed-borne); *Primary spread:* From seed-borne.

Secondary spread: Through air-borne conidia and by air-blown rain splashes.

Management

- Collet seeds only from fruits free from disease.
- Collect and destroy diseased plant parts.
- Treat seeds with captan or thiram 4g/kg.
- Spray mancozeb 0.2 per cent or copper oxychloride at 0.25 per cent and carbendazim 0.1 per cent with a sticker before flowering and at fruit formation.

3. *Alternaria* Leaf Spot

Symptoms

- ☆ Irregular, brown to dark brown spots with concentric rings appear on the leaves, which coalesce and cover major area.
- ☆ Diseased leaves become yellow and fall prematurely.
- ☆ Fruits show brown irregular spots with dark brown margin.
- ☆ Seeds in affected fruits are discoloured.

Fungus

Alternaria solani Sorauer

It is a polycyclic necrotrophic pathogen, produces dark colored, multinucleate muriform and beaked conidia arise singly in simple conidiophore.

Mode of survival: In the infected crop debris and on seeds.

Primary spread: From infected seeds and crop debris.

Secondary spread: Through air-borne conidia.

Management

- ☆ Spray mancozeb or zineb or chlorothalonil 0.2 per cent or copper oxychloride 0.25 per cent.

4. *Cercospora* Leaf Spot

Symptoms

- ☆ Circular or oval spots appear on leaves.
- ☆ These spots are with white centre and brown margin and increase up to 1.0 cm in dia.
- ☆ Spots coalesce leading to yellowing, drying and premature defoliation.
- ☆ Spots are seen on the stalks and stems also.

Fungus

Cercospora capsici (Syd.) Butler and Bisby.

Favourable conditions: Rainy season with low temperature and high humidity.

Mode of survival: In volunteer plants, infected plant debris and seeds.

Primary spread: From infected plant debris.

Secondary spread: Through air-borne conidia.

Management

- ☆ Treat seeds with carbendazim 2g/kg.
- ☆ Spray carbendazim or propiconazole 0.1 per cent or mancozeb or difolatan 0.2 per cent.

5. Powdery Mildew

Symptoms

- White talcum powder- like growth is on the lower surface of leaves in patches.
- Corresponding upper surface of leaves show yellow colour.
- Very often the patches join together and cover entire leaf surface.
- Severely diseased leaves become yellow, dry and shed prematurely.
- Severe infection leads to profuse leaf shedding and plants become barren.
- Flower and fruit production is lowered.

Fungus

Leveillula taurica (Lev.) Arn.

It is *Oidiopsis* type. Mycelium is endophytic. Conidiophores emerge through stomata in clusters of 3 or 4. Conidiophores are long, hyaline and septate. Conidia are club-shaped, single celled and hyaline.

Favourable conditions: Warm and humid weather.

Mode of survival: As dormant mycelium in the infected and fallen leaves.

Primary spread: From infected, fallen leaves. *Secondary spread:* Through Air-borne conidia.

Management

- Collect and destroy the infected fallen leaves.
- Spray carbendazim 0.1 per cent or dinocap 0.2 per cent or wettable sulphur 0.3 per cent or dust sulphur 25 kg/ha.

6. *Fusarium* Wilt

Symptoms

- The leaves turn yellow, droop and dry.
- The plant wilts completely which leads to death of plant.
- The vascular system is discoloured, particularly in the lower stem and roots and can be seen after removing the bark.
- Roots are poorly developed and become brown.

Fungus

Fusarium solani (Mart.) Sacc.

Favourable conditions: Ill-drined soil and high soil temperature and presence of nematode.

Mode of survival: As chlamydospore in soil.

Primary spread: From soil-borne chlamydospores. *Secondary spread:* Through Irrigation water and implements.

Management

- ☆ Grow wilt resistant varieties.
- ☆ Treat seeds with *Trichoderma viride* 4g/kg or *Pseudomonas fluoresence* 10g/kg or carbendazim 2g/ha.
- ☆ Provide proper drainage facilities.
- ☆ Drench Bordeaux mixture 1 per cent or copper oxychloride 0.25 per cent or carbendazim 0.1 per cent.

7. Bacterial Leaf Spot

Symptoms

- ☆ Water-soaked, circular or irregular and cankerous, scattered brown spots appear on the leaves.
- ☆ The spot bulges and turns purple with black centre, surrounded by yellow halo.
- ☆ Petioles and young stems are also infected with cankerous spots.
- ☆ Round, brown, raised spots with depression in the centre are seen on fruits.
- ☆ Diseased green fruits turn brown or black completely.

Bacterium

Xanthomonas vesicatoria (ex Doidge) Vauterin *et al.,*

[Syn. *Xanthomonas campestris* pv. *vesicatoria* (Doidge) Dye].

It is a Gram-negative, rod shaped bacterium with polar flagellum.

Mode of survival: In seed and infected plant parts.

Primary spread: Through seed.

Secondary spread: Through Infected seedlings and air-blown rain splashes.

Management

- ☆ Adopt crop rotation avoiding tomato.
- ☆ Collect seeds from healthy fruits.
- ☆ Spray Bordeaux mixture 1.0 per cent or streptomycin sulphate 250 ppm (0.025g/lit) + copper oxychloride 0.25 per cent.

8. Mosaic Diseases

Symptoms

PVY: Mottling, vein banding, crinkling, smalling and interveinal chlorosis of leaves.

Fruits show mosaic mottling and deformity.

TMV

Mosaic mottling, chlorosis, stunting with necrosis.

CMV

Affected plants are severely stunted and become bushy. Mottling, ring spot, blistering, deformation and filiformity are the leaf symptoms. Infected plants bear few flowers and small fruits. Mottling on fruits is also seen.

Virus

Potato virus Y (PVY), *Tobacco mosaic virus* (TMV) and *Cucumber mosaic virus* (CMV).

Mode of survival: As Infected volunteer plants.

Primary spread: From infected volunteer plants, cultivated hosts and weed hosts.

Secondary spread: Through diseased seedlings and insect vectors. In PVY and CMV, the aphids (*Aphis craccivora, A.gossypii* and *Myzus persicae*) act as vectors. TMV is spread by mechanical means (no vector).

Management

- ☆ Avoid monoculture of chilli crop (adopt crop rotation).
- ☆ Avoid transplanting infected seedlings.
- ☆ Cover nursery beds with nylon net to protect the seedlings from viral infection by vectors.
- ☆ Raise 2-3 rows of maize or sorghum as border crop to restrict the spread of aphid vectors.
- ☆ Also grow 2 rows of maize or sorghum for every 5 rows of chilli crop against air direction.
- ☆ Install yellow sticky traps at 13 nos/ha.
- ☆ Uproot and burn or bury infected plants to avoid spread.
- ☆ Spray monocrotophos or profenophos or dimethoate 2 ml/lit or acephate 2g/lit.

9. Leaf Curl

Symptoms

- ☆ Upward (boat-like) curling of leaves with reduced leaf size accompanied by puckering and blistering of interveinal areas and thickening and swelling of veins.
- ☆ Internodes are shortened and plants are stunted.
- ☆ Fruiting is stopped and if formed they are small and deformed with few seeds.

Virus

Tobacco leaf curl virus (TLCV).

Mode of survival: In infected plants and other cultivated and wild hosts.

Primary spread: From infected chilli and other hosts.

Secondary spread: Through infected seedlings and vector, whitefly (*Bemisia tabaci*).

Management

- ☆ Avoid monoculture of chilli crop (adopt crop rotation).
- ☆ Avoid transplanting infected seedlings.
- ☆ Raise 2-3 rows of maize or sorghum as barrier crop to restrict the spread of aphid vectors.
- ☆ Install yellow sticky traps at 13 nos/ha.
- ☆ Uproot and burn or bury infected plants to avoid spread.
- ☆ Spray profenophos or dimethoate 2 ml/lit or acephate 2g/lit.

3. Cinnamon/Darchini (*Cinnamomum zeylanicum* Blume)

1. Leaf Spot and Die back

Symptoms

- ☆ Small spots develop on leaf lamina which gradually enlarge and coalesce to form irregular patches.
- ☆ Ultimately the leaves dry up and the stem resulting in die back symptoms.
- ☆ Young seedlings are killed.
- ☆ In mature trees, small specks appear which increase in size as necrotic blotches.
- ☆ Light to deep brown concentric zones develop on the spots.

Fungi

Colletotrichum gloeosporioides (Penz.) Penz. and Sacc.

Favourable conditions: Alternate dry and wet conditions.

Management

- ☆ Collect and destroy the infected plant debris.
- ☆ Spray with Bordeaux mixture (5:5:50) or copper oxychloride (0.3 per cent) or combination of mancozeb (0.25 per cent) and carbendazim (0.1 per cent) and repeat at 14 days interval.

- Spraying of chlorothalonil 75 per cent WP or thiophanatemethyl 70 per cent WP at 7-10 days intervals.
- Reducing shades specially during nursery stages may reduce the disease severity.

2. *Phytophthora Dieback* (Stripe Canker)

Symptoms

- Irregular and vertical necrotic stripes (1-5 cm) appear on affected plants.
- The cankers are sunken with zonations and separated from the healthy tissue by a black line.

Fungi

Phytophthora cinnamomi Rands.

The hyphae are broad, hyphal swellings typically spherical. Sporangiophore are thin and occasionally branched. Sporangia are broadly ellipsoidal or ovoid, papilla absent. Oogonia have smooth walls and become yellow or golden with age. Antheredia amphigenous, long and measure. Oospores are hyaline, nearly filling the oogonium and thick.

Favourable conditions: The optimum temperature for growth of the fungus is 24-28°C.

Management

- Use only disease free seed and planting material.
- If disease is present in the soil then crop should be rotated to a non-susceptible host for a period of four years.
- Avoid planting in water saturated soils.
- Improving soil drainage keeps the disease incidence at a lower level.
- Removal and destruction of affected parts and wound dressing with tar.
- Soil drenches with appropriate Bordeaux mixture 1 per cent may give a measure of control.

3. Seedling Blight

Symptoms

- Light brown patches which girdle stem of young seedlings; death of seedlings.

Fungi

Diplodia spp.

Management

- Disease can be controlled by application of 1 per cent Bordeaux mixture.

4. Clove (*Syzygium aromaticum* L.)

1. Leaf Spot, Twig Blight and Flower Bud Shedding

Symptoms

- Three types of symptoms are seen *viz.*, leaf spot, twig blight and flower bud shedding.
- On the leaves, necrotic spots of variable sizes and shapes are noticed. Severely affected leaves wither, drop and dry up.
- In the nursery seedlings, dieback symptoms are seen.
- The affected branches stand without leaves or only with young leaves at the tips.
- The flower buds are attacked by spread of infection from the twigs. Shedding of flower buds occurs during periods of heavy and continuous rainfall.

Fungi

Colletotrichum gloeosporioides (Penz.) Penz. and Sacc.

The mycelium is grayish white, thin and sparsely septate. Acervuli are glabrous, rounded and elongated or irregular in shape with setae. Conidia are cylindrical with obtuse ends, one celled and are formed on aseptate, hyaline or faintly brown cylindrical phialidic conidiophores.

Spread: Through air-borne conidia.

Favourable conditions: Heavy and continuous rainfall.

Management

- Destruction of the weed Clerodendron from the clove garden.
- Spraying Bordeaux mixture 1 per cent at 1-1.5 month intervals just prior to flower bud formation and continued till the harvest.

2. Sudden Death

Symptoms

- On mature trees, slight chlorosis occurs on the leaves which results in a sudden and rapid leaf fall followed by wilting of the plants.
- Complete death of the plants occurs within a few days.
- The roots are also affected by the pathogen.

Fungi

Valsa eugeniae Nutman and Roberts.

The perithecia are produced in large numbers.

Spread: Soil infested ascospores.

Favourable condition: Rainy season.

Management

- Drench Bordeaux mixture 1 per cent at collar region and soil based on the infection.

3 Seedling Wilt

Symptoms

- Seedling wilt is found mainly in nurseries and causes five to 40 per cent death of seedlings. Leaves of affected seedlings loose natural lustre tend to droop and ultimately die.
- The root system and collar region of the seedling show varying degrees of, discolouration and decay.

Fungi

Cylindrocladium sp., *Rhizoctonia* sp., *Colletotrichum* sp., *Fusarium* sp., and *Trichoderma* sp

Management

- Since the infected seedlings promote spread of the disease they are to be removed and destroyed and the nursery is drenched with Copper oxychloride 0.25 per cent.

3. Leaf Rot

Symptoms

- It is noticed in the nurseries as well as in the main field both at young and mature stages.
- Infection starts as dark spots at the leaf margin and spreads sometimes with no definite pattern.
- Rotting may be in the whole leaf or at the tip resulting in defoliation.

Fungi

Cylindrocladium quinquiseptatum Boedijin and Reitsma

Management

- Seedling and young plants can be sprayed with systemic fungicides Carbendazim 0.1 per cent.

4. Little Leaf

Symptoms

- The affected trees exhibit symptoms of complete repression.

- ☆ The trunk shows numerous protuberances.
- ☆ The branches grow at acute angles with the trunk giving a cone-shaped appearance to the tree.
- ☆ Leaves are reduced in size, internodes are short and primary and secondary branches are arranged closer to each other.
- ☆ Presence of short internodes and clustering of small, crinkled leaves is also observed.

Bacterium

Candidatus Phytoplasma.

Management

The disease can be kept under check by spraying tetracycline.

Other Diseases of Clove

5. **Die back: *Cryptosporella eugeniae* Nutman and Roberts.**
6. **Sumatra or die back: *Endothia* sp., *Phymatotichum* sp. and *Phytophthora* sp.**
7. **Grey blight of clove: *Pestalotia palmarum***
8. **Leaf spot of clove: *Alternaria citri***
9. **Sooty mould of clove: *Phragmocapnius* sp.**
10. **Algal leaf spot of clove: *Cephaleuros* sp.**

5. Coriander (*Coriandrum sativum* L.)

1. Stem Gall/Tumour

Symptoms

- ☆ Tumour-like swellings appear on leaf veins, leaf stalks, peduncles, stem and fruits.
- ☆ The swellings on the veins give a swollen hanging appearance to the leaves.
- ☆ The swellings are usually elongated and 9-12 mm x 3-5 mm dia.
- ☆ The size of the swellings vary according to the dimension of the part infected.
- ☆ The fungus is restricted to tumours and become rough.
- ☆ In very severe cases the plants are killed.

Fungus

Protomyces macrosporus Unger.

The hyphae are intercullar, septate and broad. Scattered cells in the hyphae swell and form ellipsoidal or globose bodies which later develop into chlamydospores. Exospore is thick and brown. Meso and endospores are thin. Chalmydospores germinate in waer and form sporangia and spores.

Mode of survival: In seed and in soil as chlamydospores.

Primary spread: From infected seeds.

Secondary spread: Through chlamydospores in the soil.

Favourable conditions: High soil moisture, shade, low temperature, abundant dew with higher day temperature and soil pH of 7.4.

Management

- ☆ Grow tolerant varieties.
- ☆ Use clean and healthy seeds.
- ☆ Adopt field sanitation and crop rotation.
- ☆ Collect and burn diseased crop refuse.
- ☆ Seed treatment with captan or thiram 4g/kg.
- ☆ Spray captafol 0.2 per cent.

2. Wilt

Symptoms

- ☆ The plants are attacked at all stages of growth.
- ☆ Drooping of the terminal portions, followed by withering and drying of leaves resulting in death of plant.
- ☆ Discolouration of vascular system of the root.
- ☆ Partial wilting is also common.

Fungus

Fusarium oxysporum f. sp. *corianderii* Narula and Joshi.

The fungus produces macro and micro conidia and chlamydospores.

Mode of survival: As survives both in seed and soil.

Primary spread: From seed.

Secondary spread: Through irrigation water.

Favourable conditions: Soil temperature of 24-27°C.

Management

- ☆ Grow tolerant varieties.
- ☆ Use disease free seeds.
- ☆ Follow summer ploughing and crop rotation.

- ☆ Treat seeds with *Trichoderma viride* 4g/kg or *Pseudomonas fluorescens* 10g/kg or carbendazim 2 g/kg.

3. Powdery Mildew

Symptoms

- ☆ Small white circular patches appear on foliage and on young stem which leads to drying of foliage.
- ☆ Affected leaves are distorted and reduced in size.
- ☆ In severe cases umbels and seeds dry up.

Fungus

Erysiphe polygoni DC.

Mycelium is ectophytic. Haustoria are saccate. Conidiophores arise vertically on the host surface, septate. Conidia are produced singly or in short chains, ellipsoid or ovate. Each cleistothecium contains 4 to 8 asci. Each ascus contains 3 to 5 ascospores hyaline, elliptical, single celled.

Favourable conditions: Dry and humid conditions.

Mode of survival: It survives in soil and in infected plant debris as cleistothecia.

Primary spread: From infected plant debris.

Secondary spread: Through air-borne conidia.

Management

- ☆ Grow tolerant varieties.
- ☆ Dust sulphur 25kg/ha or spray wettable powder 0.3 per cent.

6. Cumin (*Cuminum cyminum* L.)

1. Wilt

Symptoms

- ☆ Disease produces wilting symptoms at seedling and later stage of plant growth.
- ☆ Brownish discolouration of vascular bundles is seen when stem is cut longitudinal.

Fungi

Fusarium oxysporum f. sp. *cumini* Prasad and Patel

Spread: Both soil and seed-borne as well as in infected plant debris.

Survival: In infected plant debris as a mycelium and chlamydospores.

Favourable conditions: Irrigation water, rains splash, wind, through inter-cultural operations and Soil temperature is between 12.5°C and 14°C.

Management

- ☆ Collect and destroy crop debris.
- ☆ Judicious use of fertilizers.
- ☆ Provide irrigation at critical stages of the crop.
- ☆ Avoid water logging.

2. Powdery Mildew

Symptoms

- ☆ The powdery growth usually develops first on leaves which later can cover all succulent stems and branches including flowers.
- ☆ In severe case seed development may not take place.

Fungi

Erysiphe polygoni (Vanha) Weltzien

Primary spread: From through soil and seed.

Secondary spread: Through by dispersal of conidia through wind, rain splashes.

Favourable conditions: Cool high humid weather (20-25° C) or cloudy weather with high relative humidity and cloudy weather coincide with flowering stage.

Management

- ☆ Spray sulphur 0.751.0 kg/acre in 300-400 lit of water or sulphur 6-8 Kg/acre or dinocap 120 ml/acre in 300 lit of water.

3. Blight

Symptoms

- ☆ Alternate concentric rings were formed in the surface of the leaves at flowering stage and seed cannot mature to full size.
- ☆ Seeds become shriveled and are easily blown away during winnowing.

Fungi

Alternaria burnsii Uppal, Patel and Kamat, *A. cucumerina* Ellis and Everh

Mode of survival: As dormant mycelium in diseased debris.

Primary spread: From air-borne conidia.

Favourable conditions: Wet weather with temperature ranging from 20-32°C accompanied by high humidity and cloudy weather or warm humid weather.

Management

- ☆ Collect and destroy crop debris.
- ☆ Judicious use of fertilizers.

- ☆ Spray aureofungin 0.02 per cent in 300 lit of water, second spray after 30 days of interval or copper oxychloride 1.0 Kg/acre in 300- 400 lit of water or difenoconazole 0.05 per cent in 200 lit of water, second spray after 15 days of interval or mancozeb 0.6- 0.8 Kg/acre in 200 lit of water or zineb 0.6- 0.8 Kg/acre in 200 l of water.

4. Damping Off

Symptoms

- ☆ Damping off occurs in two stages, *i.e.* the pre-emergence and the post-emergence phase.
- ☆ In the pre-emergence phase seeds get rotted and the seedlings are killed just before they reach the soil surface.
- ☆ Once the seedling emerges out of soil line a soft water soaked lesion appears near the collar region causes constriction and results in toppling over of the seedlings.
- ☆ The young radical and the plumule are killed and there is complete rotting of the seedlings.

Chromista

Pythium aphanidermatum (Edson) Fitzp.

Mode of survival: As soil borne mycelium and oospores.

Primary spread: From soil and water.

Secondary spread: Through by dispersal of conidia through wind, rain splashes.

Favourable conditions: High humidity, high soil moisture, high dose of nitrogenous fertilizers, cloudiness and low temperatures below 24° C for few days.

Management

- ☆ Avoid water stagnation.
- ☆ Soil drenching with Copper oxy-chloride 0.25 per cent.

7. Curry Leaf [*Murraya koenigii* (L.) Spreng.]

1. *Phyllosticta* Leaf Spot

Symptoms

- ☆ Symptoms range from a few round spots or lesions.
- ☆ The irregular, round, yellowish brown lesions are produced on leaves.
- ☆ Tiny black fruiting bodies of the pathogen are produced, usually they form a circle.
- ☆ The center of these spots is dead tissue that easily breaks away leaving a hole.

Fungi

Phyllosticta murrayae Gonz. Frag. and Cif., *P. murrayicola*

Primary spread: From soil and rain splash.

Secondary spread: Through air and rain splash during wet condition.

Management

- ✰ Disease-free seed and seedlings should always be used.
- ✰ Preventive spray of tobacco decoction could be sprayed.
- ✰ Citrus oil at 1360 ppm inhibited the maximum growth of the fungus followed by lemongrass oil at 1720 ppm and peppermint at 2260 ppm, respectively.
- ✰ Spray NSKE 5 per cent.

Other Diseases of Curry Leaf

2. **Leaf Spot - *Cylindrosporium koenigii* Thirum, *Cercospora heyneana* Wall**
3. **Anthracnose – *Colletotrichum gloeosporioides* (Penz.) Penz. and Sacc.**
4. **White sap rot – *Fornes pectinatus* Klotzsch**
5. **Collar rot – *Rhizoctonia solani* Kuhn**

8. Fennel (*Foeniculum vulgare* Mill.)

1. Leaf Blight

Symptoms

- ✰ Minute angular, brown to black spots are formed which later enlarge and get covered with grayish white erumpent growth.
- ✰ The basal leaves get defoliated.
- ✰ The seeds formed on diseased plants are shriveled and black.

Fungus

Ramularia foeniculi Sibilia

Mode of survival: As In the infected host plant.

Primary spread: From infected plant parts.

Secondary spread: Through air-borne conidia.

Favourable conditions: Wet and humid condition and temperature 22-32°C.

Management

- ✰ Avoid water stagnation.
- ✰ Follow the common cultural, mechanical and biological practices.
- ✰ Spray of carbendazim 0.1 per cent.

2. Leaf Spot

Symptoms

- ✰ Brown spots with pale centers are formed on leaves, stems, petioles and pedicels.
- ✰ Light brown conidiophores are formed in clusters which emerge through stomata.
- ✰ Under severe condition the whole foliage gets blighted and the yield is highly reduced.

Fungus

Cercospora foeniculi Magn; *Cercosporidium punctum* (Delacr.) Deighton

Mode of survival: As soil borne and in infected plant debris.

Primary spread: From seed borne pathogen.

Secondary spread: Through infected plant debris.

Favourable conditions: Higher temperature and humidity favour the disease.

Management

- ✰ Avoid water stagnation.
- ✰ Follow the common cultural, mechanical and biological practices.
- ✰ Spray of carbendazim 0.1 per cent.

3. Damping-off

Symptoms

- ✰ Seedlings exhibit water soaked areas on the stem portion, the whole cortex dies and the seedlings topple down.
- ✰ Even some seedlings do not emerge but die.
- ✰ exhibiting pre-emergence mortality.

Fungus

Cercospora foeniculi Magn; *Cercosporidium punctum* (Delacr.) Deighton

Mode of survival: As soil borne and in infected plant debris.

Primary spread: From seed borne pathogen.

Secondary spread: Through infected plant debris.

Favourable conditions: Higher temperature and humidity favour the disease.

Management

- ✰ Avoid water stagnation.
- ✰ Follow the common cultural, mechanical and biological practices.
- ✰ Spray of carbendazim 0.1 per cent.

4. Wilt

Symptoms

- The plants are attacked at all stages of growth.
- Drooping of the terminal portions, followed by withering and drying of leaves resulting in death of plant.
- Discolouration of vascular system of the root.
- Partial wilting is also common.

Fungus

Fusarium equiseti (Corda) Sacc.

Mode of survival: As survives both in seed and soil.

Primary spread: From seed.

Secondary spread: Through irrigation water.

Favourable conditions: Soil temperature of 24-27°C.

Management

- Grow tolerant varieties.
- Use disease free seeds.
- Follow summer ploughing and crop rotation.
- Treat seeds with *Trichoderma viride* 4g/kg or *Pseudomonas fluorescens* 10g/kg or carbendazim 2 g/kg.

5. Powdery Mildew

Symptoms

- The powdery fungal growth usually develops first on leaves which later can cover all succulent stems and branches including flowers.
- In severe case seed development may not take place.

Fungus

Erysiphe polygoni DC., *E. heraclei* DC.

Mode of survival: As survives both in seed and soil.

Primary spread: From seed and soil.

Secondary spread: Through rains splashes and air-borne conidia.

Favourable conditions: Cool high humid weather (20-25°C) or cloudy weather.

Management

- Grow tolerant varieties.
- Dust sulphur 25kg/ha or spray wettable powder 0.3 per cent.

6. Collar Rot

Symptoms

- ☆ Disease appears in plots where water stagnation near the plant stem is more.
- ☆ Collar portion of the affected plants start decaying and the plants turn to yellow colour and die later on.
- ☆ Later collar region of the plant or seedlings get rotted and plant topple down.

Fungus

Sclerotinia sclerotiorum (Lib.) de Bary

Mode of survival: As survives both in seed and soil.

Primary spread: From dormant mycelium and fruiting bodies in the infected crops, seeds and soil.

Secondary spread: Through irrigation disperses conidia to nearby plants.

Favourable conditions: High humidity, high soil moisture and temperature.

Management

- ☆ Avoid water stagnation in the field.
- ☆ Provide proper spacing.

7. Root Rot

Symptoms

- ☆ Symptoms consist of seed decay and brown to reddish lesions on seedling stems and roots just below the soil line.
- ☆ These reddish brown lesions may become sunken and girdle the stems and kill the plant.
- ☆ Plants may often appear stunted and unthrifty and will die.
- ☆ On older plants, the pathogen causes a reddish brown dry cortical root rot that may extend into the base of the stem.
- ☆ Foliar symptoms may include yellowing or wilting of leaves.

Fungus

Rhizoctonia solani J.G. Kühn

Mode of survival: As survives as sclerotia in soil.

Primary spread: From soil borne sclerotia.

Secondary spread: Through sclerotia spread by irrigation water, implements and cultural operations.

Favourable conditions: Dry weather following heavy rains, High soil temperature (35-39° C).

Management

- Crop rotation with non-susceptible hosts will reduce populations of *Rhizoctonia* in the soil.
- Avoid crop rotations with sugar beets if there is evidence of Rhizoctonia in the field.
- Crop rotations with dry beans may also increase incidence of disease.
- Protective seed treatment and good seedbed preparation can reduce root rot.
- Earthing up of soil around stems promotes lateral root growth and lessen the effect of root rot on older plants.

9. Fenugreek (*Trigonella foenum-graecum* L.)

1. Powdery Mildew

Symptoms

- The symptoms of the disease appear as white, floury patches on both sides of the leaves as well as on other green parts of the plant.
- The floury patches consist of the mycelium, conidiophore and conidia of the fungus.

Fungi

Erysiphe polygoni DC.

Oidiopsis taurica (Lev.) E.S. Salmon [*Leveillula taurica* (Lev.) Arn.] has also been reported to be associated with this disease. The haustoria develop as lobed swellings. Conidiophores arise vertically and each bears several spores in chain. Cleistothecia appear as sharp black specks, scattered on the surface of the white mycelium.

Mode of survival: In the infected crop debris and on seeds.

Primary spread: From ascospores from infected crop debris.

Secondary spread: Through air-borne conidia.

Favourable conditions: More prevalent in dry weather and moderate temperatures.

Management

- Use resistant variety like Prabha. Lines GC-39UM-32, GC-7, GC-20 and UM-34 have also been reported as resistant and can be utilized in breeding programme.
- Use adequate spacing when planting to avoid overcrowding.

- Spray with wettable sulphur 0.2 per cent, or hexaconazole 0.05 per cent or dinocap 0.06 per cent or difenoconazole 0.04 per cent and repeat at 10-14 days interval at the initial stage of this disease.

2. Downy Mildew

Symptoms

- The adaxial surfaces of the leaves showed yellow patches or small chlorotic spots which appeared at the margins often.
- The abaxial surface of the leaf showed white cottony mycelial growth which often appears as grayish violet.
- The disease also causes stunted growth of the plant.

Chromista

Peronospora trigonellae Guam.

The conidiophores were branched dichotomously. Conidia were slightly pigmented and oblong to ellipsoid. The oospores were embedded in the older lesions.

Mode of survival: As oospore in infected debris.

Spread: Through infected plant debris.

Favourable conditions: Temperature of 18-24°C and a relatively high humidity

Management

- Use adequate spacing when planting to avoid overcrowding.
- Spray metalaxyl + mancozeb 0.2 per cent.

3 Root Rot/Collar Rot

Symptoms

- The symptoms of the disease appear in 30-45 days old plants as rotting of roots leading to yellowing of foliage.
- The affected plants wither and dry up.

Fungi

Rhizoctonia solani Kuhn.

It produces sclerotia which is small, irregular-shaped, brown to black structures.

Mode of survival: As fungus is soil borne and survives in the form of sclerotia.

Primary spread: From the mycelia overwintering in soil and crop residues

Secondary spread: Through soil-borne sclerotia.

Management

- ☆ Follow 3-4 years crop rotation.
- ☆ Soil amendment with neem cake 1 t/ha alone and in combination with *Trichoderma* sp. reduces root rot.
- ☆ Seed treatment with carbendazim 0.2 per cent followed by drenching at 0.1 per cent effectively controls this disease.

4. Wilt

Symptoms

- ☆ Initial symptoms are minor clearings in veins on the external part of young leaflets finally accompanied with downward drooping of mature leaves.
- ☆ At the seedling stage, plants may wilt and die.
- ☆ This is followed by marginal necrosis of the infected leaves, rapid defoliation and death of the plant.

Fungi

Fusarium oxysporum Schltdl.

The mycelia are pale white to pink in color. Microconidia are borne singly or in pairs and are abundant, oval to ellipsoidal. Macroconidia are sparse to abundant on branched conidiophores or on sporodochia. The chlamydospores are abundant and form terminally or on an intercalary basis.

Mode of survival: As fungus is soil borne and survives in the form of chlamydospores.

Primary spread: From mycelia overwintering in soil and crop residues

Secondary spread: Through soil-borne chlamydospores.

Management

- ☆ Soil amendment with neem cake 1 t/ha alone and in combination with *Trichoderma* sp. reduces wilt.
- ☆ Spray copper oxychloride 0.3 per cent.

5. *Cercospora* Leaf Spot

Symptoms

- ☆ The disease is devastating in nature, the size of the lesions on mature leaves increases resulting in rapid sporulation resembling a whitish dorsal surface.
- ☆ The necrotic areas are characterized by the presence of an encircling yellowish halo.
- ☆ Most severe infections are observed in the stem and the pods.

- ☆ In case of severe infection even the youngest leaves have been reported to be wilted and dead.

Fungi

Cercospora traversiana Sacc. (Syns. *C. traversiana* var. *trigonellae coeruleae* Savul. and Sandu-Ville and *C. trigonellae* Maublanc)

- ☆ Conidiophores are dark, paler towards the tip, unbranched, rarely geniculate and rarely septate.
- ☆ Conidia are acicular to almost cylindric, straight orurved, indistinctly multi-septate, basetruncate, tip subacute to subobtuse.

Mode of survival: As seed borne fungal disease.

Spread: Through infected plant debris.

Management

- ☆ Follow 3-4 years crop rotation.
- ☆ Seed treatment with carbendazim 0.2 per cent followed by spraying at 0.1 per cent effectively controls this disease.

6. Rust

Symptoms

- ☆ Rust commonly appears as a colored powder, consisting of minute on the lower leaf surface.
- ☆ Pustules are yellowish orange to dark brown, hair-like (ligulate) structures called telia develop on the leaves.

Fungi

Uromyces trigonellae Pass.

Mode of survival: As infected plant debris

Spread: Through air-borne uredospores

Management

- ☆ Remove infected leaves and crop debris to minimize local sources of infection.
- ☆ Use drip irrigation to keep leaves dry and make sure air circulation is good.
- ☆ Keep the plants well fed and watered.
- ☆ Reduce nitrogen level.

Other Diseases

7. Charcoal rot: *Macrophomina phaseolina* (Tassi) Goid.

8. Damping off: *Pythium aphanidermatum* (Edson) Fitzp.

9. **Leaf blight: *Alternaria alternata* (Fr.) Keissl.**

10. **Bacterial blight: *Pseudomonas syringae* pv. *syringae* van Hall *Xanthomonas axonopodis* pv. *alfalfa* (Riker, Jones and Davis) Vauterin**

11. **Bean yellow mosaic virus**

10. Garlic (*Allium sativum* L.)

1. Neck Rot and Bulb Rot

Symptoms

- ☆ It is a common disease on bulbs at the time of harvest.
- ☆ In the field, the infection develops at soil level in the stem (neck).
- ☆ Affected scales tissues are water soaked become soft. Dense layer of grey mould appear at the neck.
- ☆ Dark black sclerotia appear on the older, decayed tissues in the neck region.
- ☆ Enormous number of sclerotia is seen on maturing bulbs just before and during harvest.
- ☆ In storage, a gray mold with sclerotial bodies can be seen on the surface of the bulbs or between the scales.
- ☆ In the case of bulb rot caused by *Fusarium,* failure of germination and drying of leaves from the tip are noticed.
- ☆ Rotting of bulb and production of side shoots produce clump of shoots.

Fungi

Botrytis allii Munn and *Fusarium oxysporum* Schlecht ex Fr.

Favourable conditions: Cool, moist weather favour germination of sclerotia. Fusarial bulb rot is favoured by excess rainfall followed by a dry spell of 10 to 15 days. The feeding injury by the garlic mite, *Rhizoglyphus echinopus* favours bulb rot.

Mode of survival: Both the pathogens survive in the field as well as in storage. *Botrytis allii* survives as sclerotia while *Fusarium* survives as chlamydospores in the soil and also in bulbs.

Primary spread: From soil-borne sclerotia or chlamydospores.

Secondary spread: Through irrigation water.

Management

- ☆ Select seed materials from disease free area.
- ☆ Treat the bulbs with thiram, captan 4 g/kg or carbendazim 2 g/kg.
- ☆ Spray carbendazim 0.1 per cent or vinclozolin 0.1 per cent with sticking agent.
- ☆ Minimize damage to bulbs during harvest.

2. *Macrophomina* Rot/Internal Bulb Rot

Symptoms

- ☆ It is a storage rot.
- ☆ No external symptoms are observed unless the outer scales are removed and hence it is named as internal bulb rot.
- ☆ Black pinhead-like sclerotia develop over the fleshy scales which are light in weight, brown mostly shriveled but do not lose their pungency.

Fungus

Macrophomina phaseolina (Tassi) Goid.

Mode of survival: As infected bulbs.

Primary spread: From sclerotia in the infected bulbs.

Secondary spread: Through contact with diseased bulbs.

Management

- ☆ Proper drying and storage of bulbs in well ventilated rooms.

3. *Aspergillus* Rot/Black Mould Blemish

Symptoms

- ☆ It is a storage rot.
- ☆ Rotten garlic cloves are black, brown, pink or white.
- ☆ It manifests itself with the copious growth of a dust-like fungal mass which remains concealed between the scales.
- ☆ The whole tissues are transformed into a black powdery mass of spores.
- ☆ The individual bulbs shrivel and become light in weight.
- ☆ Infected bulbs lose their pungency and smell.

Fungi

Aspergillus niger Van Tiegh., *A. alliaceus* Thom and Church and

A. sclerotiorum G.A. Huber.

Aspergillus infection may be followed by entry of other pathogens, including soft rot bacteria. The fungus is known to produce mycotoxins in infected tissue.

Mode of survival: As *Aspergillus* spp. survives in the soil, on dead plant tissue and in bulbs.

Primary spread: From spores produced in diseased plant debris or in soil cause primary infection.

Secondary spread: Through air borne conidia.

Management

- ☆ Discard diseased bulbs.

- ☆ Proper drying of bulbs.
- ☆ Fumigate storage house with formation 0.03 per cent.
- ☆ Store bulbs in well ventilated store houses.
- ☆ Treat seeds with captan or thiram 4g/kg.
- ☆ Avoid bruising and mechanical injury to bulbs at harvest and when storing.

4. Pink Root

Symptoms

- ☆ The roots turn pink.
- ☆ Later they turn red or purple and finally turn black and die.
- ☆ The newly formed roots are also turn pink or red and die.
- ☆ Diseased plants can be easily pulled.
- ☆ The leaves of infected plants become yellow and dry from tip downwards.
- ☆ The plants are stunted with undersized bulbs because of poor root system.

Fungus

Pyrenochaeta terrestris (Hansen) Gorenz *et al.* (Syn. *Phoma terrestris* Hansen). *Favourable conditions:* Temperature of 24-28°C, continuous cropping of garlic or onion and poor fertilization.

Mode of survival: As mycelium in soil.

Primary spread: From infected bulbs/seedlings.

Secondary spread: Through irrigation water and implements.

Management

- ☆ Adopt long rotation.
- ☆ Apply adequate amount of fertilizers to maintain the plant vigour.
- ☆ Avoid infested field for raising nursery.

11. Ginger (*Zingiber officinale* Roscoe)

1. Soft Rot/Rhizome Rot

Symptoms

- ☆ Most destructive disease in field and storage.
- ☆ The infection starts at the collar region and spreads upwards as well as downwards.
- ☆ The collar region becomes water-soaked, brown and exhibit rotting.
- ☆ Rotting spreads to the rhizome resulting in soft rot of inner tissue leaving the rind.
- ☆ At a later stage roots also become soft and rot.

- ☆ Foliar symptoms appear as light yellowing on the tips of lower leaves which gradually spreads to entire leaf blade.
- ☆ The yellowing spreads to all leaves with drooping and withering and drying of pseudostem.
- ☆ New rhizomes will not develop.

Chromista

Pythium aphanidermatum (Edson) Fitzp., *P. myriotylum* Drechsler and

P. vexans (de Bary) Abad.

Mycelium is coenocytic, hyaline and branched. Sporangia are lobed and hyaline. The oospores are spherical and thick walled.

Favourable conditions: High soil moisture and temperature during monsoon, excessive irrigation, continuous turmeric cultivation and poor drainage.

Mode of survival: As survives in soil as oospores, in seed rhizomes and plant debris.

Primary spread: From diseased rhizomes and soil.

Secondary spread: Through irrigation water.

Management

- ☆ Adopt early planting (May-early June).
- ☆ Grow tolerant varieties.
- ☆ Select seed rhizomes from disease free areas since the disease is rhizome-borne.
- ☆ Grow bhendi as intercrop.
- ☆ Provide proper drainage before planting.
- ☆ Dip seed rhizomes in mancozeb 0.3 per cent or chlorothalonil 0.2 per cent for 30 min before storage and once again before planting.
- ☆ Remove diseased clumps with rhizome and drench the affected area and surrounding plants with copper oxychloride 0.25 per cent or Bordeaux mixture 1.0 per cent.

2. *Colletotrichum* Leaf Spot

Symptoms

- ☆ Big, elliptic or oblong spots (4-5 cm × 2-3 cm) with greyish white centre and brown margin appear on leaves surrounded by yellow halo.
- ☆ Spots coalesce and cause drying to give a burnt up appearance.
- ☆ Older spots show many black dot-like acervuli arranged in concentric rings.
- ☆ Black stroma appears on scales covering the rhizomes.

Fungus

Colletotrichum zingiberis (Sundar.) Butler and Bisby

[*Glomerella cingulata* (Stonem.) Spauld. and Schrenk.].

Acervuli (asexual fruiting body) with black setae, conidiophores are hyaline, septate, branched. Conidia are hyaline, aseptate, oval to elliptical or straight cylindrical and 11.0 to 17.0 × 3.0 to 6.0 µm.

Favourable conditions: High humidity and low day temperature

Mode of survival: As dormant stromata between the rhizome scales.

Primary spread: From rhizomes and plant debris.

Secondary spread: Through air-borne conidia and rain.

Management

- Select seed rhizomes from healthy areas.
- Use disease tolerant varieties.
- Avoid excess shade and adopt mixed cropping.
- Spray Bordeaux mixture 1 per cent or mancozeb 0.2 per cent or carbendazim 0.1 per cent immediately after appearance of the disease.

3. Leaf Blotch

Symptoms

- Numerous small spots (1-2mm) appear on both the surfaces of the leaves and are very closely arranged.
- Yellow coloured spots turn to dirty yellow and then brown with chlorotic halo.
- Spots coalesce to form large irregular necrotic blotches.
- Leaves dry up and give a scorched appearance
- Less number of functional leaves lead to poor yield.

Fungus

Taphrina maculans (Syd.) Butler.

Hyphae are intercullar. Branched or lobed haustoria are present. Asci are formed on both sides of the leaf. Asci contain eight ovoid, hyaline unicellular ascospores. These ascospores multiply by budding within the ascus to form sprout conidia.

Favourable conditions: Temperature 21-23°C and a relative humidity of 80 per cent.

Mode of survival: As infected plant debris.

Primary spread: From collateral host and diseased plant debris.

Secondary spread: Through Air-borne ascospores and sprout conidia.

Management

- Grow disease tolerant varieties.
- Cut and remove diseased and dried leaves.
- Spray Bordeaux mixture 1 per cent or copper oxychloride 0.25 per cent or mancozeb 0.2 per cent with a sticking agent.

4. Yellow Disease/Wilt

Symptoms

- Yellowing of the margins of lower leaves which gradually spread to the entire leaf.
- Affected plant wilts and dries but does not fall down.
- Rhizomes show creamy discolouration of vascular system and a soft and watery rot.
- Shoots can be easily pulled out from the rhizomes.
- Roots rot and rhizome formation stops.

Fungus

Fusarium oxysporum f.sp. *zingiberi* Schlecht.

Favourable conditions: Temperature of 23-29°C, relative humidity of 87-95 per cent and nematode (*Meloidogyne incognita*) infestation.

Mode of survival: As chlamydospore in soil and in infected rhizomes.

Primary spread: From apparently healthy rhizomes.

Secondary spread: Through irrigation water and injuries to rhizomes made by nematodes or by cultural operations.

Management

- Adopt crop rotation
- Plant disease-free rhizomes
- Provide adequate drainage
- Apply FYM at 25 tonnes/ha.
- Treat rhizomes with *Trichoderma viride* 4g/kg or *Pseudomonas fluorescens* 10g/kg or carbendazim 0.1 per cent and drenching root region with carbendazim 0.1 per cent.
- Remove and destroy diseased plants with rhizomes and roots.

5. Leaf Spot

Symptoms

- Small, oval, elongated spots appear on the leaves.

- Matured spots are dark brown margin, papery white centre with yellow halo.
- Spots coalesce and cause extensive blighting and premature leaf fall.

Fungus

Phyllosticta zingiberi Ramakr.

Pycnidia are erumpent, amphigenous, subglobose, dark brown, ostiolate short-necked. Pycnidiospores are hyaline, unicellular, and oblong with rounded ends.

Favourable conditions: Growing without natural shade.

Mode of survival: As pycnidiospores survive in plant debris (14 months)

Primary spread: From infected plant debris.

Secondary spread: Through rain splashes.

Management

- Grow tolerant varieties.
- Spray mancozeb 0.2 per cent or copper oxychloride 0.3 per cent or Bordeaux mixture 1 per cent.

6. Bacterial Wilt

Symptoms

- Water-soaked spots at the collar region of the pseudostem which progress upwards and downwards.
- Drooping and curling of leaf margins of lower leaves which spread upwards.
- Yellowing starts from the lowermost leaves and gradually progresses to the upper leaves.
- In the advanced stage, the plants exhibit severe yellowing and wilting.
- When cut longitudinally, vascular tissues of the pseudostem show dark streaks.
- The affected pseudostem and rhizome are slimy and when pressed gently exudes a milky ooze from the vascular strands.

Bacterium

Ralstonia solanacearum (Yubucchi *et al.*)

[Syn. *Pseudomonas solanacearum* (Smith), *Burkholderia solanacearum* (Smith)].

The bacterium is rod-shaped, gram negative, lophotrichus bacteria. (more than one polar flagellum), multiplication occurs by bacterial fission.

Favourable conditions: Ill-drained soil, high soil moisture and relative humidity, low temperature and infestation of root-knot nematode (*Meloidogyne incognita*)

Mode of survival: In rhizome and in soil (for 20 months). It survives also on turmeric and many solanaceous hosts.

Primary spread: From infected rhizomes.

Secondary spread: Through irrigation water.

Management

- ✰ Treat seed rhizomes with streptocycline 500 ppm for 30 min and shade dry before planting.
- ✰ Drench the soil with Bordeaux mixture 1 per cent or copper oxychloride 0.25 per cent.

12. Nut Meg (*Myristica fragrans* Houtt)

1. Die Back

Symptoms

- ✰ The disease is characterized by drying up of mature and immature branches from the tip downwards

Fungi

Diplodia sp.

Spread: Through air-borne conidia.

Favourable conditions: Heavy and continuous rainfall.

Management

- ✰ Pruning of the affected branches.
- ✰ Application of Bordeaux paste immediately after pruning.

2. Shot Hole

Symptoms

- ✰ Necrotic spots develop on the lamina, which are encircled by a chlorotic halo.
- ✰ In advanced stages the spots become brittle and fall off resulting in shot holes.

Fungi

Colletotrichum gloeosporioides (Penz.) Penz. and Sacc.

The mycelium is grayish white. Acervuli are elongated or irregular in shape with setae. Conidia are cylindrical with obtuse ends.

Spread: Through air-borne conidia.

Management

- ☆ A prophylactic spray with Bordeaux mixture (5:5:50) 2-3 times at 14 days interval is effective against the disease.

3. Thread Blight

Symptoms

- ☆ The fungus causes two types of blights.
- ☆ **White thread blight:** Fine white hyphae aggregate to form fungal threads that traverse along the stem underneath the leaves in an irregular manner causing blight in the affected portions.
- ☆ **Horse hair blight**: Fine black silky threads of the fungus form an irregular, loose network on the stems and leaves. These strands cause blight of leaves and stems.

Fungi

Marasmius pulcherima; Marasmius equicrinus; Corticium spp.

Spread: As mycelium in dried leaves.

Favourable condition: Severe under heavy shade conditions.

Management

- ☆ Follow proper shade regulation practices.
- ☆ Spray with Bordeaux mixture (5:5:50) at initial stage of infection.

4. Fruit Rot

Symptoms

- ☆ The infection starts from the pedicel as dark lesions and causing brown discolouration of the rind resulting in rotting.
- ☆ In advanced stages, the mace also rots emitting a foul smell have been isolated from affected fruits.

Fungi

Diplodia natalensis Pole-Evans

Chromista

Phytophthora sp.

Management

- ☆ Spraying of bordeaux mixture 1 per cent or 0.2 per cent copper oxychloride is recommended when the fruits are half mature to reduce the incidence of the disease.

5. Seedling Wilt

Symptoms

- ✰ The leaves of the affected plants loose natural lusture, tend to droop and collar of seedlings show discolouration and decay.

Fungi

Cylindrocladium sp., *Fusarium* sp., *Colletotrichum* sp., *Rhizoctonia bataticola.*

Chromista

Phytophthora sp.

Other Diseases of Nutmeg

6. **Leaf Spot of Nutmeg: *Neofusicoccum parvum* (Pennycook and Samuels) Crous *Alternaria citri* Ellis and pierce, *Cladosporium oxysporum* Berk.**
7. **Dry rot: *Stigmina myristicae* (Stein.) Mand.-Sum**
8. **Root rot: *Fomes noxius* Corner, *F. lamaoensis* (Murrill) Sacc. and Trotter**
9. **Sooty mould: *Phragmocapinus betle***
10. **Algal leaf spot: *Cephaleuros* sp.**

13. Onion (*Allium cepa* L.)

1. Purple Blotch/Scald Disease

Symptoms

- ✰ It is an important disease prevalent in all the onion growing areas.
- ✰ The symptoms occur on leaves and flower stalks as small, sunken, whitish spots with purple coloured centres.
- ✰ These spots enlarge, become eye shaped under moist conditions, turn to purple and are surrounded by a broad chlorotic margin. The leaves break at the point of infection and hang down.
- ✰ These spots coalesce and cover entire leaf area.
- ✰ Older spots are black with fructifications seen in concentric rings.
- ✰ The lesions may girdle leaves/stalk and cause drooping.
- ✰ Premature drying of the foliage results in poor development of bulbs.
- ✰ Bulbs become dry and papery.
- ✰ The intensity of disease varies from season to season.

Fungus

Alternaria porri (Ell.) Neergaard

Favourable conditions: Temperature of 21-30°C, relative humidity of 80-90 per cent, cloudy weather with slight drizzling and dew formation. Onion thrips infestation predisposes the plant to purple blotch and the disease severity is also more in thrips infested crop compared to uninfested crop.

Mode of survival: As dormant mycelium in the infected plant debris and seed bulbs.

Primary spread: From infected crop debris.

Secondary spread: Through air-borne conidia.

Management

- ✰ Adopt crop rotation.
- ✰ Use disease free seed bulbs for planting.
- ✰ Provide good drainage.
- ✰ Early planting to avoid coincidence dew formation during crop period.
- ✰ Spray mancozeb or zineb 0.2 per cent or tebuconazole 0.15 per cent or tebuconazole + trifloxystrobin 0.06 per cent (0.6 ml/lit).

2. Basal Rot/Bulb Rot

Symptoms

- ✰ The disease incidence is more in the area where onion crop is grown continuously.
- ✰ The disease is seen at early growth stage of crop (up to 40 days).
- ✰ The leaves turn yellow with stunted growth.
- ✰ The plant shows wilting and complete drying of foliage from tip downwards.
- ✰ In advanced stage, the bulb starts decaying from lower ends along with roots and ultimately whole plant die.
- ✰ Whitish mouldy growth on the bulb scales. The bulbs decay in storage.

Fungus

Fusarium oxysporum f.sp. *cepae* (Hans) Snyd. and Hans.

The fungus produces many chlamydospores which are thick walled resting spores and microconidia which are one celled and thin walled.

Mode of survival: As chlamydospores in soil and infected seed bulbs.

Primary spread: From seed bulbs and soil.

Secondary spread: Through irrigation water.

Favourable conditions: Temperature of 28-32°C, low soil moisture, poor drainage, nematode infestation and injuries made during cultural operations.

Management

- Follow mixed cropping and crop rotation to reduce the disease incidence.
- Solarize the soil by spreading polythene sheet of 250 gauges in summer season for 30 days reduces the infectious propagules, which in turn reduces the disease.
- Select healthy seed bulbs for planting.
- Treat seed bulbs with *Pseudomonas fluorescens* 5 g/kg + *Trichoderma viride* 5 g/kg or carbendazim 2g/kg.
- Apply *Pseudomonas fluorescens* 1.25 kg/ha + *Trichoderma viride* 1.25 kg/ha as basal along with FYM 50kg/ha.
- Provide proper drainage.
- Avoid injuries to bulb during cultural operations.
- Spot drench with carbendazim or thiophanate methyl at 0.1 per cent is effective in the controlling the disease.
- Cure properly the seed bulbs to required moisture level.

3. Downy Mildew

Symptoms

- The disease often starts in patches.
- In systemic infection (when the plants are grown from a diseased bulb) the plants remain stunted, distorted and pale green.
- In local infections, the first sign of downy mildew is a purple-grey, velvety growth on green leaves and flower stalks (most easily seen in the early morning).
- Usually the older leaves are attacked first and the infection spread to the sheath.
- Diseased leaves quickly turn pale-green, then yellow, and collapse and die.
- Bulbs are often infected and sprout prematurely or shrivel in store.

Chromista

Peronospora destructor (Berk.) Casp.

Favourable conditions: Cool and humid weather with a temperature of less than 22°C, damp conditions (high moisture), late planting, application of higher doses of fertilizers and frequent irrigation.

Mode of survival: As oospores in soil and infected seed bulbs.

Primary spread: From seed bulbs and soil.

Secondary spread: Through irrigation water and rain splashes.

Management

- Adopt crop rotation for five years.
- Use disease free bulbs.
- Avoid overcrowding of plants and damp conditions.
- Maintain good weed control to ensure airflow through the crop.
- Provide proper drainage in the field.
- Spray mancozeb or metalaxyl + mancozeb 0.2 per cent.
- Remove and dispose affected plants and bulbs in the harvested field.

4. Smudge

Symptoms

- It appears at all growth stages but it is more common on bulbs approaching maturity and during storage.
- The disease affects the scales and lower portions of the unthickened leaves (neck of the bulb).
- In most cases the infection remain confined to one or two of the outer dry scales.
- Masses of minute black stromata are formed beneath the cuticle of the outer bulb scales (smudge).
- They may be scattered or arranged in concentric rings of 1-2 cm dia.
- Under moist conditions, the stromata produce acervuli, spore-bearing structures.
- In severe cases, infection can penetrate the entire bulb causing shrinkage and complete collapse of the fleshy scales.

Fungus

Colletotrichum circinans (Berk.) Vogl.

Favourable conditions: Bulbs stored with high moisture level.

Mode of survival: In infected bulbs and crop debris in soil.

Primary spread: From infected bulbs.

Secondary spread: Through air-borne conidia.

Management

- Follow crop rotation.
- Provide good drainage in the field.
- Onion with red scales are resistant (Pusa Red, Pusa Ratna). Chemical responsible for resistance in scales of coloured onions is protocatechuic acid (3,4-dihydroxybenzoic acid).

- ✰ Spray mancozeb 0.2 per cent or carbendazim 0.1 per cent.
- ✰ Avoid harvesting during wet weather.
- ✰ Dry the bulbs properly before storage by hot air at 37-48°C (sun drying is not recommended).

5. Pink Root

Symptoms

- ✰ The roots turn pink.
- ✰ Later they turn red or purple and finally turn black and die.
- ✰ The newly formed roots are also turn pink or red and die.
- ✰ Diseased plants can be easily pulled.
- ✰ The leaves of infected plants become yellow and dry from tip downwards.
- ✰ The plants are stunted with undersized bulbs because of poor root system.

Fungus

Pyrenochaeta terrestris (Hansen) Gorenz *et al.* (Syn. *Phoma terrestris* Hansen).

Favourable conditions: Temperature of 24-28°C, continuous cropping of onion and poor fertilization.

Mode of survival: As mycelium in soil.

Primary spread: From infected bulbs/seedlings.

Secondary spread: Through irrigation water and implements.

Management

- ✰ Adopt long rotation.
- ✰ Apply adequate amount of fertilizers to maintain the plant vigour.
- ✰ Avoid infested field for raising nursery.

6. White Rot

Symptoms

- ✰ Initial symptoms are yellowing and dying back of leaf tips.
- ✰ Roots are normally destroyed and there is a semi watery decay with abundance of superficial white fluff y mycelium.
- ✰ Brown, mustard-seed like sclerotia develop on the surface or within tissue.
- ✰ The affected bulb become watery and the outer scales crack as they dry and shrink.

Fungus

Sclerotium cepivorum Sacc.

Favourable conditions: High soil moisture and a temperature of 24°C.

Mode of survival: As sclerotia in soil.

Primary spread: From soil-borne sclerotia.

Secondary spread: Through irrigation water.

Management

- ☆ Follow crop rotation.
- ☆ Follow soil solarisation in the nursery and wherever feasible.
- ☆ Select healthy seed bulbs for planting.
- ☆ Treat seeds with *Trichoderma viride* 4g/kg or *Pseudomonas fluorescens* 10 g/kg or carbendazim 2g/kg.
- ☆ Treat seed bulbs with *Pseudomonas fluorescens* 5 g/kg + *Trichoderma viride* 5 g/kg or carbendazim 2g/kg.
- ☆ Apply *Pseudomonas fluorescens* 1.25 kg/ha +*T. viride* 1.25 kg/ha as basal along with FYM 50 kg/ha.
- ☆ Provide proper drainage.
- ☆ Remove infected plants and drench the soil around these plants.
- ☆ Avoid injuries to bulb during cultural operations.
- ☆ Cure properly the seed bulbs to required moisture level.

7. Black Mold

Symptoms

- ☆ The disease occurs in the region having high temperature ranging from 35-45°C.
- ☆ It causes postharvest blemishes and rotting of onion grown and stored in hot climate.

Fungi

Aspergillus niger Van Tiegh., *A. fumigates* Fresen and *A. alliaceus* Thom and Church

Favourable conditions: High temperature in the field and in storage, wounds made during harvesting and handling during storage.

Mode of survival: In the diseased bulbs.

Primary spread: From contaminated store houses and diseased bulbs.

Secondary spread: Through air-borne conidia.

Management

- ☆ Store the bulbs after proper drying.
- ☆ Avoid bruises on bulbs during harvest, transit and storage.
- ☆ Fumigate the storage godowns.

14. Tamarind (*Tamarindus indicus* L.)

1. Powdery Mildew

Symptoms

- Symptoms occur as white, powdery patches on upper surfaces of leaves.
- The patches coalesce and cover the entire surface, resulting in drying of the leaf.
- The fungus then spreads to surrounding plant parts.

Fungi

Erysiphe polygoni DC *Oidium* sp., and *Oidium tamarindi* Sharma and Khare

Hyphae are hyaline with smooth cell wall, and produce short and erect conidiophores. Conidia are hyaline, unicellular, cylindrical, in chains or singly.

Mode of survival: As in soil as oospores.

Primary spread: From infected plant debris.

Secondary spread: Through air-borne conidia.

Favourable conditions: Continuous rainfall.

Management

- Provide proper field sanitation.
- Spray wettable sulphur 3-4 times at 15 days interval or karathane 0.1 per cent.

Other Diseases of Tamarind

2. **Leaf spot - *Cercospora tamarindi* Khan, *Axhaetomium globosum, Bartalinia robillardes* Tassi, *Exosporium tamarindi* Syd., *Hendersonia tamarindii* Syd., *Meliola tamarindi* Syd., *Phomopsis tamarindii* Arya, *Phyllostica tamarindina* Rao, *P. tamarindicola* Rao, *Prathigada tamarindi* Muthappa, *Sphaceloma* sp., *Stigmina tamarindi* Syd., *Xanthomonus tamarindii* Cook.**
3. **Bark parasite - *Myriangium tamarindi* Tendulkar**
4. **Black pod - *Aspergillus niger* Tiegh**
5. **Brownish saprot - *Polyporus calcuttensis***
6. **Collar rot - *Phytophthora nicotianae* var. *nicotianae* Breda de Haan**
7. **Sap rot - *Xylaria euglossa***
8. **Seedling blight - *Macrophomina phaseolina* (Tassi) Goid., *Rhizoctonia solani* Kuhn**
9. **Sooty mould - *Meliola tamarindi* Syd.**

10. **Stem canker - *Hypoxylon nectrioides* Speg**
11. **Stem disease - *Fracchiaea indica* Talde**
12. **Stem rot - *Pholiota gollani* Hennings**
13. **Trunk and root rot - *Stereum nitidulum* Berk.**
14. **White root and wood rot - *Ganoderma lucidum* (Lyess) Karst.**
15. **White rot - *Trametes floccosa***

15. Turmeric (*Curcuma longa* L.)

1. Rhizome Rot

Symptoms

- ☆ Affected plants become pale, leaf tips turn yellow and infection gradually spreads down to the leaf blade and leaf sheath along the margin.
- ☆ Often the middle portion of the lamina remains green while margins turn yellow.
- ☆ Leaves droop, plants wither and dry.
- ☆ Infected plants can be pulled out from the soil easily.
- ☆ Root system is completely rotten. Rhizome small become brown, mis-shapened, soft and form a putrefying mass.

Chromista

Pythium aphanidermatum (Edson) Fitzp. and *P. graminicolum* Subram.

Mode of survival: As oospores in soil.

Primary spread: From infected plant debris.

Secondary spread: Through irrigation water.

Favourable conditions: High soil moisture and ill drained soil.

Management

- ☆ Adopt crop rotation with non-host crops.
- ☆ Provide proper drainage facilities.
- ☆ Grow tolerant varieties.
- ☆ Use disease-free seed rhizomes.
- ☆ Dip seed rhizomes in metalaxyl 0.2 per cent or copper oxychloride 0.25 per cent solution for 30 min prior to storage and at the time of sowing.
- ☆ Rouge out diseased plants.
- ☆ Drench the bed with copper oxychloride 0.25 per cent or metalaxyl 0.2 per cent when the disease is noticed in the field.

2. Leaf Spot

Symptoms

- ✰ Big, elliptic or oblong spots (4-5 cm x 2-3 cm) with greyish white centre and brown margin appear on leaves surrounded by yellow halo.
- ✰ Spots coalesce and cause drying to give a burnt up appearance.
- ✰ Older spots show many black dot-like acervuli arranged in concentric rings.
- ✰ Black stroma appears on scales covering the rhizomes.

Fungus

Colletotrichum capsici (Syd.) Butler and Bisby.

Mode of survival: As dormant stromata between the rhizome scales.

Primary spread: From rhizomes and plant debris.

Secondary spread: Through air-borne conidia and rain.

Favourable conditions: High humidity and low day temperature

Management

- ✰ Select seed rhizomes from healthy areas.
- ✰ Use disease tolerant varieties.
- ✰ Avoid excess shade and adopt mixed cropping.
- ✰ Spray Bordeaux mixture 1 per cent or mancozeb 0.2 per cent or carbendazim 0.1 per cent immediately after appearance of the disease.

3. Leaf Blotch

Symptoms

- ✰ Numerous small spots (1-2mm) appear on both the surfaces of the leaves and are very closely arranged.
- ✰ Yellow coloured spots turn to dirty yellow and then brown with chlorotic halo.
- ✰ Spots coalesce to form large irregular necrotic blotches.
- ✰ Leaves dry up and give a scorched appearance.
- ✰ Less number of functional leaves lead to poor yield.

Fungus

Taphrina maculans (Syd.) Butler.

Hyphae are intercullar. Branched or lobed haustoria are present. Asci are formed on both sides of the leaf. Asci contain eight ovoid, hyaline unicellular ascospores. These ascospores multiply by budding within the ascus to form sprout conidia.

Mode of survival: As infected plant debris.

Primary spread: From collateral host and diseased plant debris.

Secondary spread: Through air-borne ascospores and sprout conidia.

Favourable conditions: Temperature 21-23°C and a relative humidity of 80 per cent.

Management

- Grow disease tolerant varieties.
- Cut and remove diseased and dried leaves.
- Spray Bordeaux mixture 1 per cent or copper oxychloride 0.25 per cent or mancozeb 0.2 per cent with a sticking agent.

16. Vanilla (*Vanilla planifolia* Andrews)

1. Root and Stem Rot

Symptoms

- Browning of roots leading to rotting and decay.
- Yellowing of leaves.
- Leaves become flaccid shrivelled and giving a drooping appearance.

Fungi

Fusarium oxysporum f. sp. *vanillae*, (Syn: *Fusarium batatatis* var. *vanillae* Tucker)

Primary spread: From infected plant debris

Secondary spread: Through soil borne chlamydospore.

Favourable conditions: Heavy and prolonged precipitation, deficient soil drainage, excess shade, poor ventilation, drought stress, low nutrients, and high plant density.

Management

- Use land with good drainage.
- Use healthy and vigorous cuttings.
- Maintain a 10 cm cover of mulch over the roots.
- Keep plants well-nourished.
- Avoid overcrowding plants by maintaining appropriate distance between plants after pruning.
- Avoid excess shade and excess sunlight.
- Prune plants to remove infected parts.
- Soil drenching with *Trichoderma harzianum, Bacillus* spp., *Paecilomyces* sp. and *Pseudomonas fluorescens.*

2. Black Rot

Symptoms

- ✫ Water-soaked green to black rot of stems, leaves and/or pods.
- ✫ Thin white mycelium may be visible in infected tissues.
- ✫ Disease usually begins at the apical part of the plant and spreads to leaves, stems and all other parts of the plant.

Chromista

Phytophhora meadii McRae

Primary spread: From infected plant debris.

Secondary spread: Through soil borne oospore.

Favourable conditions: Disease emergence favored by prolonged wet weather, poorly draining soil, excessive shading of plants, overcrowded plants and lack of weed control in the plantation.

Management

- ✫ Use land with good drainage.
- ✫ Plant vanilla using recommended spacing.
- ✫ Control weeds around plants.
- ✫ Remove infected parts of plants and destroy by burning.
- ✫ Monthly application of the following mixtures: Fosetyl-Al (2.5 g/L of water) Carbendazim (2.0 g/L of water) and the mixture of Metalaxyl (2.5 g/L of water) Benomyl (2.0 g/L of water).
- ✫ Spray potassium phosphonate 0.4 per cent.

3. Anthracnose

Symptoms

- ✫ Small, sunken, dark brown spots on leaves, fruits, stems and/or flowers.
- ✫ Infected fruits dropping from plants before they reach maturity.
- ✫ Damage to fruit is more pronounced of the growing season.

Fungi

Colletotrichum anilla Scalia

Primary spread: From infected plant debris.

Favourable conditions: Warm and humid periods.

Management

- ✫ Use healthy and vigorous cuttings.
- ✫ Spraying any fungicide that contains copper oxychloride or mancozeb of 2 g/lit with water or Bordeaux mixture 1 per cent.

4. Rust

Symptoms

- ☆ Yellow to orange pustules on undersides of leaves which enlarge and coalesce causing the entire leaf to dry out.
- ☆ Finally plant becomes unproductive, defoliates and dies.

Fungi

Uromyces joffrini Delacr

Primary spread: From infected plant debris.

Favourable conditions: Overcrowding, excessive shading of plants and excessive moisture.

Management

- ☆ Removal of some leaves from the plants allows better penetration of sunlight.
- ☆ Spray of Bordeaux mixture, copper fungicides, of 2.5 g/lit of water weekly.

5. Root and Stem Rot

Symptoms

- ☆ Brown lesions on roots which turn brown and dry out.
- ☆ Plants begin to rot at the apical tip and stop growing.
- ☆ Plant begins to produce new roots from apical tissue.
- ☆ Stems dry out and crack longitudinally, cracks will eventually cover the whole stem and the plant will die.

Fungi

Fusarium incarnatum-equiseti species complex and *Colletotrichum* sp.

Primary spread: From Infected plant debris.

Favourable conditions: Excessive soil moisture, too much shading of plants, inadequate fertilization, over-crowding and drought stress.

Management

- ☆ Plant vanilla in well-draining soils and avoid overcrowding the plant.
- ☆ Prune out infected plant parts.
- ☆ Plant tolerant/resistant to the disease.

6. Cymbidium Mosaic Disease

Symptoms

- ☆ The plants infected with the virus are generally asymptomatic, but occasionally they exhibit mild chlorosis.

Virus

Cymbidium mosaic virus (CYMV)

Transmission

Sap and propagative materials.

Management

- ☆ Plants with virus symptoms must be removed from the plantation and burned.
- ☆ The movement of infected cuttings from one region to another must be avoided.

Other Diseases of Vanilla

7. **Black spot - *Colletotrichum orchidophilum* Damm**
8. **Brown spot - *Cylindrocladium quinquiseptatum* Boedijin**
9. **Leaf spot - *Colletotrichum gleosporoides* (Penz.) Penz. and Sacc.**
10. **Leaf blight - *C. gleosporoides, F. oxysporum***
11. **Leaf scar - *C. gleosporoides***
12. **Foot rot - *Sclerotium rolfsii, Verticillium* sp**
13. **Bean rot - *C. gleosporoides, F. oxysporum***
14. **Leaf rot/Bean rot - *Mucor racemosus* Fresen**
15. **Thread blight - *Pellicularia filamentosa* (Pat.) D.P. Rogers**
16. **Red rust - *Cephaleuros parasiticus* Karsten**

Viral Diseases

17. **Cucumber mosaic virus (CMV)**
18. **Odontoglossaum ringspot virus (ORSV)**
19. **Vanilla mosaic potyvirus (VanMV)**
20. **Vanilla necrosis potyvirus (VanNV)**
21. **Bean common mosaic virus (BCMV)**
22. **Bean yellow mosaic virus (BYMV)**

Chapter 5

Plantations Crops

1. Arecanut (*Areca catechu* L.)

1. *Mahali* Disease/*Koleroga*/Fruit Rot

Symptoms

- ☆ Characteristic symptoms include rotting and excessive shedding of immature nuts.
- ☆ The first visible symptom appears as water-soaked lesions on the surface of affected nuts. Infected nuts lose their lustre.
- ☆ The lesions gradually spread and cover the entire nut which later rot and shed from the bunches.
- ☆ The fruit bunches infected towards the end of rainy season may remain mummified on the palm.
- ☆ White mycelial mat develops on the entire surface of fallen nuts.
- ☆ Fruit stalks and rachis of inflorescence are also affected.
- ☆ Nuts are light in weight, deteriorate and become unsuitable for chewing.
- ☆ When infection occurs later in the season, it leads to rotting and drying up of nuts without shedding (known as 'Dry Mahali').

Fungus

Phytophthora arecae (Coleman) Peth.

Sporangia are pear-shaped and papillate.

Mode of survival: As oospore in affected and fallen nuts.

Primary spread: From mummified nuts on the palm and the fallen nuts.

Secondary spread: Through heavy air and rain splashes.

Favourable conditions: Continuous heavy rainfall coupled with low temperature (20 to 23°C), and high humidity (>90 per cent) favour the growth

Management

- ☆ Collect all the infected nuts and other plant parts and destroy.
- ☆ Follow field sanitation.
- ☆ Spray Bordeaux mixture 1 per cent with adhesive or vegetative oil.

2. *Anabe Roga*/Foot Rot

Symptoms

- ☆ Pale discolouration of the leaflets in the outer whorl of the leaves.
- ☆ Discolouration spreads to the whole leaf and the entire crown becomes yellow with the outer whorl drooping down and covering the stem.

- Later the leaves in the inner whorl also become yellow.
- Subsequently the leaves dry up, droop and fall off leaving the base stem.
- Development of the inflorescence and nuts is arrested.
- Infected palms exhibit symptoms similar to drought.
- Then the stem becomes brittle and easily broken by heavy air.
- The base of the stem shows brown discolouration and oozing of dark fluid.
- On cutting open the affected trunk, dark brown discolouration of internal tissues up to one metre from the ground level can be seen.
- Bracket-shaped fructification of the fungus called *anabe* appears at the base of the trunk.
- Roots are brittle, discoloured and dried.

Fungus

Ganoderma lucidum (Lyess) Karst.

Hard, leathery brackets are on the basal stem portion with brown polished upper surface and dirty white lower surface with numerous pores seen.

Mode of survival: As basidiospores in soil or in fruiting bodies.

Primary spread: From soil-borne basidiospores.

Secondary spread: Through irrigation water and root contact and by air-borne spores.

Favourable conditions: Soil pH of 3-9 and soil moisture of 40-80 per cent ; neglected, ill-drained and overcrowded plantations; hard black loamy and acid soils with high iron and low calcium; palms of 5-10 years age.

Management

- Collect and burn dead palms along with roots and fungal fruiting bodies.
- Avoid repeated ploughing and digging in the diseased gardens.
- Avoid closer spacing.
- Avoid flood irrigation and water flowing from infected palms to healthy palms.
- Improve drainage facility.
- Digging deep isolation trenches to avoid root contact from diseased to healthy.
- Follow balanced fertilizer application.
- Drench Bordeaux mixture 1 per cent at frequent interval with a sticking agent.

3. Inflorescence Die back and Button Shedding

Symptoms

- Brown patches of discolouration appear on the rachillae of the male flowers and then on the main rachis.
- It spreads from tip downwards covering the entire rachis and cause wilting.
- Affected female flowers are shed from the rachis and the whole inflorescence shows die back symptom.
- The embryo of female flowers shrivels and become brown.
- Concentric rings of pink conidial mass are seen on discoloured tissues.
- Button shedding is common during monsoon period.

Fungus

Colletotrichum gloeosporioides Pat and Penz.

Conidia are single celled, subhyaline, cylindrical, straight with obtuse ends with one or two oil globules and in mass looks pink.

Mode of survival: In infected plant parts *viz.*, inflorescence and buttons.

Primary spread: From infected plant parts.

Secondary spread: Through air-borne conidia and rain splashes.

Favourable conditions: Dry periods (February-May).

Management

- Remove fully affected inflorescence and destroy them to prevent spread.
- Spray carbendazim 0.1 per cent with a sticking agent.

4. Bud Rot/Crown Rot

Symptoms

- Initial symptoms include discolouration of spindle leaf from light green to yellow and then to brown followed by withering.
- Infection spreads inside the bud and such buds rot and the surrounding young leaves also rot.
- The affected young leaf whorl can be drawn out with a gentle pull.
- Infection spreads to the adjacent outer whorl of leaves, which become yellow, droops and drop off one by one leaving a bare stem.
- Secondary organisms colonize these affected tissues leading to a slimy mass emitting a foul smell.
- Severe fruit rot infection, especially on bunch and inflorescence stalks, which spreads to the stem and outer leaf sheaths.

- Inner portion of the leaf sheath exhibits water soaked lesions and base of leaf sheath rots resulting in yellowing of outer leaves.
- Disease spreads to subsequent upper leaf sheath base resulting in their rotting and yellowing of middle whorls of leaves.
- Tender portion of the stem and growing bud becomes infected resulting in rotting of the internal tissues of the crown and finally death of the palm.

Fungus

Phytophthora arecae (Coleman) Pethybridge.

Mode of survival: As oospores, chlamydospores and mycelium in soil, on fallen nuts and on dried nuts and inflorescence remaining on the crown.

Primary spread: From infected plant parts in the soil.

Secondary spread: Through air and air blown rain.

Favourable conditions: Continuous heavy rainfall with intermittent bright sunshine hours, low temperature of 20-23° C, constant high relative humidity more than 90 per cent, air, cloudy weather, *etc.*

Management

- Remove and destroy dead palms, fruit rot affected dried bunches, fallen nuts and crowns of affected palms.
- Scoop out infected portion and apply Bordeaux paste.
- Spray drench the spindle leaf base and the base of other leaves with Bordeaux mixture 1 per cent.

5. Stem Bleeding

Symptoms

- Small, discoloured depressions appear on the basal portion of the stem.
- Later they coalesce and cracks develop on the stem.
- Hollowness develops inside the stem upto varying depths due to disintegration of inner fibrous tissues.
- Brown gummy exudate oozes out from the cracks.
- Crowns of affected palms get reduced in size followed by reduction in yield.

Fungus

Thielaviopsis paradoxa (de Seyenes) Hohn.

Mode of survival: As chlamydospores in soil.

Primary spread: From soil-borne chlamydospores.

Secondary spread: Through irrigation and rain water.

Favourable conditions: Younger and middle-aged palms, poor drainage, poorly fertilized garden and cracks in the palm.

Management

- ✰ Improve drainage facility.
- ✰ Paint with Bordeaux paste after scooping out dead tissues.
- ✰ Apply recommended fertilizers.

6. Bacterial Leaf Stripe

Symptoms

- ✰ Dark green, water-soaked linear lesions appear on the leaflets parallel to the midrib.
- ✰ These lesions are covered with creamy-white bacterial exudates on the lower surface.
- ✰ Affected midrib and veins of the leaflets turn black.
- ✰ The entire leaflet in a frond is affected resulting in complete or partial blighting.
- ✰ In severe cases, entire crown may be affected.
- ✰ When growing buds are affected, death of palm occurs.

Bacterium

Xanthomonas campestris pv. *arecae* (Rao and Mohan) Dye

Mode of survival: In affected palm.

Primary spread: From affected plant parts.

Secondary spread: Through air blown rain.

Favourable conditions: Younger palms (3-5 years old).

Management

- ✰ Spray or inject the stem with tetracycline group of antibiotics at 500 ppm conc.

7. Yellow Leaf Disease

Symptoms

- ✰ Yellowing of leaves begin in the inner whorl, gradually spreading to the outer parts of the crown.
- ✰ Chlorosis is observed on all leaves in the whorl from the edges of the leaflets to the midrib region.
- ✰ Withering of the tips starts and gradually spreads to the older portions of the leaf.
- ✰ The freshly formed leaves are smaller, unequal and become flaccid.

- In a few cases, wilting and shedding of leaves are also observed.
- Stem becomes spongy and friable and in advanced stages, the stem breaks off at the top.
- Rotting of the roots is also observed.
- The nuts are small, shriveled with black kernel.

Bacterium

Candidatus Phytoplasma asteris

Mode of survival: In affected palm.

Primary spread: From affected plant parts.

Secondary spread: Through plant hopper (*Proutista moesta*).

Favourable conditions: Water logged condition; lack of balanced nutrition; Oct-Dec months.

Management

- Remove and destroy the diseased palms in the mildly affected areas to prevent the spread.
- Provide adequate drainage especially during monsoon season.
- Apply recommended dose of NPK fertilizers along with lime and zinc at 8.5 g each per palm.
- Adopt phytosanitary measures and plant protection measures to control anabe, bud rot, spindle bud and mite infestation.
- Avoid sun scorching of the stem by covering with arecanut leaves or painting with lime slurry.

2. Cashew (*Anacardium occidentale* L.)

1. Seedling Blight

Symptoms

- Water-soaked, irregular brown coloured spots on the young leaves which turn yellow and lose their turgidity.
- Wilting, withering and falling of leaves.
- The collar region then starts rotting and becomes fibrous.
- The roots exhibit discolouration and the finer roots rot.

Fungi

Cylindrocladium scoparium Morgan., and *Fusarium* spp.

Chromista

Pythium spp. and *Phytophthora palmivora* (E. J. Butler).

Primary spread: From soil-borne inoculum.

Secondary spread: Through splash irrigation.

Favourable conditions: Ill drained conditions.

Management

- The nursery beds or bags should be drenched with Bordeaux mixture 1.0 per cent.

2. Anthracnose/Brown Leaf Spot

Symptoms

- The disease affects tender leaves, twigs, inflorescence, shoots, nuts and apples.
- Initially reddish brown, shiny water-soaked lesions appear on shoots followed by resin exudation.
- Lesions grow longitudinally and radially killing the affected shoots.
- On leaves, brown circular to irregular spots with dark coloured margins are seen. In severe cases coalescence of spots takes place resulting in leaf fall.
- Acervuli are greyish, slightly raised and found on the spot.
- Tender leaves when attacked become crumpled.
- Unfertilized flowers when attacked turn black and fall off.
- Affected nuts and apples decay and shriveled.
- Successive die back of terminal shoots for a couple of years induce stag-headed symptoms and kill diseased plant.

Fungus

Colletotrichum gloeosporioides (Penz).

[Sexual stage: *Glomerella cingulata* (Stoneman)].

Acervuli (asexual fruiting body) with black setae, conidiophores are hyaline, septate, branched. Conidia are hyaline, aseptate, oval to elliptical or straight cylindrical and 11.0 to 17.0 × 3.0 to 6.0 μm.

Favourable conditions: Temperature of 25°C, high relative humidity and monsoon period.

Mode of survival: In affected dead plant parts.

Primary spread: From infected plant parts.

Secondary spread: Through air-borne conidia.

Management

- Plant air breaks like *Casuarina* or *Eucalyptus* checks the spread of the disease from one plantation to another.

- Remove and burn infected branches, fruits and leaves.
- Spray Bordeaux mixture 1.0 per cent.

3. Pink Disease/Die back

Symptoms

- White or pinkish growth on the bark of the branches penetrates into the deeper tissues and causes gradual death of the shoots from apex downwards and hence the name die back.
- Towards the cessation of monsoon, a film of silky thread of the fungus is seen on the branches.
- The mycelium is silvery white in the beginning but changes to pink at later stages.
- In advanced stages the bark splits and peels off. In a tree either one branch or many branches are affected.
- On the affected branches, the leaves turn yellow and fall off giving a barren appearance.

Fungus

Corticium salmonicolor Berk. and Broome.

[Syn. *Pellicularia salmonicolor* (Berk. and Br.) Dastur]

It is a basidiomycetous fungus, produces the silvery white, later pink and finally dirty septate mycelium. Basidial fructification with scattered hymenia sparingly produced basidia bearing basidiospores.

Mode of survival: As dormant mycelium in infected branches.

Primary spread: From infected branches.

Secondary spread: Through air-borne spores.

Management

- Prune the dried affected branches and protect the cut end by application of Bordeaux paste.
- Spray Bordeaux mixture 1.0 per cent, one in May-June before onset of South West Monsoon and the second in October.

4. Powdery Mildew

Symptoms

- Greyish-white growth on leaves, young twigs, inflorescence and young nuts.
- In severely affected trees the leaves become shriveled and dry.
- Fruit setting is reduced.

Fungus

Oidium anacardii Noack

Conidia is hyaline, cylindrical-to-slightly doliform, single-celled conidia (10 to 17.5 × 2.5 to 5 μm) borne in chains.

Mode of survival: In affected plant parts.

Primary spread: From infected plant parts.

Secondary spread: Through air-borne conidia.

Favourable conditions: Dry humid weather and rainy season.

Management

- ☆ Spray dinocap 0.2 per cent or quinomethionate 0.1 per cent.

5. Shoot and Inflorescence Blight

Symptoms

- ☆ Small oval to irregular brown water soaked spots are formed on the tender shoots, leaves and inflorescence.
- ☆ Many spots coalesce resulting in drying up of twigs.
- ☆ Badly affected trees present a scorched up appearance.

Fungus

Colletotrichum gloeosporioides Pat and Penz. and

Phomopsis anacardii Early and Punith.

Mode of survival: In affected plant parts.

Primary spread: From infected plant parts and tea mosquito bug.

Secondary spread: Through air-borne conidia.

Favourable conditions: Injury caused by the tea mosquito bug, *Helopeltis antonii* paves the way for the entry of pathogens.

Management

- ☆ Prune the shoots just below the affected portion.
- ☆ A combination spray of a fungicide copper oxychloride 0.2 per cent and an insecticide imidacloprid 0.05 per cent at 1.5 ml per litre is recommended.

3. Cocoa (*Theobroma cacao* L.)

1. Seedling Die back

Symptoms

- ☆ This disease is more severe on one month old cocoa seedlings though it is noticed in 1 to 4 month old seedlings.

- Infection starts from tip of the stem or cotyledon stalk or from the collar region.
- Dark brown to black, water soaked and linear lesions appear and they extend to leaves through the petiole resulting in wilting and subsequent defoliation of the seedlings.
- In advanced stages, defoliation and die back of the seedlings are noticed.

Fungus

Phytophthora palmivora Butler.

It is an oomycete possess aseptate mycelia, zoospore are the asexual spores produced in the sporangium where as oospores are sexual spores or resting/ dormant spores borne in oogonium.

Mode of survival: As oospores in infected plants.

Primary spread: From oospores in soil.

Secondary spread: Through zoospores and sporangia.

Favourable conditions: Cool weather; acidic pH; temperature of 18 to 25°C.

Management

- Remove and destroy the infected seedlings in the nursery.
- Provide adequate drainage.
- Drench the soil with Bordeaux mixture 1.0 per cent or copper oxychloride 0.3 per cent.
- Apply *Trichoderma viride* in the soil along with FYM.

2. White Thread Blight

Symptoms

- The young branches of the infected plants contain white mycelial threads which spread longitudinally and irregularly along the surface of the stem and leaf lamina.
- Later, the fungus invades the cortical tissues which eventually turn dark brown to black.
- The diseased leaves also turn dark brown and the dead leaves eventually get detached from the stem but are found suspended by the mycelial thread in a row.
- The extensive death of the young branches and suspended leaves in a row are the common field symptoms of white thread blight.

Fungus

Marasmius scandens Massee

Mode of survival: As mycelial mat on diseased leaves and stems.

Primary spread: From dead leaves with the mycelial mat are easily carried by air to the healthy plants.

Secondary spread: Through plant to plant and different branches of the same plant through the mycelium.

Favourable conditions: High humidity, less aeration and sunlight due to thick shade are the pre disposing factors.

Management

- ☆ Prune and remove dead materials and affected parts.
- ☆ Regulate the shade by structural pruning of branches to reduce humidity.
- ☆ Apply biocontrol agent, *Trichoderma viride* in the soil.
- ☆ Paste with copper oxychloride or Bordeaux paste at cut ends.

3. Black Pod/Pod Rot

Symptoms

- ☆ It is the most destructive fungal disease of cocoa.
- ☆ Pods are attacked at any stage of pod development and the infection may be seen at stalk end, distal end or in laterals.
- ☆ Brown discolouration of pod spreads rapidly over the whole pod with a line of demarcation.
- ☆ Under humid conditions, white mould appears on the surface of pod.
- ☆ Internal tissues of pods become brown and the infected beans are discoloured.
- ☆ Infected pods shrivel to form mummified pods.
- ☆ The beans become infected internally about 15 days after the initial infection and are soon of no commercial value.

Fungus

Phytophthora palmivora Butler.

It is an oomycete possess aseptate mycelia, zoospore are the asexual spores produced in the sporangium where as oospores are sexual spores or resting/ dormant spores borne in oogonium.

Mode of survival: As mycelium, oospores and chlamydospores in diseased pods.

Primary spread: From diseased pods.

Secondary spread: Through air-borne sporangia and zoospores; rain splashes; insects and rodents; contact between diseased and healthy pods; contaminated pruning materials.

Favourable conditions: High rainfall and long periods of high humidity; closer planting and cool damp weather.

Management

- ☆ Prune the trees properly to minimize the shade.
- ☆ Remove infected pods at regular intervals.
- ☆ Provide drainage facilities.
- ☆ Follow frequent and complete harvesting of pods and disposal of pod mummies and infected pods and pod husks. Ripe healthy pods should be regularly harvested, often daily.
- ☆ Modify the canopy by pruning to provide optimum shade, light and aeration through appropriate spacing and pruning to reduce surface wetness.
- ☆ Avoid intercropping with areca.
- ☆ Apply *Trichoderma viride*.
- ☆ Spray copper oxychloride 0.25 per cent or Bordeaux mixture 1.0 per cent.

4. Charcoal Pod Rot

Symptoms

- ☆ The infection appears as dark brown to black spot on any part of the pod surface and spreads rapidly.
- ☆ The beans may not develop fully and get mummified.

Fungus

Lasiodiplodia theobromae Pat. [*Botryodiplodia theobromae* (Pat.) Griff. and Maubl]

Mode of survival: In diseased pods.

Primary spread: From diseased pods.

Secondary spread: Through air-borne pycnidiospores.

Management

- ☆ Avoid intercropping with areca palm.
- ☆ Adopt proper spacing.
- ☆ Destroy infected pods.
- ☆ Adopt proper pruning to maintain adequate shade.
- ☆ Spray Bordeaux mixture 1.0 per cent or mancozeb 0.2 per cent.

5. Cherelle Rot or Anthracnose

Symptoms

- ☆ The young shoots, leaves and stems exposed to the sun in the crown region of the tree are affected.

- ☆ The lesions with yellow edges spread from the edge to the center of the leaves leading to defoliation and leaving the branches bare.
- ☆ The emerging new shoots from such branches are also affected and give a small broom-like appearance.
- ☆ Similar lesions and defoliation are noticed in the seedlings.
- ☆ The lesions on the stem are deeper.
- ☆ Deep brown lesions on fruits and ultimately the diseased pods turn brown to black and remains on the tree as mummified fruit.
- ☆ White mycelia appear on the lesions, which turn pink when the fungus produces spores.

Fungus

Colletotrichum gloeosporioides Pat and Penz.

Mode of survival: In infected plant parts.

Primary spread: From diseases plant parts.

Secondary spread: Through air, rain water or irrigation, insects and tools.

Favourable conditions: Humid conditions; rainy season; wounds by insects.

Management

- ☆ Provide adequate shade in the nursery and in the plantation.
- ☆ Adopt raised beds in the nursery.
- ☆ Use disease free seedlings for planting.
- ☆ Prune affected parts of the plants and spray Bordeaux mixture 1 per cent or mancozeb 0.2 per cent or carbendazim 0.1 per cent.

6. Witches' Broom

Symptoms

- ☆ The disease affects only actively growing tissues (shoots, flowers and pods) causing cocoa trees to produce branches with no fruit and ineffective leaves.
- ☆ Vegetative brooms are characterized by proliferation of axillary buds, stipules, hypertrophied petioles with smaller, pale yellow laminae and increased number of nodes.
- ☆ The other symptoms include necrosis of leaves, swelling of stem, axillary and terminal brooms.
- ☆ Fruits are hypertrophied, distorted and ripe unevenly.
- ☆ The seeds are also infected with lesions and are not viable.

Fungus

Moniliophthora perniciosa (Stahel)

[(*Crinipellis perniciosa* (Stahel)] (*Marasmius perniciosus* Stahel)

Primary spread: From basidiospores, seeds and bud wood.

Secondary spread: Through air borne basidiospores blown rain splashes, air and human beings.

Favourable conditions: Relative humidity of 80 per cent and temperature of 20-30°C.

Management

- ☆ Adopt phytosanitary pruning.
- ☆ Remove and destroy infected plant materials.
- ☆ Use resistant varieties.
- ☆ Remove young blooms before the production of sporophore.
- ☆ Adopt proper spacing.

7. Vascular Streak Die back (VSD)

Symptoms

- ☆ The first visible symptom is the yellowing of a single leaf in the second or third flush from the twig with sets of green patches scattered over the yellowish lamina.
- ☆ Affected leaves defoliate prior to the older leaves of the same branch and the bark of such branches becomes rough.
- ☆ The auxillary buds of the fallen leaves sprout and die rapidly.
- ☆ Later, die back appears on these branches and the xylem vessels turn brownish which appear as streaking within the vascular tissues. Hence the disease is called as vascular streak disease.

Fungus

Ceratobasidium theobromae (P.H.B. Talbot and Keane)

Syn: *Oncobasidium theobromae* Pat.

Mode of survival: As basidiocarps on fresh leaf scars during wet weather.

Primary spread: From basidiospores.

Secondary spread: Through air-borne spores.

Management

- ☆ Use disease tolerant varieties.
- ☆ Raise the nursery away from infected plantations with plastic roof over it.
- ☆ Use disease free seedlings for planting.
- ☆ Follow strict quarantine measures.

- ✰ Adopt regular pruning of infected branches atleast 30 cm below the discoloured xylem tissues.
- ✰ Spray of hexaconazole or propiconazole 0.1 per cent at monthly intervals.

8. Cocoa Swollen Shoot Disease (CSSD)

Symptoms

- ✰ Red vein banding: Reddening of primary veins or banding in young leaves followed by yellow banding along the main veins of leaves. Vein clearing in leaves producing a 'fern-like pattern and chlorosis or flecking and mottling of mature leaves.
- ✰ Stem and root swellings: Development of swellings at nodes, inter-nodes, apex of stems and roots.
- ✰ Malformation of pods, usually smaller and spherical.
- ✰ Death of infected tree.

Virus

Cocoa swollen shoot virus (CSSV)

Mode of survival: As semi-persistent manner in white fly.

Transmission: Mealy bug, *Planococcoides njalensis* transported by animals, ants, air and human beings.

Management

- ✰ Use of disease tolerant or less susceptible varieties.
- ✰ Do not carry cocoa plant material such as pods, seeds, bud wood or leaves to other farms or regions since once a cocoa plant becomes infected with CSSV it can take up to 2 years before any symptoms appear.
- ✰ Use barrier crops.
- ✰ Eradicate or remove diseased plants.
- ✰ Do not remove any diseased plant material, insects or soil samples from farms unless under the supervision of the relevant research authority.
- ✰ Wash hands after handling diseased cocoa material.
- ✰ Wear rubber boots so they can be cleaned using a 10 per cent bleach solution before visiting another farm.
- ✰ Sanitize all pruning tools and knives with 10 per cent bleach solution before visiting another farm.
- ✰ Spray insecticide to control mealy bugs.

4. Coconut (*Cocos nucifera* L.)

1. Basal Stem Rot/Tanjore Wilt/Ganoderma Wilt

Symptoms

The diseased trees show the following typical symptoms in different parts of the palm *viz.,* stem, leaves, inflorescence and roots.

Stem

- ☆ Exudation of reddish brown, viscous liquid from the basal portion of the stem and it extends up to three metres height.
- ☆ Discolouration of the stem and internal rotting are noticed up to the height of bleeding (exudation).
- ☆ In advanced stages the basal portion of the stem decays completely.
- ☆ Some palms show wilting symptoms without external bleeding.
- ☆ In some trees, the bark of the stem peels off.
- ☆ In advanced phase, the fructifications of the fungus (sporophore or mushroom) is observed at the base of the tree just above the ground level as a bracket.

Leaves

- ☆ The leaflets in the outer one or two whorls show yellowing, withering and drooping.
- ☆ These leaves hang around the trunk for several months.
- ☆ The remaining leaves droop down in quick succession leaving the spindle leaf alone.
- ☆ Newer leaves are smaller and their production is delayed.
- ☆ Outer leaves fall off.
- ☆ The spindle size is reduced and do not unfold properly.

Inflorescence and Nuts

- ☆ Inflorescence development is suppressed and the fruit set is arrested.
- ☆ Button shedding is commonly noticed.
- ☆ The subtended bunches hang down and the quality of kernels from such bunches is poor and nuts are barren.
- ☆ In severely diseased palms, nut and kernel weight, water content, copra weight and oil content are decreased.

Roots

- ☆ The finer roots decay and die
- ☆ Only few new roots are produced

Fungus

Ganoderma lucidum (Fr). Karst.

Mycelium is hyaline, thin-walled, branches with frequent clamp connections. Chlamydospores are ellipsoid and slightly thick-walled. The fruit body is perennial, stipitate, lateral and sometimes sessile. It is corky at first and become woody later. Fruit bodies are upto 30 cm in dia.

Mode of survival: As basidiospores in soil.

Primary spread: From basidiospores.

Secondary spread: Through irrigation water, rain water and root contact.

Favourable conditions: Trees grown in sandy loam and sandy soil, waterlogged condition, low soil moisture during summer months, denser planting, damages caused by beetles and weevils.

Management

- Follow the cultural practices such as minimized ploughing in diseased areas, avoidance of closer planting, providing proper drainage facilities, adequate irrigation, application of mulches in the basins, *etc.*
- Apply *Pseudomonas fluorescens* at 200 g + *T. viride* 200 g/tree/year along with 200g phosphobacteria + 200 g Azotobacter + 50 kg FYM/tree/year.
- Apply FYM or compost or green manure at 50 kg/tree + 5 kg neem cake/ tree once in 6 months along with the recommended fertilizers.
- Irrigate at least once in 15 days during summer months.
- Remove and burn severely infected palms.
- Isolate severely infected tree by digging trench 1.2m away from the base of the trunk and apply sulphur dust in the trench.
- Drench the soil with Bordeaux mixture 1 per cent solution at 40 l/tree around the trunk in a radius of 1.5 m.
- Chisel out bleeding patches in the stem and protect with copper oxychloride (5 per cent) and subsequently with hot coal tar.
- Apply aureofungin sol 2 g + copper sulphate 1 g in 100 ml of water or hexaconazole 2 ml/100 ml of water through root feeding (The active absorbing root of pencil thickness must be selected and a slanting cut is made. The solution to be taken in a polythene bag and the cut end of the root is dipped in the solution).

2. Bud Rot

Symptoms

- Primary symptoms are pale colouration of leaflets, bending over, browning and breaking down of the heart leaf or central shoot.

- The affected heart leaf comes off easily when it is pulled.
- Brown, sunken spots develop at the base of the young leaves affecting a long strip of leaf tissue leading to rotting.
- When rotting extends downward the surrounding whorl becomes dull yellow and then brown.
- Young nuts cease to develop and fall off prematurely.
- The softer infected portion rots and degenerates to slimy mass emitting foul smell.
- When the growing bud is affected, the tree will be killed.

Fungus

Phytophthora palmivora Butler.

It is an oomycete possess aseptate mycelia, zoospore are the asexual spores produced in the sporangium where as oospores are sexual spores or resting/ dormant spores borne in oogonium.

Mode of survival: As oospores in infected plant debris.

Primary spread: From soil-borne oospores.

Secondary spread: Through zoospores, rain or overhead irrigation.

Favourable conditions: High rainfall, high humidity (more than 85 per cent), low temperature, dense plant population and wound caused by *Rhinoceros* beetles. Trees of all ages are affected. But young trees are severely infected.

Management

- Provide adequate drainage in gardens.
- Adopt proper spacing and avoid overcrowding in bud rot prone gardens.
- Remove all the affected tissues at the crown region and paint with Bordeaux paste (Dissolve 100 g of copper sulphate and 100 gm of quick lime each in 500ml water separately and mix to form 1 lit of Bordeaux paste) followed by spray the crown with copper oxychloride 0.25 per cent or Bordeaux mixture 1 per cent to protect the emerging shoot. The treated crown region may be protected from rain by covering with polythene sheet.
- Remove and burn badly affected trees.
- In the early stage of infection, cut, remove and burn the infected spindle leaf and apply Bordeaux paste.
- Spray copper oxychloride 0.25 per cent or Bordeaux mixture 1 per cent with sticking agent on the crown of the neighbouring palms as a prophylactic measure before and after the onset of monsoon. Palms that are sensitive (Dwarf palms) to copper containing fungicides can be protected by mancozeb. Small, perforated sachets containing 2 g

of mancozeb may be tied to the top of leaf axil. When it rains, a small quantity of the fungicide is released from the sachets to the leaf base, thus protecting the palm.

- ☆ Fill the leaf axil with carbaryl 25 g mixed with 200g of sand to red palm weevil infestation of affected palms.

3. Stem Bleeding

Symptoms

- ☆ The disease occurs on palms of all ages anywhere on the trunk but the spread is more rapid in young palms.
- ☆ The characteristic symptom of the disease is the exudation of a dark reddish brown fluid from cracks in the outer tissue mostly on the lower trunk.
- ☆ The fluid turns black as it dries up on the bark.
- ☆ The tissues beneath the bleeding patches decay and become yellow.
- ☆ In the early stages, the decay of the internal tissue is localized but as the disease progresses, these patches coalesce resulting in a general decay.
- ☆ The infected trees are not killed but their yield is reduced.
- ☆ The trees become very thin at the top and are likely to be broken by air.

Fungus

Ceratostomella paradoxa Dade

[Syn. *Ceratocystis paradoxa* (Dade) Moreau; *Ophiostoma paradoxa* (Dade) Nannf.].

Mycelium is brown. Matured endoconidia hyaline to pale brown and smooth walled. Chlamydospores are terminal in chains, oval, thick-walled and brown. Perithecia are partly immersed, light brown and ostiolate. Ascospores are ellipsoid with unequally curved sides, hyaline, non-septate and smooth.

Mode of survival: As dormant mycelium, chlamydospores/ascospores in soil.

Primary spread: From air-borne conidia.

Secondary spread: Through irrigation and rain water.

Favourable conditions: Cracks and wounds in the trunk, poor maintenance of garden and damage by beetles, July to November months, sever summer followed by sudden wetting or rain, imbalanced nutrition, salinity, *etc.*

Management

- ☆ Maintain the garden with proper fertilization.
- ☆ Scoop out the diseased or rotten tissue and swab the exposed surface with hot coal tar or Bordeaux paste 5 per cent. If cavities are found, these have to be filled up with cement for reinforcing.

- Avoid mechanical injuries since it is a wound pathogen.
- Irrigate the crop during summer months at 15 days interval or provide drainage during rainy season.
- Increase the soil organic matter by applying FYM 50 kg + neem cake 5 kg/tree.
- Apply *Trichoderma viride* or *Pseudomonas fluorescens* 200 g each/palm/year during September.

4. Grey Blight

Symptoms

- The disease symptoms develop on the outer whorl of leaves.
- Minute yellow spots encircled by a greyish band on the leaflets.
- The spots increase gradually up to 5 cm. The centre of these spots turn greyish white with brown margin.
- Several such spots coalesce into irregular and grey necrotic patches with black, minute dot-like acervuli are formed on them.
- In advanced stages, the tips and the margins and at times the whole leaflet dry and shrivel giving the leaf a burnt-up appearance.

Fungus

Pestalotiopsis palmarum Cooke. (Syn. *Pestalotia palmarum* Cooke).

The fungus produces acervuli. Conidia are 5-celled straight or curved. The three median cells are coloured and the terminal cells are hyaline. Apical cell tapering, crowned with three setulae.

Mode of survival: As dormant mycelium.

Primary spread: From infected palms and seedlings.

Secondary spread: Through air-borne conidia.

Favourable conditions: Dry weather, excessive nitrogen and water logging.

Management

- Remove and burn infected and dried leaves periodically.
- Apply heavy doses of potash to give resistance.
- Improve drainage condition of the soil.
- Spray the foliage with copper oxychloride 0.25 per cent or Bordeaux mixture 1.0 per cent or propiconazole 0.1 per cent before the onset of rain.

5. *Lasiodiplodia* Leaf Blight (LLB)

It is an emerging serious problem in Pollachi tract of Tamil Nadu. At present, the disease is spreading at a faster rate in adjoining districts

Symptoms

- Though the incidence is noticed throughout the year, the maximum incidence is observed during summer months.
- The disease attacks both seedlings and adult palms. Both leaves and nuts are affected.
- In adult trees, outer 3 to 4 whorls of leaves are affected.
- The affected leaflets start drying from the tip downwards and exhibit a charred or burnt appearance.
- Dark grey to brown lesions with wavy margins appear at the stem end of the nuts.
- The pathogen enters into the kernel through mesocarp resulting in the decay of the endosperm.
- The affected nuts desiccate, shrink, deform and drop prematurely results in loss in nut yield up to 10 to 25 per cent.

Fungus

Lasiodiplodia theobromae Pat.

[Syn. *Botryodiplodia theobromae* (Pat.) Griff. and Maubl]

The asexual fruiting body of the fungus is pycndia. The pycnidium produces and releases two-celled, dark brown, striated conidia.

Mode of survival: As pycnidia/resting spores in infected nuts and leaves.

Primary spread: From infected leaves and nuts.

Secondary spread: Through air-borne conidia.

Favourable conditions: Temperature of 25°C favours the disease.

Management

- Remove and burn the severely affected leaves to avoid further spread.
- Apply *Pseudomonas fluorescens* 200g along with 50 kg of FYM + 5 kg of Neemcake/palm/year.
- Spray Bordeaux mixture 1.0 per cent or copper oxychloride 0.25 per cent two times at 45 days interval during summer months.
- Root feeding of carbendazim 2 g or hexaconazole + 100 ml water 3 times at 3 months interval.
- Application of 2 kg of muriate of potash in addition to the recommended fertilizer dosage.

6. Leaf Rot

Symptoms

- Leaf rot disease commonly occurs on coconut palms already affected by root wilt disease.

- ☆ The first symptom is the appearance of water-soaked brown lesions in the spear leaves.
- ☆ Gradually they enlarge and coalesce resulting in extensive rotting.
- ☆ Each new leaf of the diseased tree gets infected and all the leaves show disease symptoms.
- ☆ As the leaf unfurls the rotten portions of the lamina dry and get blown off in air, giving a 'fan' shape to the leaves.
- ☆ In acute cases, the spear fails to unfurl.
- ☆ The reduction in leaf surface adversely affects the yield.

Fungus

Bipolaris incurvata Ellis

The mycelium was septate, branched and brown in colour. Conidiophores mainly arose in small groups and conidia were typically curved, fusiform, septate, pale straw coloured.

Mode of survival: As infected leaf debris.

Primary spread: From infected leaf debris.

Favourable conditions: Dry humid weather.

Management

- ☆ Remove the rotten portions from the spear and the adjacent leaves.
- ☆ Spray the crown and the leaves with hexaconazole 2 ml or mancozeb 3 g in 300 ml of water per palm or Bordeaux mixture 1 per cent or copper oxychloride 0.3 per cent to soak the base of spindle leaf.

7. Root (Wilt) Disease/Kerala Wilt

Symptoms

- ☆ It is a non-lethal debilitating disease and the affected palms survive for a long period giving a reasonably good yield.
- ☆ Important diagnostic symptom is the flaccidity of leaflets *i.e.*, they curve abnormally inwards, resembling the ribs of mammals.
- ☆ In the beginning yellowing is restricted from the leaf tips to the middle of the leaves followed by necrosis of leaflets.
- ☆ Reduction of leaf size and tapering of terminal portion of the trunk.
- ☆ Wilting of leaves from middle to outer whorl.
- ☆ Flowering is delayed and also yield is considerably reduced.
- ☆ Abnormal shedding of buttons and immature nuts.
- ☆ Matured nuts are small with thin kernel.
- ☆ Rotting of roots from tip backwards.

☆ Accelerated transpiration and reduced absorption by the roots are the causes of wilting. This disease predisposes the plant to leaf rot (*Bipolaris halodes*) infection and pests like rhinoceros beetle and red palm weevil.

Bacterium

Candidatus Phytoplasma oryzae'-related strain

Mode of survival: In infected trees.

Transmission: Infected tree sap and lace wing bug (*Stephanitis typica*) and plant hoppers (*Proutista moesta*).

Favourable conditions: Sandy loam soil, severe flood, abundance of lace wing bug, (*Stephanitis typicus*), poor maintenance of plantation.

Management

☆ Growing of tolerant varieties Chowghat Green Dwarf and West Coast Tall.

☆ Remove all infected uneconomic trees yielding less than 10 nuts per palm per year.

☆ Grow green manure crops - cowpea, Sunnhemp (*Crotalaria juncea*), *Mimosa invisa, Calapagonium mucanoides, Pueraria phaseoloides etc.* in coconut basins and incorporate. The ideal green manure crops for the sandy and alluvial soils are cowpea and sesbania respectively.

☆ Raise intercrops like tapioca, yam, fodder crops, *etc.*,

☆ Apply balanced dose of fertilizers along with 50 kg of FYM + 5 kg neemcake/palm/year.

☆ Irrigate coconut palms with atleast 250 lit of water per tree per week.

☆ Adopt strictly all the prescribed prophylactic measures against leaf rot disease, red palm weevil, rhinoceros beetle *etc.*

☆ Initial stage of the disease, spray the crown and drench the basin with 500 ppm tetracycline.

8. Lethal Yellowing

It is a pandemic disease in South Florida of USA.

Symptoms

☆ Premature dropping of most of the nuts.

☆ Necrosis of new inflorescence with blackening of tips.

☆ Most of the male flowers die and thereby no fruit set.

☆ The lower fronds turn yellow and spread gradually to younger leaves.

☆ The yellow leaves become turgid, brown, desiccate and hang down.

☆ Finally, the newly emerged spear leaf collapses.

☆ The terminal bud dies, crown wither and falls within 6 months leaving a bare trunk.

Bacterium

'Candidatus Phytoplasma palmae'

Mode of survival: As infected leaf and plants.

Primary spread: From infected plants and other palms such as *Phoenix* species.

Secondary spread: Through leafhopper (*Myndus crudus*).

Favourable conditions: Neglected plantation, high vector population and summer season.

Management

- ✰ Remove and destroy diseased palms.
- ✰ Grow resistant cultivars.
- ✰ Spray insecticide like diazinon or dimethoate.
- ✰ Injection of tetracycline-type products or oxytetracycline HCl into the trunk.

9. Cadang-Cadang

Symptoms

- ✰ The name Cadang-Cadang means 'dead or dying' which reflects the slow but persistent lethal progress of the disease. It is also known as yellow mottle decline.
- ✰ This is a slow killer takes 5-8 years to die and thus only the oldest palms are more susceptible.
- ✰ In the initial stage, the nuts become smaller, more rounded and have equatorial scarifications. The fibrous layer becomes thinner and thinner.
- ✰ Chlorotic leaf spots begin to appear and inflorescences become stunted.
- ✰ Later, the spathe, inflorescence and nut production decline and then cease.
- ✰ Leaf spots become more numerous and they coalesce.
- ✰ The leaflets become brittle and the fronds decline in size and number and remain in upright position in the crown.
- ✰ In advanced stages, the central fronds become bronze-yellow, main bud dies and falls showing crownless.
- ✰ The roots deteriorate and rot.

Viroid

Coconut cadang-cadang viroid (CCCVd)

Mode of survival: In infected trees. The African oil palm is the other natural host.

Primary spread: From infected seed nuts.

Secondary spread: Through pollination and harvesting scythe.

Management

- ✰ Cut and burn the diseased palms.
- ✰ Replant with resistant cultivars.
- ✰ Follow strict enforcement of quarantine measures.
- ✰ Follow good soil and fertilizer management methods.

10. Pencil Point Disorder

Symptoms

- ✰ The stem will taper towards its tip (pencil point) with lesser number of leaves.
- ✰ The leaf size will be greatly reduced and the leaves will be pale and yellow in colour.
- ✰ The axillary inflorescences do not develop properly in advanced cases.

Cause

Micronutrient deficiency.

Management

- ✰ Dissolve 225 g each of borax, zinc sulphate, manganese sulphate, ferrous sulphate, copper sulphate and 10 g of ammonium molybdate in 10 lit of water and pour in the basin of 1.8 m radius along with the recommended fertilizer dose.
- ✰ Remove severely affected palms and replant with new seedlings.

11. Crown Choking

Symptoms

- ✰ It is characterized by the emergence of short, wrinkled leaves manifested as sharply bent leaflet tips commonly called hook leaf. Leaves have a serrated zigzag appearance.
- ✰ The leaflets show severe tip necrosis and fail to unfurl and stick together or unseparated like a fan. In many cases, it gives a choked appearance to the frond.
- ✰ The inflorescence and nuts are become necrotic.
- ✰ Ultimately the affected palm dies.

Cause

Boron deficiency

Management

- ✰ Apply 50 g of Borax at half-yearly intervals (Feb-Mar and Sept-Oct) along with recommended fertilizer in the basins. In root wilt affected areas, a dosage of 200-300g per palm per year is recommended.

5. Coffee (*Coffea arabica* L.)

1. Collar Rot/Damping Off

Symptoms

Pre-emergence Damping Off

- ✰ Embryo and endosperm are invaded by the fungus before germination and radicle during germination.
- ✰ Seeds rot and disintegrate.

Post-emergence Rot

- ✰ Seedlings show brownish discolouration on the stem near the ground level leading to rotting of the tissue.
- ✰ Growing apex wilts and the seedlings collapse and dies.

Fungus

Rhizoctonia solani Kuhn.

Mode of survival: As sclerotia in soil.

Primary spread: From infected plant refuses in soil.

Secondary spread: Through irrigation water.

Favourable conditions: Overcrowding of seedlings, thick overhead shade, hot and humid climate, excessive soil moisture in the nursery beds

Management

- ✰ Expose the nursery soil to the sun light for 2-3 months.
- ✰ Prepare raised seed beds and drain excess moisture.
- ✰ Avoid excessive watering.
- ✰ Avoid overcrowding of seedlings by adopting recommended seed rate.
- ✰ Provide filtered overhead shade using green leaves/coir mats/nylon mats.
- ✰ Remove and destroy the affected seedlings by burning to prevent further spread.
- ✰ Treat coffee seeds with carbendazim 2 g/kg and drench the nursery beds with captan 0.2 per cent or carboxin 0.1 per cent.
- ✰ Remove the affected seedlings and spray with any of the fungicides *viz.*, carbendazim 0.1 per cent or captafol 0.2 per cent to protect the seedlings.

2. Leaf Rust/Oriental Leaf Disease

Symptoms

- ✰ Small, yellow translucent oily spots appear on upper surface of the young leaves and they enlarge into round shape of 10 to 15 mm in dia.

- Spots later turn into orange yellow colour.
- The corresponding lower surface is covered with orange brown, dusty, powdery mass of uredospores.
- Severe attack results in coalescence of spots and complete defoliation.
- Severe defoliation leads to die back, debilitation of the bushes which results in poor cropping in the succeeding seasons.
- Die back results in death of whole branch or stem and eventual death of entire plant.
- Berries from diseased plants ripen prematurely and the beans are of poor quality and light in weight.

Fungus

Hemileia vastatrix Berk. and Broome

It is a hemicyclic fungus with theurediniosporic life cycle. The dusty powder consists of uredospores, which are orange-segment shaped. The convex surface is echinulated and the concave surface is smooth. Teleutospores are thick walled, smooth and turnip-shaped mostly seen from January to April. Aecial and pycnial stages are not noticed.

Mode of survival: In infected plants.

Primary spread: From infected plants.

Secondary spread: Through air-borne uredospores and teliospores; insects, animals, birds and workers.

Favourable conditions: Intermittent rain and sun shine, mist or rain during dry weather from December to March. Thin or no overhead shade during the monsoon is favourable.

Management

- Grow resistant/tolerant cultivars.
- Prune dead and dried twigs including die back affected twigs.
- Apply balanced nutrients to maintain plant vigour and to balance the plant biomass.
- Maintain good overhead shade in plantations.
- The antagonistic fungi, *Verticillium lecanii* and *Darluca* sp. found on rust pustules may be useful for a short period during winter months as biocontrol agent.
- Prophylactic spray of Bordeaux mixture 0.5 per cent or copper oxychloride 0.25 per cent before and during monsoon period.
- Spray of triadimefon 0.1 per cent or hexaconazole 0.1 per cent during August-September is effective.

- Spray oxycarboxin 0.15 per cent with caution since it leads to defoliation sometimes.

3. Anthracnose/Nilgiri Twig Disease

Symptoms

- Greyish spots are observed on leaves, twigs and berries.
- Spots are necrotic on the leaves and branches.
- Spots are circular or irregular, greyish and 2 to 3 cm dia on leaves.
- Elongated brown or black lesions are seen on the branches, which result in blackening of twigs and drying from tip downwards.
- Spots on berries are black, sunken and irregular; Beans become black in colour.
- Berries fail to mature and form *katte kai.*
- Root system decays and feeder roots die.

Fungus

Colletotrichum gloeosporioides Pat and Penz. (Sexual stage*: Glomerella cingulata*).

Acervuli (asexual fruiting body) with black setae, conidiophores are hyaline, septate, branched. Conidia are hyaline, aseptate, oval to elliptical or straight cylindrical and 11.0 to 17.0 × 3.0 to 6.0 mm.

Mode of survival: As saprophyte on dead tissues.

Primary spread: From infected plant parts.

Secondary spread: Through air-borne conidia, rain water, rain splashes, birds, machinery and coffee pickers.

Favourable conditions: Dry period following the monsoon rain, inadequate overhead shade, prolonged drought and soil moisture stress.

Management

- Provide drainage and apply balanced fertilizer.
- Conserve soil moisture by providing shade and mulch.
- Spray Bordeaux mixture 1 per cent.

4. Brown Eye Spot/Berry Blotch

Symptoms

- Disease is severe on seedlings and on young crop.
- Reddish-brown or brown spots with greyish centre occur on upper leaf surface and are surrounded by yellow halo.
- After sporulation the centre of the spots become olive green.

- ☆ Berries turn dark brown or black with irregular, sunken blotch with a purple halo (berry blotch).
- ☆ Berries shrink, dry and fall off. The skin of the fruits becomes dry and hard, making it difficult to pulp.

Fungus

Cercospora coffeicola Berk. and Cooke.

[Sexual stage*: Mycosphaerella coffeicola* Cooke.]

Mode of survival: In seed.

Primary spread: From infected plant refuses and from seeds.

Secondary spread: Through air-borne conidia.

Management

- ☆ Maintain adequate shade to avoid sun scorching on berries.
- ☆ Spray Bordeaux mixture 1 per cent or captan or mancozeb or ziram or captafol 0.2 per cent.

5. Sooty Mould

Symptoms

- ☆ Leaves, stem and berries are covered with a black fungal growth (mycelium and spores) resulting in reduced photosynthetic activity.
- ☆ Wherever heavy attack of aphids and scale insects are found sooty mould occurs; fungus feeds on the secretions of these insects.

Fungus

Capnodium braziliense Puttemans

Mode of survival: In infected plants.

Primary spread: From infected plants.

Secondary spread: Through air-borne spores and rain splashes.

Management

- ☆ Control aphids and scales with fish oil rosin soap 1kg + 1 kg starch for 200 lit of water.
- ☆ Boil maida 1 kg with 5 lit of water, cool, dilute to 20 lit with water (5 per cent) and spray. In due course the mouldy growth will peel off after drying.

6. Black Rot/*Koleroga*/Thread Blight

Symptoms

- ☆ Blackening and subsequent rotting of young leaves berries and shoots (Black rot).

- Diseased leaves defoliate, get detached from branches and hang out by means of slimy fungal strands.
- The under surface of the affected leaves is covered by a thin, hyaline, spreading web of fungal mycelium.
- Berries become brown, finally black and drop off from affected branches.

Fungus

Corticium koleroga (Cooke) Höhn. (Syn. *Pellicularia koleroga* Cooke).

Mycelium is hyaline, septate, profusely branched with numerous basidia, which arise in groups of 4 or 5 on short branches. Basidiospores are hyaline and thin-walled.

Mode of survival: In infected plant parts.

Primary spread: From infected fallen leaves and berries.

Secondary spread: Through air-borne spores.

Favourable conditions: Heavy rainfall with saturated atmosphere with 95-100 per cent humidity, thick overhead shade and continuous mist.

Management

- Maintain uniform, medium intensity overhead shade.
- Centering the bushes to allow sunlight and air.
- Prune and burn dead twigs.
- Collect infected leaves and berries and bury them.
- Spray Bordeaux mixture 1 per cent or carbendazim 0.1 per cent before and during monsoon.

6. Large/Black Cardamom (*Amomum subulatum* Roxb.)

1. Seedling Rot and Wilt

Symptoms

a. Seedling Rot

- Maximum damage to the nursery occurs in February and March.
- Leaves turn pale and their tips become yellow.
- The infection spreads to entire leaf extending to leaf sheath resulting in wilting of seedlings.
- The collar portion decays and the entire seedlings die.
- Infection spreads in the nursery beds resulting in death of seedlings in small patches.

- In grown up seedlings, rotting extends from the collar region to the rhizomes resulting in decay and death of the plant.

b. Wilt

- The intensity of the disease is severe in plantations from October to February.
- Sudden wilting of the plant or individual leaf is the characteristic symptom of the disease.
- The first symptom is chlorosis of the older leaves and progresses to the younger ones.
- The emerging heart leaf shows necrosis and the base of the pseudostem may split.
- All the leaves eventually collapse leaving the pseudostem exposed.
- Brown to black vascular discolouration is seen in the outer leaf sheath throughout the pseudostem.
- The roots become black and the bark can be easily peeled off.
- Finally the whole clump dries up.

Fungus

Fusarium oxysporum Schlecht ex Fr.

Mode of survival: In infected crop debris in soil.

Primary spread: From diseased plant debris in soil.

Secondary spread: Through irrigation water.

Favourable conditions: Crowded seedlings, dampness due to high rainfall, poor drainage, high humidity, high soil moisture, cloudiness and low temperatures below 24° C for few days.

Management

- Avoid planting in swampy or dry areas.
- Provide proper drainage.
- Collect and destroy the affected plants by burning.

2. Leaf Spot and Blight

Symptoms

- Water-soaked lesions appear on the leaf margin or on the tips of the leaves which rapidly enlarges, coalesce and cover the entire leaf lamina giving blighted appearance.
- The affected leaves become necrotic and dry.
- Leaf sheath covering the pseudostem shows blackish-brown discolouration which progress down to the rhizome.

- Gradually the pseudostem becomes brittle and breaks in the middle or at the collar region.
- The affected clump dries up completely giving the entire plant a burnt appearance.

Fungus

Pestalotiopsis versicolor (Speg.) Stey.

Mode of survival: In infected crop debris in soil as acervuli and dormant mycelium.

Primary spread: From diseased plant parts.

Secondary spread: Through air-borne conidia.

Favourable conditions: Cool and moist weather, inadequate shade, lack of field sanitation, poor nutrient management and monocropping.

Management

- Spray carbendazim 0.1 per cent or zineb 0.2 per cent.

3. Chirkey Disease

Symptoms

- In young leaves, discrete pale green to yellow longitudinal stripes of 2.5 mm are seen running parallel to veins giving characteristic mosaic symptoms.
- In mature leaves, the symptoms are masked.
- Later, chlorotic and necrotic streaks appear along the lateral veins of leaves.
- The affected plants produce flowers and spikes but without capsules.

Virus

Large cardomom chirke virus (LCCV)

Mode of survival: In infected crop debris in soil as acervuli and dormant mycelium.

Primary spread: From planting infected suckers.

Secondary spread: Through insect vector, corn aphid *Ropholosiphum maydis* Fitch.

Favourable conditions: Lack of phyto-sanitation to destroy the insect vector, presence of infected old plants in the garden.

Management

- Removal of volunteer plants.
- Use healthy planting material.

- ☆ Control aphid vectors.
- ☆ Plant healthy planting materials.

4. Foorkey Disease/Bushy Dwarf Virus

Symptoms

- ☆ The affected plants produce profuse stunted shoots which fail to produce flowers.
- ☆ The leaves become small, lightly curled and pale green in colour.
- ☆ The inflorescence becomes stunted, thereby producing no flowers and fruits.
- ☆ The capsule size reduced and chaffy without seeds.

Virus

Cardomom bushy dwarf virus (CBDV)

Mode of survival: In infected crop debris in soil as acervuli and dormant mycelium.

Primary spread: From planting infected suckers.

Secondary spread: Through insect vector, aphids *viz., Pentalonia nigronervosa* Cog. and *Micromyzus kalimpongensis.*

Favourable conditions: Lack of phyto-sanitation to destroy the insect vector, presence of infected old plants in the garden.

Management

- ☆ Do not raise nurseries in the vicinity of infected plantations.
- ☆ Uproot infected clumps and drench the surrounding area with systemic insecticides to control movement of aphids on the main host as well as collateral hosts such as corn, banana, peach and squash.
- ☆ Destroy all the collateral host plants of the aphids in and near the plantation.
- ☆ Use healthy disease free planting material preferably seedlings.
- ☆ Spray insecticides to control the aphid vectors with methyl demeton or dimethoate.

7. Pepper (*Piper nigrum* L.)

1. Quick Wilt/*Phytophthora* Foot Rot

Symptoms

Rotting of foot (basal portion of vine) region is named as foot rot. It is also known as quick wilt because of death of vines in 2-3 weeks. All parts of the vine are vulnerable for infection and the pathogen produces four types of symptoms.

Leaf and Spike Infection

- Dark brown or black water-soaked spots with fimbriate margin appear on the leaves (starting from leaf tip) which enlarge causing concentric zonation with grey centre.
- Infection of spikes and berries leads to spike shedding.
- Affected leaves and spikes shed.

Die back

- Infection on aerial branches causes rotting of stem and die back.
- A dark patch appears on the stem.
- Stem breaks off at the nodes and the leaves turn yellow flaccid and droop.
- The vine dies in 2-3 weeks (Quick wilt).

Foot Rot or Collar Rot

- Infection at collar region is fatal and it causes death of vines within 3 weeks.
- Affected portion emits bad odour.
- Infection spreads to root system also.

Root Rot

- Root symptom appears without foot rot symptom.
- Roots weakened due to death of vine particularly during Oct -Nov.
- These vines may recover after the rain and survive for more than two seasons till the root infection culminates in collar rot and death of the vine.

Chromista

Phytophthora capsici Leonian

Sporangia are produced in infected leaves of runner shoots and are carried to upper leaves by rain splashes, thereby spreading in a ladder like fashion.

Mode of survival: In infected plant debris and in soil as oospores and chlamydospores.

Primary spread: From infected plant debris and soil.

Secondary spread: Through irrigation water, termites and snails (chlamydospores and oospores).

Favourable conditions:

- The leaf infection is serious in irrigated areca-pepper mixed plantations.
- Areas with heavy rainfall and poor drainage recorded more disease.
- Soil temperature of 20-24°C and relative humidity of above 90 per cent favour disease incidence.

Management

Phytosanitation

- Remove and destroy dead vines along with root system.
- Select planting material from disease free gardens.
- Raise the nursery preferably in fumigated or solarized soil.

Cultural Practices

- Provide adequate drainage to reduce water stagnation.
- Avoid injury to the root system.
- Do not allow freshly emerging runner shoots to trail on the ground. They must either be tied back to the standard or pruned off.
- Prune the branches of support trees with the onset of monsoon to avoid buildup of humidity and for better penetration of sunlight.

Chemical Control

Any of the following chemical control measures can be adopted.

- Drench the soil (45-50 cm radius) with copper oxychloride 0.25 per cent or potassium phosphonate 0.3 per cent or metalaxyl + mancozeb 0.2 per cent at 5-10 lit/vine during May–June followed by foliar spray with Bordeaux mixture 1 per cent or potassium phosphonate 0.3 per cent or metalaxyl + mancozeb 0.2 per cent.
- Repeat drenching and spray once again during August-September and during October if the monsoon is prolonged.
- Apply *Trichoderma viride* or *Pseudomonas fluorescens* around the base of the vine at 50g/vine at the onset of monsoon (May-June) and again during August - Sept.

2. Pollu Disease/Anthracnose

Symptoms

- Spots on leaves are brown to dark brown which coalesce to form bigger lesions with concentric zonations and are surrounded by yellow halo.
- Water-soaked dirty brown sunken areas appear on the berries.
- Discoloured berries occur in groups of two to five.
- Berries split at the lesion site and dry up resulting in hollow and chaffy berries. The word pollu in Malayalam means a hollow.

Fungus

Colletotrichum gloeosporioides Pat and Penz.

Mode of survival: In infected plant debris.

Primary spread: From through infected plant debris.

Secondary spread: Through air and rain splashes.

Management

- ☆ Spray Bordeaux mixture 1 per cent or carbendazim 0.1 per cent.
- ☆ Regulate shade in the plantation.
- ☆ Spray profenophos or dimethoate 2 ml/lit + mancozeb 2g/lit.

3. Basal Wilt

Symptoms

- ☆ Grey lesions appear on stem and leaves.
- ☆ White mycelial growth is seen at the edges of leaf lesions.
- ☆ The pathogen later girdles the stem resulting in drooping of leaves beyond the point of infection.
- ☆ In advanced stages, the rooted cuttings dry up.
- ☆ Small, whitish grain-like sclerotial bodies appears on lesions.

Fungus

Sclerotium rolfsii Sacc. [Sexual stage*: Athelia rolfsii* (Curzi)].

It is a soil living saprophytic pathogen, produces ribbon like hyphae with clamp connections. Basidia are club-shaped, bearing four smooth, ellipsoid basidiospores. Small, brownish sclerotia (hyphal propagules) are also formed.

Mode of survival: In infected plant debris and in soil.

Primary spread: From infected plant debris.

Secondary spread: Through irrigation water and rain

Management

- ☆ Remove and burn affected cuttings along with fallen leaves.
- ☆ Spray carbendazim 0.1 per cent or Bordeaux mixture 1 per cent.

4. Stunt

Symptoms

- ☆ The vines exhibit shortening of internodes to varying degrees.
- ☆ The leaves become small and narrow, deformed and appear leathery, puckered and crinkled.
- ☆ Chlorotic spots and streaks also appear on the leaves occasionally. The yield of the affected vines decrease gradually.

Virus

Tomato bushy stunt virus (TBSV)

Mode of survival: In infected plants.

Primary spread: From infected planting materials.

Secondary spread: Through Vector.

Management

- ✰ Remove and burn affected vines.
- ✰ Plant only disease-free cuttings.

5. Phyllody

Symptoms

- ✰ The floral buds are transformed into narrow leaf-like structures.
- ✰ Malformed spikes show leafy structures exhibiting phylloid symptoms.
- ✰ The vines become unproductive within 2 to 3 years.

Bacterium

Candidatus Phytoplasma [Aster yellows group of Phytoplasma]

Mode of survival: In infected plants.

Primary spread: From infected planting materials.

Secondary spread: Through cotton leafhopper or jassid (*Hishimonus phycitis*).

Management

- ✰ Remove and destroy affected vines.

6. Red Rust

Symptoms

- ✰ Several reddish brown cushiony spots appear on older leaves and stem.
- ✰ Minor disease not causing any economic loss. However, severe infection leads to defoliation.

Algae

Cephaleuros mycoides Sood and *C. virescens* Kunze ex E.M.Fries

Mode of survival: In infected plant parts.

Primary spread: From infected planting materials.

Secondary spread: Through rain splashes.

Favourable conditions: Continuous rainfall.

Management

- ✰ Spray Bordeaux mixture 1 per cent along with wetting agent.

7. Slow Decline – Nematode and Fungal Complex

Symptoms

- The affected vines exhibit varying degrees of root degeneration due to infestation by plant parasitic nematodes.
- The diseased vines exhibit foliar yellowing coinciding with depletion of soil moisture and recover with the onset of monsoon.
- Defoliation and die back are the other aerial symptoms of the disease.
- Symptoms reappear in subsequent seasons after the cessation of the monsoon and the diseased vines gradually lose their vigour and productivity.
- The affected vines show varying degrees of loss of feeder roots, necrosis and presence of root knots due to infestation by plant parasitic nematodes such as *Radopholus similis* and *Meloidogyne incognita* respectively leading to rotting of feeder roots.

Nematodes

Radopholus similis and *Meloidogyne incognita*

Chromista

Phytophthora capsici either independently or together.

Management

- Remove and destroy the affected vines.
- Treat planting pits with carbofuran at 50 g/pit at the time of planting.
- Plant only nematode-free rooted cuttings raised in fumigated or solarized nursery mixture.
- Apply carbofuran 100 g/vine during May/June (with the onset of southwest monsoon) and September/October. Along with carbofuran, the basins should be drenched with copper oxychloride 0.25 per cent or potassium phosphonate 0.3 per cent or metalaxyl + mancozeb 0.2 per cent.
- Provide sufficient soil moisture at the time of nematicide application.
- Apply *Trichoderma viride* or *Pseudomonas fluorescens* 2.5 kg/ha + FYM 50kg/ha in soil.

8. Rubber (*Hevea brasiliensis* Mull. Arg.)

1. Powdery Mildew

Symptoms

- Newly formed tender flush during the refoliation period of January to March are infected with white powdery growth on both the surfaces.

- ☆ Diseased leaves shrivel, fall off leaving petioles on the stem for sometime giving a broom-stick appearance.
- ☆ Diseased older leaves with powdery growth gives distorted appearance with restricted brittle patches.
- ☆ Severe attack leads to extensive defoliation, resulting in serious retardation of rate of growth and bark renewal.
- ☆ Infected shoots dry up and cause die back.
- ☆ Infection on flowers and tender fruits cause premature drop.

Fungus

Oidium heveae Steinm.

Mycelium is hyaline and septate. The conidiophores bear the conidia in chain. Conidia are chain, barrel-shaped or cylindrical. The youngest conidium is at the base of the chain and the mature one at the tip.

Mode of survival: As dormant mycelium in barks.

Primary spread: From infected trees.

Secondary spread: Through air-borne conidia.

Favourable conditions: Cloudy days with light rain and or cool misty nights with dew formation during refoliation favour serious outbreak; shaded conditions.

Management

- ☆ Dust sulphur (325 mesh) 11-14 kg/ha or sulphur dust and talc powder at 7:3 ratio for plantation.
- ☆ Spray wettable sulphur 0.25 per cent or carbendazim for nurseries and young plants.

2. Abnormal Leaf Fall

Symptoms

- ☆ It induces shedding of leaves during June-August, while general leaf fall occurs in December.
- ☆ First the pods rot and the leaves shed prematurely either green or after turning coppery red within a fortnight.
- ☆ A black lesion develops on the petiole with a drop of latex, often coagulated in the centre.
- ☆ The young leaf buds and fruits rot result in reduced seed viability.
- ☆ Heavy defoliation leads to considerable lose of yield and die back of terminal twigs.
- ☆ A thick carpet of rotting foliage fallen on the ground emit foul odour.
- ☆ Rotting of bark near tapping cut results in reduced latex yield.

Fungi

Phytophthora palmivora Butler, *P. meadi* McRae and *P. nicotianae* var *parasitica* and *P. botryosa* Cke. are also causes the disease.

The fungus produces both inter-and intracellular hyphae. The mycelium is non-septate, hyaline and produces sporangiophores. The sporangia are oval or pear shaped, hyaline.

Mode of survival: In rubber and other hosts like coconut, palmyra, jack, *citrus* spp. Survives during dry months in mycelial stage inside affected tissues (fruits or fruit shoots); also survives as oospores and chlamydospores in affected fruits.

Primary spread: From affected fruits and shoots.

Secondary spread: Through air-borne rain splashes.

Favourable conditions: High rainfall, high relative humidity (more than 80 per cent) low temperature of 15.6-20 °C (than 29°C).

Management

- ✰ Grow resistant varieties.
- ✰ Prophylactic sprays of the foliage with Bordeaux mixture 1 per cent through high volume sprayer or oil-based copper oxychloride through low volume air blaster or aerial application.

3. *Colletotrichum* Leaf Disease/Secondary Leaf Fall

Symptoms

- ✰ Disease is observed during April to October.
- ✰ It usually attacks tender and immature leaves at the leaf tip region.
- ✰ Numerous, minute, circular, brown spots with yellow halo appear on leaves and these spots coalesce and the leaves dry up.
- ✰ Infected leaves get crinkled, distorted and shed.
- ✰ Lesions are also observed on green stems leading to shoot die back.

Fungi

Colletotrichum gloeosporioides Pat and Penz; (Asexual stage: *Glomerella cingulata* Stoneman); *C. acutatum* Simmonds is also responsible for the disease.

The mycelium is septate and hyaline at first and later become light brown. Conidia are single celled, oblong. The perithecia are pherical, brown and are isolated or grouped. Asci are oblong and each ascus contains eight curved unicellular ascospores.

Mode of survival: Survive saprophytically in infected leaves and dried twigs during dry season.

Primary spread: From infected rubber plants and also from hosts like mango, papaya, anona and citrus.

Secondary spread: Through air-borne conidia and rain splashes.

Favourable conditions: Weak plants under water logged conditions.

Management

- ✰ Apply recommended dose of fertilizers in addition to adequate drainage.
- ✰ Spray Bordeaux mixture 1 per cent or copper oxychloride 0.25 per cent or mancozeb 0.2 per cent or carbendazim 0.1 per cent.

4. Bird's Eye Spot

Symptoms

- ✰ A hot weather disease serious mainly in nurseries and in young plants.
- ✰ Small, circular, necrotic spots appear on leaves with dark brown/reddish brown margins and pale/papery centre.
- ✰ These spots look similar to eyes of birds and hence named as bird's eye spot.
- ✰ Matured leaves are generally not infected.
- ✰ Severe infection leads to premature defoliation and die back.

Fungus

Drechslera heveae (Petch.) Ellis. [Syn. *Helminthosporium heveae* Petch.].

Conidiophores are solitary, straight to flexuous sometimes geniculate. Conidia are curved, navicular or fusiform, pale to mid-golden brown or reddish brown and smooth.

Mode of survival: In infected leaves.

Primary spread: From infected plant parts.

Secondary spread: Through air-borne conidia, irrigation/rain splashes and movement of workers in the nursery.

Favourable conditions: Weaker plants and plants grown under exposed situation.

Management

- ✰ Adopt balanced fertilization to get vigorous seedlings.
- ✰ Provide shade to seedlings in the nursery.
- ✰ Spray Bordeaux mixture 1 per cent or copper oxychloride 0.25 per cent or mancozeb 0.2 per cent or carbendazim 0.1 per cent.

5. Pink Disease

Symptoms

- ✰ It is a most serious stem disease affecting mainly the young rubber trees (2-8 years).

- The first indication is the exudation of drops of latex from branches followed by the appearance of white silky cob-web like mycelium on woody branches, usually at the fork, where several branches join the main trunk.
- Later, the stem assumes a characteristic pink colour.
- When the pathogen damages the latex vessels, the latex exudes and the affected bark gets rotten and dries up.
- The bark rots, dries up and cracks.
- The stem translocation is affected and the leaves turn yellow and dry up and remain attached to the dead branches.
- The sprouts develop from below the affected portion of the stem.
- The pink encrustations on the bark contain the basidia of the fungus on which basidiospores are produced.

Fungus

Corticium salmonicolor Berk. and Br.

Mode of survival: As dormant mycelia and basidiospores on the infected barks.

Primary spread: From infected rubber plants and other hosts like cocoa, citrus and coffee.

Secondary spread: Through air-blown basidiospores and rain splashes.

Favourable conditions: Wet weather; high relative humidity; 2-8 years old trees.

Management

- Prune infected branches and protect the cut ends with Bordeaux paste.
- Spray Bordeaux mixture 1 per cent or copper oxychloride 0.25 per cent.
- Apply thiram (10g/kg) incorporated in a wound dressing compound in 1 per cent ammoniated field latex or thiride 0.75 per cent or propiconazole 0.1 per cent in pidyvyl, china clay and water (1:2:4 by volume) as rain-wash is prevented.

6. Leaf Spot

Symptoms

- The disease is noticed in nursery during November to May. On mature trees, disease is observed during the refoliation period from February to May.
- Large spots with brown margins and pale centre appear on leaves and the centre of the spot fall off leaving a shot-hole symptom.
- Several lesions coalesce to form large blighted area.
- Disease spreads along the veins leading to a brownish "railway track" or "fish bone" like appearance.

- ☆ The localised infection on midrib or base of a leaf causes leaf abscission. Defoliation leads to die back of branches.

Fungus

Corynespora cassiicola (Berk. and Curt.)

[Syn. *Helminthosporium cassiicola, H. papaya* and *H. vignae*].

Mode of survival: In infected leaves.

Primary spread: From infected leaves.

Secondary spread: Through air-borne conidia.

Favourable conditions: Weak plants.

Management

- ☆ Maintain seedlings in vigorous condition through adequate balanced nutrition.
- ☆ Provide shading.
- ☆ Spray Bordeaux mixture 1 per cent or mancozeb 0.2 per cent or carbendazim 0.1 per cent in the nursery.
- ☆ Spray mancozeb 0.2 per cent or carbendazim 0.1 per cent or copper oxychloride 0.25 per cent during refoliation in the mature plantation.

7. White Root Disease

Symptoms

- ☆ It causes damage both to immature and mature rubber trees.
- ☆ The fungus penetrates the roots and causes damage to the water conducting tissues with general discolouration of the foliage producing symptoms similar to water deficient during a drought period.
- ☆ Later, leaves fall off and the branches show die back followed by death of thc trce.
- ☆ Typically thick, ramifying white mycelial strands (rhizomorphs) are seen on the affected roots. They become brown with age and the wood is usually cream in colour and later becomes soft and friable.

Fungus

Rigidoporus (*Fomes*) *lignosus*

Fruiting bodies are bracket shaped, with a yellow margin. They are concentrically banded reddish brown on the upper surface with a pore-studded red brown lower surface.

Mode of survival: In diseased old stumps and large roots.

Primary spread: From infected stumps and roots.

Secondary spread: Through root contact/rhizomorphs/basidiospores.

Management

- ☆ Remove the old and dead infected stumps and burn *in situ*.
- ☆ Apply 100 g of sulphur dust around the plant on the surface of soil while replanting. It should not touch the sprout since it causes scorching.
- ☆ Drench the soil with Bordeaux mixture 1 per cent.

8. Brown Root Disease

Symptoms

- ☆ The symptoms appear as loss of shiny appearance of leaves, drooping, yellowing and buckling (curving) of leaflets and retardation in growth of trees.
- ☆ The leaves later turn to reddish yellow and dry up and the trees show die back.
- ☆ On the infected roots, yellowish brown patches of rhizomorphs are visible.
- ☆ Cortex appears as brown, mottled with white patches.
- ☆ Wood shows brownish discolourations and brown lines.
- ☆ On old infected stumps large, hard, brownish purple brackets of fungus with grey underside develop.

Fungus

Phellinus (*Fomes*) *noxius* Corner.

Mode of survival: In infected roots and stumps.

Primary spread: From infected stumps and roots.

Secondary spread: Through root contact/rhizomorphs/basidiospores.

Management

- ☆ Remove and burn the affected and dead trees.
- ☆ Drench the nursery bed with propiconazole 0.1 per cent
- ☆ Apply lime at 2.5t/ha in the field.
- ☆ Drench the soil with Bordeaux mixture 1 per cent.

9. Dry Rot/Stump Rot/Collar Rot/Charcoal Rot

Symptoms

- ☆ The disease affects roots, collar, trunk and branches with copious exudation of latex from the lesions.
- ☆ Affected roots show dry rot.
- ☆ Affected bark and wood become soft and powdery with double black lines in the wood.

- ☆ Later, soft, grey, ear lobe- like fruiting bodies appear and turn black on the bark.
- ☆ As the fungus penetrates the wood, it becomes soft and the trees or branches are killed or blown over.

Fungus

Ustulina zonata (Lev.) Sacc. (syn. *Ustulina deusta*)

Mode of survival: In infected roots and stumps.

Primary spread: From infected stumps and roots.

Secondary spread: Through root contact/rhizomorphs/air-borne basidiospores and boring beetles.

Management

- ☆ Remove infected and destroy dead trees and stumps.
- ☆ Scrape off the fructifications, affected bark and wood showing black lines.
- ☆ Apply wound dressing compound in which thiram incorporated.
- ☆ Avoid accumulation of latex at the base of the tree.

9. Small Cardamom (*Elettaria cardamomum* Maton)

1. Nursery Leaf Spots

Symptoms

***Phyllosticta elettariae* Choudhary**

- ☆ It is most common in primary nurseries.
- ☆ It appears mostly during February-April with the receipt of summer showers.
- ☆ The disease appears as small, dull white, round or oval spots on the leaves.
- ☆ These spots become necrotic and the central portion of the spot withers off leading to the formation of shot hole.

***Cercospora zingiberi* Rathaiah**

- ☆ It is observed in secondary nurseries.
- ☆ Yellowish to reddish brown rectangular patches are seen on the lamina which are almost parallel to the side veins.

Fungi

Phyllosticta elettariae S. Chowdhury and *Cercospora zingiberi* Togashi et Katsuki

Mode of survival: In infected plants.

Primary spread: From diseased plant parts.

Secondary spread: Through air-borne conidia.

Management

- ☆ Maintain seedling vigour so that they develop tolerance to the disease.
- ☆ Avoid exposure to direct sunlight.
- ☆ Avoid raising nurseries continuously on the same site.
- ☆ Spray mancozeb 0.2 per cent as prophylactic measure.

2. Clump Rot/Rhizome Rot/Damping Off

Symptoms

- ☆ The disease appears mainly in nurseries during rainy season.
- ☆ As a result, the infected seedlings die and collapse in masses.
- ☆ Yellowing of leaves and decay of tillers starting from collar region. Decay extends to rhizomes and roots.
- ☆ Rotten rhizomes become soft, dark brown and decay resulting in death of the clumps.
- ☆ Affected tillers fall off by a slight disturbance.

Chromista

Pythium vexans de Bary.

Rhizoctonia solani Kuhn and *Fusarium* sp. are also responsible in some cases)

Mode of survival: As oospores in soil.

Primary spread: From diseased rhizomes

Secondary spread: Through irrigation water

Favourable conditions: Heavy and continuous rainfall, excess soil moisture, overcrowding of plants.

Management

- ☆ Practice thin sowing in primary nurseries to avoid overcrowding of seedlings.
- ☆ Provide adequate drainage to prevent water stagnation.
- ☆ Remove infected and dead seedlings and maintain proper phytosanitary measures.
- ☆ Drench nursery beds with Bordeaux mixture 1 per cent or copper oxychloride 0.25 per cent (3-5 lit/m^2).
- ☆ Treat the seeds with *Trichoderma* or *Pseudomonas* followed by soil application of *Trichoderma* or *Pseudomonas* to the nursery bed (100 g/m^2).

3. Capsule Rot/Fruit Rot/*Azhukal* Disease

Symptoms

- The name *azhukal* denotes rot.
- The disease affects leaves, tillers, panicles and capsules.
- The disease appears as water soaked lesions on tender leaves and capsules, which later form dead areas surrounded by yellow halo.
- As a result, the leaves rot and shred along the veins.
- In the advanced stages, the affected leaves break at the base of the petiole and remain hanging.
- On the immature capsules, water soaked discoloured areas develop, which later turn brown.
- Such capsules decay, emit a foul smell and subsequently drop off in 3-5 days.
- Matured capsules shrivel on drying.
- Panicle infection starts at the tip and proceeds downward. Such diseased portion later dry off, break and fall off at the collar region.

Fungus

Phytophthora meadii McRae.

Mode of survival: As oospores in soil (soil-borne).

Primary spread: From soil-borne oospores.

Secondary spread: Through air-borne sporangia and rain splashes.

Favourable conditions: Temperature of 22-24°C and relative humidity (85-92 per cent), heavy and continuous rainfall, excess soil moisture, thick shade, overcrowding of plants, south west monsoon period (August-September). Malabar variety is highly susceptible.

Management

- Destroy disease affected portions and plant debris.
- Provide adequate drainage to avoid dampness.
- Do trashing and cleaning the plant base before onset of monsoon.
- Regulate thick shade by gentle lopping of tree branches.
- Apply *Trichoderma viride* or *T. harzianum to* plant basins at 1 kg/clump. If the soil is drenched with copper oxychloride or other fungicides, *Trichoderma* should be applied only after 15 days.
- Spray Bordeaux mixture 1 per cent or Fosetyl-Al 0.2 per cent three times.
- Drench the plant base (2-3 lit/clump) with Bordeaux mixture 1 per cent or copper oxychloride 0.25 per cent.

4. Chenthal Disease/Leaf Blight

Symptoms

- The disease assumes severity during the post-monsoon period.
- Elongated water-soaked lesions appear on the lower surface of the young leaves.
- They turn yellow to reddish brown with a pale yellow halo.
- In severe cases, the leaves dry up and give a burnt appearance.
- Later the leaves wither and the pseudostems wilt.

Fungus

Colletotrichum gloeosporioides Pat and Penz.

Mode of survival: In infected plants.

Primary spread: From diseased plant parts.

Secondary spread: Through air-borne conidia.

Favourable conditions: Insufficient shade.

Management

- Provide adequate shade and adopt phytosanitary measures.
- Burn affected shoots and leaves.
- Spray carbendazim 0.1 per cent or copper oxychloride 0.3 per cent or Bordeaux mixture 1.0 per cent.

5. Rust

Symptoms

- The disease is observed as minute, brown uredosori on the lower leaf surface.
- These uredosori are surrounded by chlorotic halos.
- Later, the whole leaf gets covered by uredosori resulting in premature drying.

Fungus

Uredo elettariae Racib

Uredospores sessile, obovoid to ellipsoidal, echinulate, yellowish brown and become cinnamon brown with age. Teliospores are one-celled, regularly arranged in 2-7 layers, oblong to ellipsoidal.

Mode of survival: In infected plants.

Primary spread: From diseased plant parts.

Secondary spread: Through air-borne uredospores.

Favourable conditions: Climatic conditions during May and June.

Management

- Spray mancozeb 0.2 per cent or tebuconazole 0.05 per cent.

6. Katte Disease/Mosaic/Marble Mosaic

Symptoms

- Mosaic disease is locally known as *Katte* meaning a disorder.
- It affects the plants of all ages.
- Spindle-shaped chlorotic flecks (2.5mm dia) appear on the youngest leaves and these flecks progress into light green, discontinuous stripes parallel to the veins from the midrib to leaf margin.
- Subsequent leaves also show mosaic symptom.
- Mosaic mottling is also observed on the leaf sheath and young pseudostem.
- The disease spreads to all tillers of the clump and the affected clumps produce slender tillers with few shorter panicles.
- The clump size is reduced after one or two years of infection.
- The affected plants do not die but give poor yield.
- The loss is total when young seedlings are infected.

Virus

Cardamom mosaic virus (CdMV)

Mode of survival: In infected clumps

Primary spread: From infected planting materials (rhizome).

Secondary spread: Through aphid vector (*Pentalonia nigronervosa* var. *caladii*).

Favourable conditions: Cardamom aphid is more efficient than banana aphid. Single alate and viviparous female can transmit the virus. Nymphs also transmit disease. The vector is a non-persistent type and loses it infectivity immediately after feeding a cardomom or banana plant.

Management

- Use only healthy seedlings (raised from seeds) instead of using rhizomes.
- Avoid clones from diseased gardens for gap filling/replanting.
- Uproot and destroy the infected plants regularly.
- Destroy wild plants like *Amomum, Alpinia, Colocasia, Curcuma, etc.* if they show symptoms of katte disease.
- Do not raise nursery near disease affected plantation.
- Spray systemic insecticides to control vector population.

7. Other Viral Diseases

a. Cardamom Vein Clearing or *Kokke Kandu* - *Cardamom vein clearing virus*

Symptoms

- This is an endemic disease in some areas of Karnataka.
- The disease *Kokke Kandu* means hook-like tiller.
- Plants of all stages are affected from seedlings to capsule stage.
- The disease appears as continuous or discontinuous intraveinal clearing, stunting, rosetting of leaves, loosening of leaf sheath, shredding of leaves and mottling on the pseudostem.
- The tillers are stunted and their tips form hook-like structures after the younger most leaf roll gets hooked up in the leaf sheath.
- Panicle length and number of capsules are much reduced.
- The light green patches on the immature capsules because cracking of fruits.
- Seeds become partially sterile.
- The disease is primarily transmitted by infected rhizomes and the Secondary spread is by the aphid, *Pentalonia nigronervosa* f. sp. *caladii.*

b. Cardamom Necrosis Disease or Nilgiri Necrosis Disease - *Nilgiri necrosis virus* (NNV)

Symptoms

- It was first noticed in severe forms in Nilgiris, Tamil Nadu and hence the name Nilgiri Necrosis Disease (NND).
- Alternate light green and white to yellow streaks appear on the leaves as mosaic.
- Later, these streaks become necrotic, turn reddish brown leading to shredding of the leaves.
- The infected leaves are reduced in size with distorted margins.
- Early infected plants produce few panicles and capsules.
- In advanced stages, tillers are highly stunted, narrow and produce small and brittle leaves and fail to bear panicles and capsules.
- It is transmitted through planting of infected rhizomes and no vector transmission.

C. Chlorotic Streak Disease - *Banana bract mosaic virus* (BBrMV)

Symptoms

- The disease appears as spindle shaped yellow or light green streaks on leaves intravenously along the midrib.

- ☆ Later they coalesce and the veins turn yellow or light green.
- ☆ Discontinuous spindle shaped mottling appears on the pseudostem and on the petioles.
- ☆ It is transmitted through planting of infected suckers.

Integrated Management of Viral Diseases

- ☆ Inspect regularly and rogue out virus infected clumps.
- ☆ Use virus-free planting materials.
- ☆ Raise seedlings and clonal nurseries in isolated sites.
- ☆ Remove infected volunteer plants in the replanted area and totally avoid the presence of volunteers from nursery area.
- ☆ Spray neem based products at 0.1 per cent to control aphid vector.

10. Tea [*Camellia sinensis* (L.) Kuntze.]

1. Blister Blight

Symptoms

- ☆ The disease attacks seedlings in the nursery and in planted crops. Repeated attacks cause death of seedlings.
- ☆ The first visible symptom is circular, oily, yellowish, translucent spot on the tender leaf and it turns into deep red, shiny blisters.
- ☆ The circular spot gradually enlarges to 3 to13 mm dia. bulged on the under surface of the leaf with concave trough-like depression on the upper surface forming a classic blistered lesion.
- ☆ Leaves become curled and distorted.
- ☆ The disease attacks 2 to 3 young leaves of first flush and kills the young shoots and buds.
- ☆ Matured leaves are rarely affected.

Fungus

Exobasidium vexans Massee.

Mycelium is septate. It produces two kinds of spores *viz.*, conidia and basidiospores. The conidia are borne singly at the tips of long stalks. They are hyaline, elliptical, straight or slightly curved. Basidia are long, club-shaped and thin walled. The basidiospores are ovate or oblong and hyaline.

Mode of survival: As spores and mycelia in living bushes as well as in abandoned nurseries and badly kept plantations.

Primary spread: From infected bushes.

Secondary spread: Through air-borne spores.

Favourable conditions

- Cool day and nights with wet or humid conditions.
- Bushes in moist, poorly drained and shady localities.
- Closer spacing and excess N application.
- Cool, moist, relatively still air favours infection.
- Medium pruned tea which provides vigorous succulent shoots and large tender leaves.

Management

- Prune during November/December and remove affected shoots.
- Avoid entry of workers from diseased plantation to healthy areas.
- Avoid planting broad-leaved varieties.
- Spray 2-3 rounds of 5-10 per cent aqueous extracts of *Cassia alata/Acorus calamus/Adhatoda vasica* at 15 days intervals.
- Spray Bordeaux mixture 1 per cent or copper oxychloride 0.25 per cent or hexaconazole 0.1 per cent or chlorothalonil or mancozeb 0.2 per cent or bitertanol 0.1 per cent or triadimefon 0.1 per cent or nickel chloride 210g/ha or propiconazole 0.1 per cent.

2. Grey Blight

Symptoms

- It is a weak parasite which normally enters through wounds.
- Mature leaves, young shoots and bare stalks are commonly affected by the pathogen.
- Infection on young shoots result in die back symptom.
- Irregular brown spots appear on older leaves and turn grey and coalesce to form irregular patches.
- The spots have fine concentric lines in which acervuli appear as black dots in the older spots on the upper surface.
- Young leaves are blackened even before unfolding and distorted.
- The fungus attacks plucking points and causes die back.

Fungus

Pestalotiopsis theae (Sawada) Steyaert

Acervuli, sub-epidermal erumpent and dehisce rupturing epidermis by a pore. Conidia are cylindrical and five celled with three filiform setulae.

Mode of survival: In infected plant parts.

Primary spread: From infected bushes.

Secondary spread: Through air-borne conidia.

Favourable conditions: Weak bushes, K deficiency, sun scorch, insect puncture, pruning and plucking wounds. The incidence is more serious during July-December.

Management

- ☆ Spray potash fertilizer to improve resistance.
- ☆ Spray carbendazim 0.1 per cent or mancozeb 0.2 per cent or thiophanate methyl 0.1 per cent or Bordeaux mixture 1 per cent or copper oxychloride 0.25 per cent.

3. Pink Disease

Symptoms

- ☆ Number of fine silky threads united into a thin film which appears only on the stem, and not found on leaves.
- ☆ The fungus forms pink fructifications over affected stems.
- ☆ Young branches on the outside bush lose their leaves and show die back symptom.
- ☆ Bark tissues die in patches.
- ☆ Pink tissues of the affected plant parts become white when old.

Fungus

Pellicularia salmonicolor (Berk. and Br.) Dastur

[Syn. *Corticium salmonicolor* Berk. and Br.]

Mode of survival: In infected plant parts.

Primary spread: From infected plants.

Secondary spread: Through air-borne spores.

Management

- ☆ Remove and destroy the affected plant parts.
- ☆ Apply recommended dose of potash which promotes resistance.

4. Black Rot

Symptoms

- ☆ The disease appears normally at the end of May or early June.
- ☆ Small, dark-brown irregular spots on leaves coalesce to form dark brown patches, which eventually cover the whole leaf.
- ☆ Before the leaves turn black, the lower surface assumes a white powdery appearance.
- ☆ Diseased leaves fall off.

Fungi

Corticium invisum Petch and *C.theae* Bernard

Mode of survival: In infected plant parts.

Primary spread: From infected plants and infected plant parts in soil.

Secondary spread: Through basidiospores carried by workers through clothes.

Favourable conditions: High temperature and humid air.

Management

- ✰ Prune the affected branches and leaves and burn them immediately.
- ✰ Spray copper oxychloride 0.25 per cent during third week of April.

5. Twig Die back and Stem Canker

Symptoms

- ✰ Browning and drooping of affected leaves and the affected shoots dry.
- ✰ The entire branch dies from the tip downward.
- ✰ The drying branches often have cankers which have shallow lesions lesions surrounded by a thick area of bark.

Fungus

Macrophoma theicola Petch.

Mode of survival: In infected plant parts.

Primary spread: From infected plants.

Secondary spread: Through rain splashes

Favourable conditions: Rainy weather favours its spread and dry conditions promote its development.

Management

- ✰ Prune the affected branches and leaves and burn them immediately.
- ✰ Spray copper oxychloride 0.25 per cent during third week of April.

6. Red Rust

Symptoms

- ✰ Most algal spots develop on the upper leaf surface.
- ✰ Leaves develop lesions that are roughly circular, raised, and purple to reddish-brown and it spreads from leaves to branches and fruit.
- ✰ Older infections become greenish-grey and look like lichen.
- ✰ Infection on new stem reduces vigour or causes death.

Algae

Cephaleuros virescens Kunze. ex E.M.Fries. (syn. *C. parasiticus*)

It consist of branched filaments that comprise a thallus possess haustoria. Gametangia are terminal or intercalary. Sexually, zygote produces a dwarf sporophyte.

Mode of survival: As survive in spots on leaves or stems of bushes and in fallen plant host debris.

Primary spread: From infected plant parts.

Secondary spread: Through air and rain splashes

Favourable conditions: Frequent rain, warm weather, poor nutrtion, poor drainage, still air, severe pruning of young plants, weakening by red spider mite attack and drought.

Management

- ✰ Improve soil nutrition and maintain good vigour.
- ✰ Provide proper drainage facilities and good air circulation by maintaining plant canopy.
- ✰ Remove and burn badly diseased bushes plant parts.
- ✰ Spray Bordeaux mixture 1 per cent or copper oxychloride 0.25 per cent immediately after pruning.

6. Sooty Mould

Symptoms

- ✰ A black superficial fungal growth (mycelia and spores) appear on leaves, stem and branches.
- ✰ The infection is associated with the presence of scale insects and aphids.
- ✰ The disease reduces photosynthetic activity and vigour of the plant and quality of leaves.

Fungus

Capnodium theae Boedijn and *C. footii* Harv. ex Berk. and Desm.

Mode of survival: In infected plant parts.

Primary spread: From infected plants.

Secondary spread: Through air-borne spores and rain splashes.

Management

- ✰ Spray dimethoate 2 ml/lit + wettable sulphur 2g/lit or copper oxychloride 2.5g/lit.

7. Root Diseases

a. Charcoal Stump Rot - *Ustulina zonata* (Lev.) Sacc.

- ☆ Tea bushes die suddenly in patches as if suffering from drought.
- ☆ Leaves wither, turn brown and drop off and the infected bushes dry up.
- ☆ Brown or white mycelia with fan-shaped fructifications are seen on the wood when bark is removed in root but the roots do not show any external mycelium.
- ☆ Charcoal like encrustations appear on bark in advanced stages.

b. Brown Root Rot - *Fomes lamoenis*

- ☆ It is slow spreading and quick killing disease.
- ☆ The mycelial mat at the root surface produces tawny-brown cushions and deep-brown honey comb-like reticulations on the wood.
- ☆ Fructifications are seen on stumps which are bracket shape, irregular and hard.
- ☆ Wood turns soft and spongy.
- ☆ Soil encrustations on the roots cannot be easily washed off.
- ☆ The alternative hosts are bread fruit, cocoa, coffee and rubber.
- ☆ The disease is more common in low elevation areas and the disease spreads through root contact.

c. Black Root Rot - *Rosellinia arcuata* Petch

- ☆ Black wooly mycelial growth appears on root surface and at collar region.
- ☆ White and star-shaped mycelium seen on wood surface below the root bark.
- ☆ Girdling and formation of cankers at collar region.
- ☆ The disease spreads rapidly in damp weather.
- ☆ High humidity, frequent rains and a temperature of 24-32°C favours the development of disease

d. *Armillaria* Root Rot - *Armillaria mellea* (Vahl) P. Kumm.

- ☆ Due to root infection the foliage begins to wilt, turns chlorotic and falls.
- ☆ The fungus spreads up and reaches collar region.
- ☆ The root bark cracks and shows sheets of creamy mycelium on the wood accompanied by brown rhizomorphs.
- ☆ Mushroom-like basidiocarps are formed in advanced stages.

e. Red Root Rot - *Poria hypolateritia* Berk. ex Cooke

- ☆ It is a fast spreading and slow killing pathogen.

- ✰ Whitish mycelia which turn later to red and in advanced stages become black.
- ✰ When root bark is removed, blood -red mycelial strands become visible.
- ✰ The bark gets softened and wood becomes discoloured, spongy and sodden.
- ✰ It spreads mostly by root contact and plants of 1-2 years old are more vulnerable for infection.
- ✰ The alternative hosts are Coffee, Grevillea, Albizia and Erythrina.

Mode of survival: In infected roots and stumps.

Primary spread: From infected plant refuses.

Secondary spread: Through root contact/rhizomorphs/basidiospores.

Management

- ✰ Uproot infected, dead bushes and burn them.
- ✰ Isolate the infected bushes by making trenches (1.2 m, 0.5 m) to avoid root contacts with healthy bushes.
- ✰ Manure adequately to promote vigorous growth in bushes.
- ✰ Drench dug out soil and pits with Copper oxychloride 0.25 per cent
- ✰ Fumigate soil with vapam and metham sodium where ever possible with 20-25 cm away from the collar region of the living plant to avoid phytotoxicity. They are injected into drilled holes to a depth of 20 cm at a distance of 30 cm.

Chapter 6

Aromatic Plants

1. Camphor/Camphor Laurel (*Cinnamomum camphora* L.)

1. *Verticillium* Wilt

Symptoms

- ☆ Wilting and drying of twigs, branches or entire tree occurs.
- ☆ Partial or complete wilting is seen.
- ☆ Wilting in twigs or branches show brown discolouration of xylem vessels.
- ☆ Leaves are smaller and show interveinal and marginal yellowing.

Fungi

Verticillium albo-atrum Reinke and Berthold and *V. dahliae* Kleb.

Mode of survival: As microsclerotia in infected root tissues or mycelium or conidia.

Primary spread: From infected seedlings.

Secondary spread: Through irrigation water and wind-blown soil.

Favourable conditions: Temperature of 21-27°C, clay soil.

Management

- ☆ Select disease free areas planting.
- ☆ Follow crop rotation.
- ☆ Irrigate the field to required moisture level.
- ☆ Maintain optimum soil fertility.
- ☆ Follow deep ploughing to bury infected plant materials.
- ☆ Destroy affected trees by burning.
- ☆ Dip the seedlings in carbendazim 0.1 per cent solution.
- ☆ Drench root region of tree with carbendazim 0.1 per cent.

2. Powdery Mildew

Symptoms

- ☆ White powdery growth on upper leaf surfaces which become grey later.
- ☆ White powdery growth on petioles also.
- ☆ Chlorosis, cupping and leaf dip.

Fungi

Erysiphe cinnamomi Sawada and *Microsphaera alphitoides* Griffon and Maubl.

Mode of survival: As cleistothecia in plant debris.

Primary spread: From ascospores in infected plant debris.

Secondary spread: Through wind-borne conidia.

Favourable conditions: Warm humid conditions.

Management

- ✰ Collect and destroy diseased and fallen leaves.
- ✰ Maintain fertility level of soil.
- ✰ Spray wettable sulphur 0.3 per cent or carbendazim 0.1 per cent.

3. Leaf Blight

Symptoms

- ✰ Minute dots with irregular chlorotic lesions appear at the tips and base of the leaves.
- ✰ Circular to oblong concentric, black, velvety rings appear in chlorotic area.
- ✰ Lesions have yellow halo, increase in size, coalesce and causes blighting.
- ✰ Affected leaves dry and die from tip downwards.

Fungus

Alternaria sp.

Mode of spread: As dormant mycelium in infected plant debris.

Primary spread: From infected plant debris.

Secondary spread: Through wind-borne conidia.

Favourable conditions: Warm weather and humid conditions due to rain or heavy dead, poor nutrient status of soil.

Management

- ✰ Use disease free planting materials.
- ✰ Apply optimum fertilizer dose.
- ✰ Spray mancozeb 0.2 per cent.

4. Leaf Spot

Symptoms

- ✰ Water-soaked, round, oval or irregular spots occur on leaves.
- ✰ These spots coalesce to cause blighting and drying up of plants.

Bacterium

Pseudomonas sp.

Mode of Spread: Infected seeds and plant debris.

Primary spread: From infected seeds and plant debris in soil.

Secondary spread: Through self-sown plants, irrigation water.

Favourable conditions: Sandy loam and sandy clay soil, high soil moisture, free moisture on leaves, temperature of 24-27°C and relative humidity 85-92 per cent.

Management

- ☆ Use clean and disease free seeds.
- ☆ Treat seeds with streptomycin 0.05 per cent solution.
- ☆ Provide good drainage.
- ☆ Adopt crop rotation.
- ☆ Destroy infected plants and weeds.
- ☆ Avoid irrigation from infected fields to healthy fields.

5. *Botryosphaeria* Canker

Symptoms

- ☆ Weakened trees due to drought or over-pruning are highly susceptible.
- ☆ Trees on sandy soils are easily attacked.
- ☆ Die back of branches occur.
- ☆ Fungus kills sap wood tissues causing cankers (areas of dead tissues) on stem.
- ☆ The disease produces inward until the entire branch is killed and finally the whole trees die.

Fungus

Botryosphaeria dolhidea (Moug.) Ces.

Management

- ☆ Cut the infected branches along with a small length of healthy stem.
- ☆ Sterilize the cutting tools before using every tree.
- ☆ Avoid over pruning and mechanical root damage.

2. Citronella (*Cymbopogon winterianus* Jowitt.)

1. Leaf Blight/Blotch

Symptoms

- ☆ Circular, reddish, brown spots appear on margins and tips of leaves.
- ☆ Spots enlarge, coalesce and form elongated, longitudinal, reddish brown, necrotic lesions resulting in drying and premature leaf fall.
- ☆ Profuse conidial production on the spots and black dots.
- ☆ Reduction in leaf and twig production.

Fungus

Curvularia andropogonis (Zimm.) Boedijn and *C. eragrostidis* Henn.

Mode of survival: In infected plant debris in soil.

Primary spread: From dormant mycelium in infected plant debris.

Secondary spread: Through wind-borne conidia.

Favourable conditions: Rainy season, nutritionally poor soil.

Management

Spray zineb 0.2 per cent or mancozeb 0.2 per cent.

2. Collar Rot and Wet

Symptoms

- ☆ Yellowing, drooping and wilting of leaves.

Fungus

Fusarium moniliforme J. Sheld.

Mode of survival: As chlamydospores in infected plant debris.

Primary spread: From infected plant debris.

Secondary spread: Through irrigation water.

Favourable conditions: Rainy season, nutritionally poor soil.

Management

- ☆ Follow crop rotation for 3-4 years.
- ☆ Follow deep summer ploughing.
- ☆ Spray captan or carbendazim 0.1 per cent.

3. Sheath Rot and Blight

Symptoms

- ☆ Sheath near soil surface is affected and gradually spreads to sheath tissues.
- ☆ Concentric spots cover large area of leaf and leaf sheath.
- ☆ Rotting of leaf and leaf sheath in the basal portion with a weft of greyish mycelium and small, round or irregular, dark-brown sclerotia are formed inside the leaf sheath.

Fungus

Rhizoctonia solani Kuhn.

Mode of survival: As sclerotia in infected plant materials.

Primary spread: From infected plant materials.

Secondary spread: Through irrigation water.

Management

- ☆ Follow deep summer ploughing.
- ☆ Regulate irrigation water.

4. Lethal Yellowing

Symptoms

- ☆ Disease is seen in patches in the field.
- ☆ Root infection spreads to stem.
- ☆ Leaves become chlorotic and yellow and roots rot.
- ☆ Poor growth and death of plants occur in irregular patches

Chromista

Pythium aphanidermatum (Edson) Fitzp.

Mycelium is hyaline, septate. Sporangia are lobed. Oospores are motile and biflagellate. Sexual spores are oospores.

Mode of survival: As oospores in soil.

Primary spread: From soil-borne oospores.

Secondary spread: Through irrigation water.

Favourable conditions: Ill-drained soil conditions.

Management

- ☆ Improve soil drainage.
- ☆ Adopt crop rotation.
- ☆ Treat the slips with carbendazim 0.1 per cent.

Other Diseases of Citronella

5. **Wilt - *Fusarium moniliforme, Fusarium oxysporum.***

6. **Leaf spot - *Monochaetiella cymbopogonis.***

3. Davana (*Artemisia pallens* Walker)

1. Damping-off

Symptoms

- ☆ Pre-and post-emergence damping off of seedlings cause poor stand in the nursery.

Fungus

Rhizoctonia solani Kuhn

Mode of survival: As sclerotia in infected plant parts.

Primary spread: From soil-borne sclerotia.

Secondary spread: Through irrigation water.

Favourable conditions: High soil moisture and seed rate; High humidity and cloudy weather.

Management

- ✰ Grow seedlings in raised seed bed.
- ✰ Regulate irrigation.
- ✰ Provide good drainage.
- ✰ Treat the seeds with carbendazim 2g/kg and drench soil with carbendazim.

4. East Indian Rosebay (*Tabernaemontana coronaria* R. Br.)

1. Anthracnose

Symptoms

- ✰ It produces irregular, brownish to greyish spots on the foliage.

Fungus

Colletotrichum dematium (Pers. ex Fr.) Grove and

Colletotrichum tabernaemontanae Pavgi

Favourable conditions: Moist weather.

Management

Spray carbendazim 0.1 per cent or zineb 0.2 per cent is effective.

5. Garden Geranium (*Pelargonium × hortorum* L.H. Bailey)

1. Root Rot/Wilt

Symptoms

- ✰ Yellowing and drying of leaves.
- ✰ Rotting of roots with numerous, minute, black sclerotia are seen.
- ✰ Death of the plant at the end.

Fungus

Rhizoctonia solani Kuhn.

Mode of survival: As sclerotia in infected plant debris.

Primary spread: From soil-borne sclerotia.

Secondary spread: Through irrigation water.

Management

Dip the cuttings in carbendazim 0.1 per cent before planting.

2. Leaf Spot

Symptoms

- ☆ Small, light reddish brown with dark brown margins.
- ☆ Spots increase in size, coalesce and cause premature leaf fall.

Fungus

Cercospora brunkii Ellis and Galloway

Mode of survival: As dormant mycelium in plant debris.

Primary spread: From infected plant debris.

Secondary spread: Through wind-borne conidia.

Management

Spray carbendazim 0.1 per cent or copper oxychloride 0.3 per cent or zineb 0.2 per cent.

6. Ginseng

(Asian Ginseng - *Panax ginseng* Baill.)

(American Ginseng - *Panax quinquefolium* L.)

1. *Phytophthora* Root Rot and Foliar Blight

Symptoms

Root Rot

- ☆ Root rot is a serious symptom primarily on 3 and 4 year old plants and it occurs in patches.
- ☆ Symptoms appear first on the crown and extend down the tap root or develop further down on the root system.
- ☆ Infected roots show light brown discolouration of the smooth exterior and creamy white soft rubbery interior texture.
- ☆ Squeezing infected roots by hand causes soft inner tissues to extrude like tooth paste from a tube and exhibit pungent odours due to secondary colonizers.
- ☆ Hyphae ramify throughout the affected tissues.

Foliar Blight

- Water-soaked, dark green lesions with white centre appear on foliage.
- Affected foliage collapse downward from the base of the petiole.
- Leaves turn yellow or red.
- Stems become hollow and brown

Chromista

Phytophthora cactorum (Leb. and Cohn)

Mode of survival: As oospores in infected rotten roots.

Primary spread: From oospores in soil.

Secondary spread: Through zoospores on leaf lesions spread by wind or rain.

Favourable conditions: Cool, wet or humid weather, warming soil temperature, heavy soil, excessive rain and poor drainage.

Management

- Remove and destroy affected plants.
- Avoid planting in excessively heavy or poorly drained soils.
- Avoid driving heavy machinery in wet soil.
- Grow ginseng adopting ridges and furrows method.
- Use seeds from disease free areas.
- Avoid continuous cultivation of ginseng in an area.
- Spray bleaching powder 10 per cent onto boots, shovels used to dig up infested areas.
- Pressure wash tractors that traverse a root rot infested area.
- Apply pre- and postharvest treatment of metalaxyl or Fosetyl-Al to reduce root rot.

2. Damping Off, Root Rot and Seed Decay

Symptoms

- Pre-emergence decay of seeds and seedlings and post-emergence rotting of stems and roots near soil surface (*Pythium* and *Fusarium*).
- Poor germination and poor stand.
- Stem above soil level causes tip over when a water-soaked brown lesion weakens the stem and it falls over and and the roots become brown (*Rhizoctonia solani*).
- Brown stem lesions above soil line and girdling of stem in 1-2 year old plants (*Allernaria panax*).

Fungi

Alternaria panax Whetzel, *Fusarium* sp., *Rhizoctonia solani* Kuhn.

Chromista

Pythium sp., *Phytophthora cactorum* (Leb. and Cohn)

Mode of survival: As infected plant debris and straw mulch.

Primary spread: From seed and soil-borne.

Secondary spread: Through irrigation water and farm equipments.

Favourable conditions: Wet cool weather, excessive soil moisture and dense saw dust mulch.

Management

- ☆ Select field with good drainage.
- ☆ Fallow the land for one year preceding new planting.
- ☆ Use only clean, treated disease free seeds.
- ☆ Adopt raised seed bed.
- ☆ Use light mulch as dense mulch harbour pathogens and slow ginseng plant emergence.
- ☆ Avoid heavy soils and low wet fields.
- ☆ Avoid excessive fertilization.
- ☆ Sterilize seed beds with formaldehyde well before planting.
- ☆ Sterilize seeds with 10 per cent bleaching powder.
- ☆ Pre-planting soil application with metalaxyl- Al and post planting treatment with metalaxyl or azoxystrobin is also effective.

3. Disappearing Root Rot

Symptoms

- ☆ Attacks at all ages of the crop.
- ☆ Small, discoloured, gold to brown lesions appear on root surface.
- ☆ Lesions enlarge rapidly, deepen producing a reddish-brown, spongy rot below the level of periderm.
- ☆ Affected root becomes dark brown.
- ☆ Tap root is distorted and lateral rootlets are also affected.
- ☆ Rot extends from root into crown and stem. When diseased plant is pulled stem alone gets separated from the plant.
- ☆ Damage to roots cause foliage inciting one side of the plant. Foliage turn red to brown after wilting.
- ☆ Harvest of roots after 2-4 years from infected plants show partial rotting and presence of conidia of the pathogen.

Fungus

Cylindrocarpon destructans (Zinssm.) Scholten (*Nectria radicicola* Gerlach and L. Nilsson).

Favourable conditions: High plant density, coniferous forest soils, soils with pH above 6.5. *Mode of survival:* As chlamydospores in infected plant debris in soil.

Primary spread: From infected plant debris in soil.

Secondary spread: Through irrigation water, farm implements.

Management

- ☆ Follow crop rotation.
- ☆ Avoid denser planting.
- ☆ Minimize wounding of roots.
- ☆ Soil drench with Fludioxonil.

4. *Sclerotinia* White Rot

Symptoms

- ☆ The stem and roots rot.
- ☆ Roots become soft and watery without any discolouration.
- ☆ Foliage wilt, discoloured, dry and shrivel.
- ☆ Black sclerotia are observed on infected plant parts.

Fungus

Sclerotinia sclerotiorum (Lib.) de Bary

Mode of survival: As sclerotia in infected plant debris.

Primary spread: From soil borne sclerotia.

Management

- ☆ Follow crop rotation avoiding beans.
- ☆ Avoid dense plant canopy.
- ☆ Provide good drainage.
- ☆ Remove and destroy infected plant parts.
- ☆ Disinfect soil with copper sulphate or thiophanate methyl.

5. *Verticillium* Wilt

Symptoms

- ☆ Usually a disease of older gardens.
- ☆ Leaves wilt and droop parallel to the stem.
- ☆ Wilting kills the plant.
- ☆ Roots are firm, vascular tissue is conspicuously discoloured and yellow.

Fungus

Verticillium dahlia Kleb.

Mode of survival: As microsclerotia in infected plant debris.

Primary spread: From infected plant debris.

Secondary spread: Through irrigation water.

Favourable conditions: Temperature below 20°C.

Management

- ✰ Follow crop rotation.
- ✰ Remove and destroy diseased plants.

6. Rusty Root Rot

Symptoms

- ✰ All ages of plants are affected.
- ✰ Rusted roots exhibit slightly raised reddish brown lesions which girdle the stem.
- ✰ Lesions are superficial and rough in appearance and may cover most of the root surface.
- ✰ Plants are stunted and mature early and not killed.

Fungi

Cylindrocarpon destructans (Zinssm.) Scholten, *Fusarium* sp., *Rhexocercosporidium panacis* Reeleder

Management

Spray of fludioxanil and azoxystrobin

7. *Alternaria* Blight

Symptoms

- ✰ Both seedlings and mature plants of 2-4 years old are attacked.
- ✰ Young seedlings are directly killed.
- ✰ Circular, water-soaked spots turn brown with target board appearance and dark yellow-brown margin. Centre of the lesion dries leaving shot-hole symptom.
- ✰ More spots cause blighting of leaves.
- ✰ Premature defoliation due to rapidly spreading foliar infection causes reduced flower (stigma) yield.
- ✰ Reddish to dark brown lesions appear on the stem, petiole and peduncles.
- ✰ Stem lesions cause the leaves to turn to reddish brown.

- Stem collapse, bends over and the shoot dies.
- Leaves are killed when the lesions girdle the stem.
- Sporulation of fungus on the surface of lesions gives velvety-brown appearance.
- Premature defoliation leads to reduced yield of flowers (stigmas).
- Abortion of developing berries.
- Injured roots are infected.
- Roots become dark brown or black, remains firm.
- Roots rot slowly without any odour.
- Seed head infection causes shelling of berries.

Fungus

Alternaria panax Whetzel

Mode of survival: As diseased leaves and stems as conidia or mycelia and on straw mulch.

Primary spread: From seed-borne.

Secondary spread: Through air-borne conidia.

Favourable conditions: Warm wet humid weather, temperature of 20-25°C or rainy or humid weather and dense canopy.

Management

- Adopt proper spacing (1-2 plants/sq.ft.) to avoid high plant density and to reduce relative humidity to increase air circulation.
- Remove and destroy infected plant parts.
- Apply fertilizer judiciously.
- Avoid moving in the field when leaves are wet.
- Keep foliage as dry as possible.
- Spray Bordeaux mixture 0.6 per cent with 0.4 per cent calcium arsenate (600 g copper sulphate + 600 g stone lime in 100 times of water + calcium arsenate 400 g). It is sprayed four times when plants emerged through soil, when leaflets are fully spread, just before blossoming and after fruit set.
- Spray mancozeb 0.2 per cent at 15 days interval or iprodione or chlorothalonil at maturity.
- Spray Iprodione + copper hydroxide at 7days interval with a sticker to avoid development of Iprodione resistant *Alternaria.*

8. *Botrytis* Blight

Symptoms

- Plants of 3 years old are generally affected.
- Leaves, flowers and fruits are attacked leading to premature defoliation.
- Rapidly enlarging water soaked concentric lesions starts at leaf tip and spreads back along the midrib, giving the lesion a "V" shape.
- Spots shows target-load appearance.
- Affected leaves turn reddish to brown prematurely.
- Lesions are covered with fuzzy mycelial growth with *sclerotia.*
- Infection on flowers causes them to abort and the aborted flowers desiccate and persistent pedicels turn reddish to brown.
- Fruit development is uneven.
- Immature green berries turn brown with sporulation of the fungus.
- Mature red fruits are covered with dense grey fungal growth.
- Stem infection cause wilting of plants. In root infection rotting is not seen.

Fungus

Botrytis cinerea Pers.

[Sexual: *Botryotinia fuckeliana* (de Bary) Whetzel; Syn. *Sclerotinia fuckeliana* (de Bary) Fuckel].

Mode of survival: As sclerotia in infected plant debris and on primary stems and surrounding susceptible hosts.

Primary spread: From infected plant debris.

Secondary spread: Through wind-borne conidia; rain splashes also help.

Favourable conditions: Temperature of less than 20°C, crowded condition, poor ventilation and continued leaf wetness.

Management

- Remove and destroy diseased crop residues.
- Avoid mechanical damage to plant parts.
- Provide good air circulation following optimum spacing.
- Avoid excessive N application as it enhances leaf growth.
- Preventive spray with chlorothalonil or fenhexamid. Iprodione + copper hydroxide spray is also effective.

9. Powdery Mildew

Symptoms

- Powdery white coating on the upper leaf surface which later becomes reddish purple.

- ✰ Leaves turn yellow and drop pre-maturely.
- ✰ Early and severe infection reduces seed yield and root yield.

Fungus

Erysiphe panax Bai and Wang.

Mode of survival: As cleistothecia in infected plant debris.

Primary spread: From infected plant debris.

Secondary spread: Through wind borne conidia.

Management

- ✰ Follow crop rotation to avoid pathogen build up.
- ✰ Avoid denser planting and improve air circulation.
- ✰ Spray azoxystrobin or pyraclostrobin or boscalid or thiophanate methyl.

10. Anthracnose

Symptoms

- ✰ Circular or irregular purplish brown spots with reddish margin appear on leaves.
- ✰ Spots increase in size with concentric rings and cause leaf fall.
- ✰ Infection spreads down to petiole to the main stalk.
- ✰ Affected central tissues in the spot fall off leaving.
- ✰ Small, black fruiting bodies called acervuli are seen on diseased leaves, petioles and stems.

Fungi

Colletotrichum dematium (Pers.) Grove

Mode of survival: As mycelium and acervuli in diseased plant debris.

Primary spread: From diseased plant debris.

Secondary spread: Through wind-borne conidia.

Management

Spray iprodione or mancozeb.

7. Gymnema (*Gymnema sylvestris* R.Br.)

1. Powdery Mildew

Symptoms

- ✰ White powdery patches appear on upper leaf surface.
- ✰ Later the fungal growth covers the entire leaf surface, become grey in colour.

- ☆ Affected leaves turn yellow and dry.
- ☆ In severe cases leaves and stem dry off results in reduced plant growth.

Fungus

Oidium sp.

Primary spread: From infected plant debris.

Secondary spread: Through air-borne conidia.

Favourable conditions: Warm humid weather.

Management

- ☆ Spray wettable sulphur 0.3 per cent or carbendazim 0.1 per cent or or propiconozole 0.1 per cent or hexaconazole 0.1 per cent or thiophanate methyl 0.1 per cent.

8. Haritaki/Chebulic Myrobalan (*Terminalia chebula* Retz.)

1. Leaf Rust

Symptoms

- ☆ Light brown to yellowish brown pustules seen on lower leaf surface.
- ☆ Corresponding upper leaf surface appear as greyish brown.
- ☆ Severe infection causes premature defoliation.

Fungus

Uredo terminaliae P.Henn.

Mode of survival: As infected leaves.

Primary spread: From diseased leaves.

Secondary spread: Through wind borne uredospores.

Management

Spray wettable sulphur.

9. Henbane (*Hyoscyamus niger* L. and *H. muticus)*

1. Seedling Blight

Symptoms

- ☆ Yellowish water-soaked lesions appear on hypocotyl region.
- ☆ Infection spreads upward and down to root region.
- ☆ Infected stem and root become dark brown with soft rot.

Chromista

Pythium butleri Subraman

Mode of survival: As survives as oospores in infected material in soil.

Primary spread: From soil-borne oospores.

Secondary spread: Through irrigation waters.

Management

Drench the soil with copper oxychloride 0.3 per cent.

2. *Henbane* Mosaic

Symptoms

- ✰ Yellow mosaic mottling.
- ✰ Vein clearing and vein banding.
- ✰ Stunting of plant.

Virus

Henbane mosaic virus (HMV)

Transmission: Sap transmissible.

Management

Remove and destroy diseased plants.

10. Indian Ipecac/Nanjaruppan [*Tylophora indica* (Burm. f.) Merr.]

1. Leaf Web Blight

Symptoms

- ✰ Premature defoliation.

Fungus

Rhizoctonia solani Kuhn.

Mode of survival: As sclerotia.

Primary spread: From soil-borne sclerotia.

Secondary spread: Through irrigation water and farm implements.

Management

- ✰ Reuse the plant with support instead of allowing to creep on the ground.
- ✰ Rogue out diseased plants and burn.
- ✰ Foliar spray with Bayleton 0.06 per cent twice at 15 days interval from June.

11. Khus Khus/Vetiver
[*Chrysopogon zizanioides* (L.) Roberty]

1. Leaf Blight

Symptoms

- ☆ Tan to dark coloured spots on leaves turn black in colour.

Fungus

Curvularia trifoloi (Kauffman) Boedijn

Primary spread: From infected plant debris.

Secondary spread: Through air-borne conidia.

Management

Spray copper oxychloride 0.3 per cent.

Other Disease in Khus Khus

2. **Anthracnose – *Gloecospora sorghi***
3. **Leaf spot – *Helminthosporium* sp.**

12. Lemongrass
[*Cymbopogon flexuosus* (Nees ex. Steud.)]

1. Leaf Blotch

Symptoms

- ☆ Small, circular, reddish brown spots appear on margins and tips of leaves.
- ☆ Spots enlarge and coalesce to form longitudinal, elongated necrotic, reddish brown lesion causing blighting and premature leaf fall.

Fungi

Curvularia andropogonis (Zimm.) Boedijn and

C. verruciformis G.P. Agarwal and V.P. Sahni

Mode of survival: In infected plant debris in soil.

Primary spread: From dormant mycelium in infected plant debris.

Secondary spread: Through wind-borne conidia.

Favourable conditions: Rainy season, nutritionally poor soil.

Management

Spray zineb 0.2 per cent or copper oxycholoride 0.3 per cent or mancozeb 0.2 per cent.

2. Leaf Spot

Symptoms

- ☆ Oval, brown spots with dark purple borders appear on leaves.
- ☆ Spots enlarge, coalesce and cause necrotic leaf blight.

Fungus

Drechslera sacchari (E.J. Butler) (Syn: *Helminthosporium sacchari* E.J. Butler)

Mode of survival: As infected plant debris in soil.

Primary spread: From dormant mycelium in infected plant debris.

Secondary spread: Through wind-borne conidia.

Management

- ☆ Burn the stubbles in summer.
- ☆ Spray mancozeb 0.2 per cent or Bordeaux mixture 1.0 per cent or copper oxychloride 0.3 per cent.

3. Red Leaf Spot

Symptoms

- ☆ Scattered, small, brown spots on the ventral surface of older leaves.
- ☆ Disease spreads to midrib and leaf sheath.
- ☆ Spots coalesce to turn larger necrotic spots causing drying of leaves.
- ☆ Presence of small, black acervuli on the leaf spots.

Fungus

Colletotrichum graminicola (Ces.) G.W. Wilson

Mode of survival: As infected plant debris and stubbles.

Primary spread: From dormant mycelium in infected leaf and leaf sheath.

Secondary spread: Through wind-borne conidia.

Management

Spray carbendazim 0.1 per cent or mancozeb 0.2 per cent.

4. Rust

Symptoms

- ☆ Discrete, linear cinnamon brown uredinia are seen on both upper and lower leaf surface cause chlorotic streaks.
- ☆ Severe infection leads to coalescence of uredia and extensive necrosis of leaves.
- ☆ Telia also appear but no aecia.

Fungi

Puccinia nakanishikii Dietel and *P. cymbopogonis* Massee

Mode of survival: As infected fallen leaves.

Primary spread: From infected plant parts and fallen leaves.

Secondary spread: Through air-borne uredospores.

Favourable conditions: High rainfall, high humidity and warm air temperature.

Management

- ✰ Apply optimum fertilizer dose for vigorous growth.
- ✰ Adopt intercropping with non-host crops.
- ✰ Prune or cut back diseased plants periodically.
- ✰ Grow in well-drained soil.
- ✰ Keep weeds under check to reduce relative humidity.
- ✰ Avoid overhead irrigation.
- ✰ Spray oxycarboxin 0.1 per cent or zineb 0.2 per cent.

5. Long Smut

Symptoms

- ✰ Flowers in inflorescence are transformed into slender, tubular or conical, cream coloured sori.
- ✰ Sori flake off at maturity, starting from tip and hang in shreads.
- ✰ Leaf and oil yield is reduced.

Fungus

Tolyposporium christensenii Ragunath

Spread: Through wind-borne spores.

Management

Spray zineb 0.2 per cent at flower initiation.

6. Grey Blight

Symptoms

- ✰ Brown spots appear on leaves.
- ✰ Spots increase in size and become dark brown with black dots (fruiting bodies are present on leaf spots).

Fungus

Pestalotiopsis mangiferae Henn.

Mode of survival: As dormant mycelium in infected plant tissues.

Primary spread: From infected plant refuse.

Secondary spread: Through wind-borne conidia.

Management

Spray wettable sulphur 0.3 per cent or carbendazim 0.1 per cent.

7. Little Leaf/Grassy Shoot

Symptoms

- The disease affects floral parts.
- Affected plants are stunted with little leaf formation.
- In the place of normal inflorescence, proliferated small, leafy structures are produced.

Fungus

Balansia sclerotica (Pat.) Höhn

Management

- Remove and destroy diseased plants.
- Avoid using seeds from diseased plants.
- Uproot and destroy infected plants.
- Follow crop rotation.

13. Liquorice (*Glycyrrhiza glabra* L.)

1. Collar Rot/Root Rot

Symptoms

- Dark brown spots on the collar region causes girdling.
- Rotting of roots.
- Stolon and roots turn soft and pulpy of stem.
- Decay and death of the plant.
- Branches and main stem show pycnidia and roots show sclerotia.
- Old and new sprouts are attacked and leaves on sprouts turn pale yellow and dry up and fall off.

Fungi

Rhizoctonia bataticola (Taubenh.) E.J. Butler, *Fusarium* sp. and *Sclerotium* sp.

Favourable conditions: Temperature above 35°C.

Management

- Remove and destroy infected sprouts/branches.

- ☆ Use soil mulch to reduce soil temperature.
- ☆ Adopt optimum irrigation.
- ☆ Drench the soil with carbendazim 0.1 per cent.

2. Leaf Spot

Symptoms

- ☆ Brown spots appear on lower surface of leaves which are restricted by veins.
- ☆ Spots turn dark brown and velvety and are covered with conidia.
- ☆ Numerous spots on the leaves cause drying and premature defoliation.

Fungus

Cercospora carotae (Pass.) Solheim

Favourable conditions: Severe from late monsoon and prevail up to onset of summer.

Management

- ☆ Foliar spray with zineb or mancozeb 0.2 per cent or carbendazim 0.1 per cent at 15 days interval.

3. Leaf Blight

Symptoms

- ☆ Pink head size brown lesions appear on leaves.
- ☆ Spots develop into circular to irregular in shape with 2-3 mm dia.
- ☆ Later the spots become dark brown in the centre and light brown in the margin.
- ☆ Dot-like pycnidia are seen on the spots.
- ☆ Light brown to black spots appear on leaves and stem.
- ☆ Spots increase in size, coalesce and cause blighting of leaves.

Fungi

Phyllosticta glycyrrhizae (Hollós), *Alternaria alternate* (Fr.) Keissl.

Management

- ☆ Spray copper oxychloride 0.3 per cent at 10-15 days interval.

4. Rust

Symptoms

- ☆ Yellowish pustules appear on the leaves.
- ☆ Stem elongation.

- ☆ Rosetting and defoliation.
- ☆ Browning of pith and formation of many buds in the crown region.

Fungus

Uromyces glycyrrhizae (Rabenh.) Magnus

Primary spread: From mycelium in infected buds.

Secondary spread: Through wind-borne conidia.

Management

Spray triadimefon 0.1 per cent at 10 days interval.

14. Malabar Nut/Vasaka (*Justicia adhatoda* L.)

1. Leaf Spot

Symptoms

Leaves of all ages are attacked; older leaves are attacked first and show small black spots. The spots enlarge and show distinct concentric rings and coalese causing drying of leaves.

Fungi

Alternaria alternata (Fr.) Keissl. *Septoria adhatodae*

Favaroble conditions: Rainy season.

Management

Spray copper oxychloride 0.3 per cent or Bordeaux mixture 1 per cent.

15. Mayapple (*Podophyllum peltatum* L.)

1. Rust

Symptoms

- ☆ Bright orange (aecia) appear on the under surface of leaves.
- ☆ Yellow or light green spots blotched appearance on the corresponding upper leaf surface.
- ☆ Affected leaves are puckered and disfigured and fall.
- ☆ Heavy aecial cups causes reduction in plant growth.
- ☆ Black telia develop just before leaf fall.

Fungus

Allodus podophylli (Schwein.) Arthur (Autoecious rust, No alternate host)

Spread: Through air-borne aeciospores.

Favourable conditions: Wet weather.

Management

- ☆ Remove diseased leaves or prune diseased branches.
- ☆ Provide proper spacing.
- ☆ Avoid overhead watering/splashing water.

16. Mint (*Mentha arvensis* L.)

1. Powdery Mildew

Symptoms

- ☆ Small chlorotic spots appear on the upper surface of the leaves.
- ☆ The corresponding lower surface showing brownish discolouration prior to the appearance of powdery patches.
- ☆ Later appearance of white or grey colour powdery growth on both upper and lower surface of leaves.
- ☆ It leads to heavy defoliation.

Fungi

Erysiphe cichoracearum DC.

The fungi produce external septate mycelia, haustoria sub epidermal. Septate mycelia produces conidiophore on which barrel shaped conidia are borne in chains. Sexually ascospores borne in ascus. Asci are situated in Cleistothecium.

Primary spread: From dormant mycelia and Short period cleistothecium.

Secondary spread: Through air borne barrel shaped conidia.

Favourable conditions: Warm weather, temperature of 28-32°C, cloudy weather.

Management

- ☆ Crop rotation with non-host crops.
- ☆ Altering the date of sowing.
- ☆ Proper nutrient management.
- ☆ Avoid dense planting.
- ☆ Prophylactic aerial spray with wettable sulphur 0.3 per cent or carbendazim 0.1 per cent at 10-15 days interval.

2. Rust

Symptoms

- ☆ Brown rusty pustules on the lower surface of leave.
- ☆ The spots initially circular slightly elevated and later coalesce to form irregular spots.

- Chlorotic streaks on the upper surface, telial stage are characterized by swelling on the upper portion of the stem.
- Defoliation and death of the leaves can be seen in severely affected plants.

Fungi

Puccinia menthae Pers.

The fungi produce septate mycelia, intercellular mycelia and intracellular haustoria. urediospores in uredium and teliospores in telium.

Spread: Air borne urediospores.

Management

- Crop rotation with non-host crop.
- Spray mancozeb 0.25 per cent and hexaconazole 0.1 per cent.
- Spray recommented doses of nickel chloride, tebuconazole, propiconazole.
- Spary dinitroroamine near soil surface.

3. Leaf Spot

Symptoms

- Small unclear brown spots scattered over the leaf lamina.
- The minute spot increase in size, forming big spherical or irregular patches.
- Severe in neglected crops.

Fungi

Curvularia lunata (Wakker) Boedijn

Others: *Corynespora cassicola* Berk., *Cercospora menthicola* Tehon and E.Y. Daniels, *Septoria menthae* Oudem.

Septate mycelia, inter and intracellular haustoria, any one of the conidia or conidiophore is coloured.

Primary spread: From dormant mycelium.

Secondary spread: Through air borne conidia.

Favourable conditions: Nutritionally poor soil, temperature of 28-32°C, cloudy weather.

Management

- Collect and destroy affected plant parts.
- Increased nutritional status.
- Reduce plant population.
- Spray biological agent *Trichoderma* spp.
- Spray mancozeb 0.25 per cent as aerial spray.

5. Leaf Blight

Symptoms

- ☆ Lower most leaves first infected and disease later develops on upper leaves.
- ☆ The infected leaves shows round to oval to irregular or slightly irregular dark brown spots on the upper surface of the leaves with concentric rings.
- ☆ Spot size increases and covers the entire leaves.

Fungi

Alternaria alternata (Fr.) Keissl.

Septate mycelia, coloured muriform conidia.

Primary spread: From dormant mycelium.

Secondary spread: Through air borne conidia.

Favourable conditions: Nutritionally poor soil, temperature of 28-32°C, cloudy weather.

Management

- ☆ Proper nutritional management.
- ☆ Reduce plant population.
- ☆ Spray mancozeb 0.25 per cent as aerial spray.

6. Stolon Rot

Symptoms

- ☆ Initially yellowing of leaves, stunted growth, in advance stages the plants wilts followed by death of above ground parts.
- ☆ Stolons exhibit pinkish brown lesions initially, which gradually turns into dark brown lesions.
- ☆ Later turns to dark brown to black patches that increase in size resulting in soft decay.

Fungi

Rhizoctonia bataticola Kuhn, and *Sclerotium rolfsii* Sacc.

Septate mycelia, sub epidermal haustoria.

Primary spread: From sclerotial bodies present in the debris.

Secondary spread: Through soil borne mycelial strands.

Favourable conditions: Warm weather, optimum moisture, neutral pH, soil temperature of 28-32°C.

Management

- ☆ Collect and destroy affected plant parts.
- ☆ Summer ploughing.
- ☆ Crop rotation.
- ☆ Soil sterilization.
- ☆ Spray biological agent *Trichoderma* spp.
- ☆ Spray carbendazim 0.1 per cent or mancozeb 0.25 per cent as aerial spray.

7. Wilt

Symptoms

- ☆ Initially lower leaves starts V shape yellowing.
- ☆ Dwarfing unilateral development of the branches.
- ☆ Etiolating of leaves leading to wilting death.
- ☆ Light tan colour discolouration of vascular bundles.

Fungi

Verticillium albo-atrum Reinke and Berthold, *Verticillium dahlia* Kleb.

Septate mycelia inter and intra-cellular haustoria, produces V shaped conidiophore on which round shaped small conidia are borne.

Primary spread: From dormant mycelia.

Secondary spread: Through soil and air borne micro and macro-conidia.

Favourable conditions: Alkaline pH, black clay soil, low moisture, temperature of 30-32°C.

Management

- ☆ Collect and burn affected plant parts.
- ☆ Use soils free from nematode.
- ☆ Crop rotation with non-host crops.
- ☆ Spray biological agent *Trichoderma* spp.
- ☆ Spray carbendazim 0.1 per cent or mancozeb 0.25 per cent as aerial spray

Other Diseases in Mint

8. Downy mildew - *Peronospora stigmaticola* Raunk

9. Typhula root rot - *Typhula itoana* S. Imai

10. Rhizome or Stem Rot - *Phoma strasseri* Moesz

11. Sclerotinia Blight - *Sclerotinia sclerotiorum* (Lib.) de Bary.

17. Opium Poppy (*Papaver somniferum* L.)

1. Seedling Blight and Damping Off

Symptoms

- ☆ The seeds rot before germination.
- ☆ Rotting of roots and at collar region is common.
- ☆ Diseased seedlings die after emergence.
- ☆ Post emergence damping off occurs mostly 2-4 weeks of sowing and the death of seedlings occur in 3-5 days of attack.

Chromista

Pythium ultimum Trow., *P. dissotocum* Drechsler

Mode of survival: As oospores in infected plant debris in soil.

Primary spread: From From oospores in soil.

Secondary spread: Through Through zoospores in irrigation water.

Favourable condition: High soil moisture, low lying areas and higher seed rate.

Management

- ☆ Sterilize the soil with formaldehyde 4 per cent up to 30 cm depth.
- ☆ Sow the seeds in raised seed bed.
- ☆ Treat the seeds with metalaxyl 0.4 per cent or captan 0.3 per cent.
- ☆ Drench the soil with propamocarb hydrochloride 0.3 per cent.

2. Wilt and Root Rot

Symptoms

- ☆ Circular or irregular, dirty black leaf spots occur first.
- ☆ Later these lesions extend to entire length of leaf resulting in rotting of tender leaves.
- ☆ Irregular grey patches with brown margin appear at base of leaf sheath.
- ☆ Basal portion of pseudostem rots and breaks away at collar region.
- ☆ Tender and green fruits show light brown lesions cause rotting and shedding within 3-6 days of infection.

Fungus

Embellisia phragmospora (Emden) E.G. Simmons

Mode of survival: As chlamydospores in soil.

Primary spread: From chlamydospores in infected plant debris in soil.

Secondary spread: Through wind-borne conidia.

Favourable conditions: Soil pH of 6-7, temperature of 21-26°C, high rainfall, high soil moisture and relative humidity.

Management

Spray Mancozeb or Zineb 0.2 per cent.

3. *Fusarium* Wilt/Root Rot

Symptoms

- Black necrotic lesions developed on the stem.
- Yellowing and drooping of leaves and rapid wilting of plant.
- Drying of leaves.
- Browning of vascular tissues.

Fungus

Fusarium oxysporum Schltdl, *F. semitectum* Berk.

Management

Drench the soil with carbendazim 0.1 per cent in the root region.

4. *Macrophomina* Root Rot

Symptoms

- Root rot is noticed in patches in a field in 2 months old plants.
- Reddish brown patches seen at the collar region cause drying of plant.
- Rotting of roots and shredding of root bark are also seen with numerous small, black sclerotial bodies.
- Black necrotic lesions at the stem base causing rotting of roots.

Fungus

Macrophomina phaseolina (Tassi) Goid.

Mode of survival: As sclerotia in infected roots in soil.

Primary spread: From soil-borne sclerotia.

Secondary spread: Through irrigation water.

Management

- Drench the soil with carbendazim 0.1 per cent.
- Apply *Trichoderma viride* or *T. harzianum* along with FYM.

5. Collar Rot

Symptoms

- Black lesions appear at the collar region.

- They extend and cause blackening of stem and stunted growth.
- Shrivelling and rotting of roots are seen.
- Black, minute, sclerotia are noticed on the root barks.

Fungus

Rhizoctonia solani Kuhn

Mode of survival: As sclerotia in infected roots.

Primary spread: From soil-borne sclerotia.

Secondary spread: Through irrigation water.

Management

- Follow 3 year crop rotation.
- Treat the seeds with thiram 4 g/kg.

6. White Mould/Sclerotinia Blight

Symptoms

- Fungus infects roots, stems, heads and seeds.
- Infected parts rot and are covered with white mycelium.
- Many sclerotia are seen on or inside the stem.

Fungus

Sclerotinia sclerotiorum (Lib.) de Bary (=*Pleospora herbarum* (Pers.) Rabenh.)

Mode of survival: As sclerotia and mycelia in plant debris.

Primary spread: From ascospores produced in apothecia in infected plant stubbles.

Secondary spread: Through irrigation water.

Management

- Adopt four year crop rotation.
- Use healthy seed for sowing.
- Apply fertilizers judiciously.
- Drench the soil with Bordeaux mixture.

7. Downy Mildew

Symptoms

- Disease appears from seedling stages to maturity stage in opium crop.
- Infected seedlings are stunted and killed outright.
- Leaves near the tips and margins show pale brown spots on the upper leaf surface.

- Corresponding lower leaf surface shows fluffy grey-white fungal growth.
- Fungal growth spread to entire leaf and cause the leaves to dry.
- Stem, branches and capsules are also affected resulting in premature death of the plant.
- Hypertrophy and curvature of stem and flower stalks are common.
- Latex production is reduced and the plants are stunted.
- Capsules show fungal growth and dry.

Chromista

Peronospora arborescens (Berk.) de Bary

Mode of survival: As oospores in infected plant materials.

Primary spread: Infected seeds, oospores in infected plant debris.

Secondary spread: Through zoospores through irrigation water or wind.

Favourable conditions: Intermittent rainfall, temperature of 40-60°C, high relative humidity (90 per cent) and many collateral weed hosts.

Management

- Adopt 4-5 year crop rotation with one year fallow.
- Grow resistant varieties.
- Sow disease-free healthy and certified seed.
- Practice wide spacing.
- Adopt balanced nutrition.
- Avoid low lying fields.
- Follow recommended spacing for air circulation.
- Remove and destroy diseased leaves and plants.
- Treat seeds with metalaxyl at 8 g/kg or mancozeb 4 g/kg.
- Spray metalaxyl or Fosetyl-Al 0.2 per cent or mamcozeb 0.3 per cent 3-4 times 20, 60 and 80 days of sowing control.

8. Powdery Mildew

Symptoms

- Plants at early stages are heavily damaged.
- Circular patches of white powdery fungal growth seen on the upper leaf surface.
- Fungal growth covers the entire leaf and cause drying of leaves.
- Stem, flowers and capsules are also covered with white fungal growth.
- Reduction in capsule size, seed and latex yield.

Fungus

Erysiphe polygoni DC

Mode of survival: As dormant mycelium and cleistothecia in infected leaves.

Primary spread: From ascospores in infected leaves.

Secondary spread: Through wind-borne conidia.

Favourable conditions: Temperature below 27°C and wet summer months.

Management

- ✰ Raise the crop before 15th October to escape from powdery mildew incidence.
- ✰ Spray wettable sulphur 0.3 per cent or Baytelon or dinocap 0.1 per cent or dust sulphur 25 kg/ha.

9. Leaf Blight

Symptoms

- ✰ Diffused yellow necrotic spots appear on leaves.
- ✰ Necrosis of leaf and stem leads to stem girdling and stunting of plants.
- ✰ Foliar and capsule blight occur which leads to premature defoliation.
- ✰ Seed do not develop properly and become small and deformed.

Fungus

Helminthosporium papaveins (*Pleospora papaveracea* (De Not.) Sacc., *Crivellia papaveracea* (De Not.) Shoemaker and Inderbitzin)., *Dendryphion penicillatum* (Corda) Fr.

Mode of survival: In seed and plant debris.

Primary spread: From infected seeds and plant debris.

Secondary spread: Through wind-borne conidia.

Favourable conditions: High rainfall, high relative humidity and acid soil.

Management

- ✰ Adopt four year crop rotation.
- ✰ Follow mixed cropping with sugarbeet.
- ✰ Avoid late sowing. Sow during April-May.
- ✰ Use healthy and certified seed.
- ✰ Treat seeds with thiram or captan.

10. Black Spot

Symptoms

- Stems and leaves are affected.
- Scattered or continuous dark olive black coatings appear on poppies.

Fungus

Stemphylium botryosum Wallr.

Mode of survival: In infected plant debris.

Primary spread: From diseased plant debris.

Secondary spread: Through wind-borne conidia.

Management

- Follow crop rotation.
- Apply fertilizers judiciously.

11. Capsule Rot

Symptoms

- Velvety black spots appear on green capsules.
- Reduction in morphine, cocaine and thebaine content.

Fungi

Alternaria alternata (Fr.) Keissl., *Dendryphion penicillatum* (Corda) Fr. *Macrosporium papavenis* and *Fusarium scirpi* var *caudatum* (Wollenw.).

Management

Spray carbendazim 0.1 per cent or mancozeb 0.2 per cent 4 times at 15 days interval.

12. Bacterial Leaf Spot

Symptoms

- Water soaked areas appear on leaves which soon become black and are surrounded by a translucent ring.
- Leaves, stems, buds, flowers and capsules are attacked.
- Stem lesions are elongated and intense black, coalesce and girdle the stem leading to collapse of the plant.

Bacterium

Xanthomonas campestris pv. *papavericola* (Bryan and McWhorter 1930) Dye

Mode of survival: In infected stubbles.

Primary spread: From infected plant materials.

Secondary spread: Through splashing water.

Favourable conditions: Hot and humid weather.

Management

- ☆ Follow crop rotation.
- ☆ Select and sow healthy, certified seeds.

13. Soft Rot

Symptoms

- ☆ External blackening and internal disintegration with discolouration of midrib of leaves.
- ☆ Disease initiates at apex and spreads downward and cover the whole plant.
- ☆ Diseased plants turn into slimy mass.

Bacterium

Erwinia papaveris (Ayyar)

Mode of survival: In seed and in infected plant debris.

Primary spread: From seed-borne inoculum and infected plant debris.

Secondary spread: Through rain splashes.

Management

Spray copper oxychloride 0.3 per cent.

14. Mosaic

Symptoms

- ☆ Green and yellow mosaic symptoms appear on leaves.
- ☆ Chlorosis and deformation of leaves followed by deformation of stalks and heads.
- ☆ Poppy heads are weak, capsules are malformed.
- ☆ Flower breaking is seen.
- ☆ Infected plants are stunted.
- ☆ Reduction in opium and seed yield.
- ☆ Diseased plants die in 4 weeks of infection.

Virus

Turnip mosaic virus (TuMV)

Transmission: Through sap.

Management

Remove and destroy diseased plants.

15. Yellow Mosaic

Symptoms

- Irregular chlorotic bands are seen along the veins.
- Chlorosis spreads rapidly in interveinal areas leaving green stripes along the veins.
- Leaves are malformed and the plants are stunted.

Virus

Bean yellow mosaic virus (BYMV)

Transmission: Through aphid vector (*Aphis fabae*).

Management

Spray methyl demeton 0.05 per cent.

16. Opium Poppy Mosaic

Symptoms

- Mottling and mosaic symptoms seen in leaves.
- Vein banding, deformation of capsules and stunting of plant are the other symptoms.

Virus

Opium poppy mosaic virus (OPMV)

Transmission: Through aphid vector.

Management

Spray dimethoate 0.03 per cent.

17. Broom Rape

Symptoms

- The affected plants grow very poor and are stunted.
- Reduction in flower production.
- Produce very few capsules containing very little latex.

Parasite

Orobanche papavens

Management

- Follow crop rotation.

- ☆ Sow catch and trap crops and plough them in September.
- ☆ Pull out parasitic broom rape plants before flowering.
- ☆ Swab spikes of parasitic plants with allyl alcohol 0.3 per cent (in diesel oil) in April.

18. Palmarosa [*Cymbopogon martini* (Roxb.) Wats.]

1. Red Leaf Spot

Symptoms

- ☆ Small, brown circular or irregular spots appear first on the ventral surface of the leaves.
- ☆ Infection spreads to midrib and leaf sheath.
- ☆ Lesions increase in size, show concentric rings and minute, black acervuli on the spots during warm weather.
- ☆ Diseased leaves dry and fall off in severe infection.
- ☆ Leaf and oil yield is reduced.

Fungi

Colletotrichum graminicola (Ces.) G.W. Wilson, *Colletotrichum caudatum*,

Mode of survival: Aas mycelium in infected stubbles.

Primary spread: From self-sown plant and infected plant materials.

Secondary spread: Through wind-borne conidia.

Favourable conditions: Warm humid conditions.

Management

- ☆ Collect and destroy infected and dried leaves.
- ☆ Spray carbendazim 0.1 per cent twice at 20 days interval or 3 sprays of mancozeb 0.2 per cent at 10-12 days interval.

2. Little Leaf/Grassy Shoot

Symptoms

- ☆ Stunting and production of little leaves.
- ☆ Small abnormal leaf like flowers are produced.
- ☆ Reduction in seed and oil production.

Fungus

Balansia claviceps Speg.

Management

- ☆ Follow crop rotation.
- ☆ Spray zineb 0.3 per cent before flowering.

3. Smut

Symptoms

- ☆ Inflorescence disease.
- ☆ Seeds are transformed into long sori which contain smut spreads.

Fungus

Tolyposporium christensenii Ragunath

Management

- ☆ Treat seeds with fungicide.
- ☆ Spray zineb 0.3 per cent before flowering.

4. Leaf Blotch and Leaf Blight

Symptoms

- ☆ Circular, oval, reddish brown spots appear on the margins and tips of leaves.
- ☆ Spots turn dark brown and surrounded by dark reddish brown border.
- ☆ Spots enlarge and coalesce to form longitudinal, elongated necrotic lesions resulting in premature drying of affected leaves.

Fungus

Curvularia trifolii (Kauffman) Boedijn and *C. andropgonis* (Zimm.) Boedijn

Mode of survival: As infected plant debris.

Primary spread: From infected plant debris.

Secondary spread: Through wind-borne conidia.

Favourable conditions: Nutritionally poor soil.

Management

- ☆ Collect and destroy infected and dried leaves.
- ☆ Apply recommended NPK dose.
- ☆ Spray mancozeb 0.2 per cent.

Minor Diseases of Palmarosa

Leaf blight – *Ellisiellina candata*

Leaf spots - *Helminthosporium sacchari, Helminthosporium leucortylum, Dreshlera victorae, Dreshlera helmi, Curvularia verruciformis,* and *Curvularia eragrositidis.*

Grey blight - *Pestalotia mangiferae.*

Leaf spot and clump rot - *Drechlera colocasia.*

19. Patchouli (*Pogostemon cablin* Benth.)

1. Leaf Blight

Symptoms

- ☆ Brown spots appear at the tip and margins of leaves.
- ☆ Spots enlarge, coalesce and cause blighting and drying of leaves.

Fungus

Cercospora sp.

Spread: Through wind-borne conidia.

Management

Spray Zineb 0.2 per cent or two sprays of Dithane Z-78 0.5 per cent, at one-month interval.

2. *Fusarium* Wilt/Root Rot

Symptoms

- ☆ Yellowing of the older leaves.
- ☆ Drying of green parts.
- ☆ Brown to black discolouration of stem and roots followed by dark brown to red discoloration of vascular tissue.
- ☆ The infected plants wilted completely and could be pulled off easily from the soil.

Fungus

Fusarium solani (Mart.) Sacc.

Chromista

Pythium aphanidermatum (Edson) Fitzp.

Spread: Through soil-borne chlamydospore.

Management

Soil drenching with copper oxychloride 0.25 per cent.

3. *Sclerotial* Rot

Symptoms

- ☆ Appear water soaked lesions near collar region followed by darkening of the tissue.
- ☆ The leaves became flaccid and dropped off.
- ☆ The sclerotial bodies were white initially, later turned brown with age.

Fungus

Sclerotium rolfsii Sacc.

Spread: Sclerotial bodies from the fallen leaves.

Management

Soil drenching with copper oxychloride 0.25 per cent.

4. Mosaic

Symptoms

- ✰ Pale green mosaic pattern and malformation of leaves are seen.
- ✰ Affected plants are stunted.

Virus

Telosma mosaic virus (TeMV) and *Passion fruit woodiness virus* (PaWV).

Management

Collect and destroy diseased plants.

20. Physicnut (*Jatropha curcas* L.)

1. Anthracnose

Symptoms

- ✰ Brown to black, necrotic lesions, irregular with yellow halo appear on margins and centre of the leaves.
- ✰ Spots enlarge, coalesce and blight the leaves reducing plant vigour.
- ✰ Dark brown lesions also appear on fruits.
- ✰ On seedlings stem canker causes death of apex.

Fungi

Colletotrichum gloeosporioides Pat. and Penz. and *C. capsici* (Syd. and P. Syd.)

The fungus produces acervuli containing setae, conidiophore and conidia. Conidia are hyaline, single celled, smooth and thin walled.

Mode of survival: As dormant mycelium in infected crop debris and in soil as saprophyte.

Primary spread: From infected seeds and seedlings.

Secondary spread: Through wind-borne conidia.

Management

Spray mancozeb 0.2 per cent.

2. Powdery Mildew

Symptoms

- White fungal patches seen on the upper leaf surface and the patches increase in size, cover the entire leaf, become grey and cause drying.
- Necrotic lesions on stem and petiole cause premature leaf fall.
- Infected flower buds die and fruits deformed.

Fungus

Pseudoidium jatrophae (Hosag., Siddappa, Vijay. and Udaiyan)

(=*Vidium heveae* and *O. jatrophas*).

Mycelim is septate and branched. Conidiophores are erect, hyaline, septate and form conidia singly. Conidia are simple, smooth, ellipsoid to avoid doliform, hyaline and single celled.

Mode of survival: As cleistothecia in infected plant debris and abundant alternative hosts.

Primary spread: From plant debris and weed hosts.

Secondary spread: Through wind-borne conidia.

Favourable conditions: As warm temperature, reduced sunlight and relative humidity of 75-80 per cent.

Management

- Spray wettable sulphur 0.3 per cent
- Destroy alternate hosts in *Euphorabiaceae*.

3. Rust

Symptoms

- Chlorotic points appear on the upper leaf surface and corresponding lower surface exhibit erumpent uredopustules.
- Reddish brown uredopustules coalesce to form irregular necrotic spots in the leaves destroying the affected leaf.
- Pustules rupture and expose orange coloured uredospores.

Fungus

Phakopsora arthuriana (Syn: *Phakopsora jatrophicola*). Uredospores are ellipsoid to ovoid, sessile, closely and finely echimulate. Paraphyses are non septate. Telia are closely turned formed around uredospustula. Teliospores are irregularly arranged, cuboid, ellipsoid to polygonal.

Mode of survival: As teliospores in infected plant debris.

Primary spread: From teliospores in leaves of living plants.

Secondary spread: Through uredospores.

Management

Spray copper oxychloride 0.3 per cent.

4. Stem Canker and Die back

Symptoms

- ☆ Cankers form around initial infection point on the twig/stem and eventually complete damage is done to vascular system.
- ☆ Necrosis and die back of the symptoms are common. Die back is characterized as dead arm.

Fungus

Lasiodiplodia theobromae Pat.

[Syn. *Botryodiplodia theobromae* (Pat.) Griff. and Maubl]

The fungus produces grey mycelium. Pycnidia produce dark brown, two celled constricted conidia. Ascospores are formed in pseudothecia.

Mode of survival: As pycnidia on diseased stem.

Primary spread: From infected plant parts.

Secondary spread: Through wind and rain splashed conidia and ascospores.

Favourable conditions: Pruning the branches.

Management

- ☆ Avoid pruning during wet weather. Pruning increases number of infection sites.
- ☆ Apply fungicides at cut ends after pruning. Use copper fungicide or thiophanate methyl.
- ☆ Remove dead woods.
- ☆ Follow balanced fertilization.

5. Collar and Root Rot

Symptoms

- ☆ Sudden death of plants is common.
- ☆ Wilting, leaf yellowing and leaf fall symptoms are seen.
- ☆ Cracks in the collar region with black fungal structures in the bark of the plant.
- ☆ Roots rot, vascular system show light brown to black discolouration.

Fungi

Fusarium solani (Mart.) Sacc., *Lasiodiplodia theobromae* Pat., *Neoscytalidium dimidiatum* (Penz.) Crous and Slippers and *Macrophomina phaseolina* (Tassi) Goid.

Mode of survival: As chlamydospores or sclerotia in soil or pycnidia on dead plant parts.

Primary spread: From infected plant parts.

Secondary spread: Through wind-borne conidia.

Favourable conditions: Prolonged dry season or drought.

Management

- ☆ Irrigate the field at regular intervals.
- ☆ Follow balanced fertilization to the plant.
- ☆ Plant healthy certified seedlings.
- ☆ Avoid injuries to seedlings and plants.

6. Yellow Mosaic

Symptoms

- ☆ Appearance of mosaic pattern, reduced leaf size, leaf distortion and blistering.
- ☆ Stunting of plant.

Virus

Indian cassava mosaic virus (ICMV).

Transmission

Through whitefly, *Bemisia tabaci* and not through seed or sap.

Management

- ☆ Plant disease-free seedlings.
- ☆ Spray insecticides to control insect vector.

21. Rose Scented Geranium [*Pelargonium capitatum* (L.) Ait., *Pelargonium graveolens* L'Herit.]

1. Leaf Spot

Symptoms

- ☆ Spots are circular, light reddish brown with dark brown borders.
- ☆ The spots coalesce and kill the leaf.

Fungus

Cercospora brunkii Ellis and Gall.

Management

Spray carbendazim 0.1 per cent or zineb 0.2 per cent or copper oxychloride 0.2 per cent.

2. Root Gall

Symptoms

- ☆ Plants in isolated pockets exhibit yellowing of leaves and wilting.
- ☆ Later the plants completely wither.
- ☆ The infection spreads in patches and old plants are commonly affected.
- ☆ On uprooting and examining the wilted plants, a large number of yellowish-brown galls are found in the root system.
- ☆ The galls are approximately 10 in each plant, oval, round or irregular with 5.0 to 7.5 cm in dia weighing approximately 20 to 25g.

Fungus

Agrobacterium tumefaciens (Smith and Townsend)

Management

Spray carbendazim 0.1 per cent or zineb 0.2 per cent or copper oxychloride 0.2 per cent.

22. Rosemary (*Rosmarinus officinalis* L.]

1. Cottony Soft Rot

Symptoms

- ☆ Plants rapidly wilting and dying, often without turning yellow, as plants dry out they may turn straw yellow in color.
- ☆ Small black fungal bodies (sclerotia) may be present on the surface of the root just below the soil line together with white fluffy mycelium.
- ☆ Water soaked lesions may be present on the stem infected tissues dry out and may become covered in white mycelium.

Fungus

Sclerotinia sclerotiorum (Lib.) de Bary

Favourable conditions: Warm and humid conditions.

Management

- ☆ Plant only disease free material.
- ☆ Adopt crop rotation with non-host crops.

2. Downy Mildew

Symptoms

Leaves yellowing; white-gray fuzzy or downy growth developing on leaves.

Fungus

Peronospora lamii A. Braun

Favourable conditions: Cool, humid weather and prolonged periods of wetness on leaves.

Management

- ✰ Promote good air circulation around plants by using adequate spacing
- ✰ Avoid wetting the foliage by watering the plant at the base.

3. Crown Gall

Symptoms

Galls of various sizes on roots and root crown below the soil line; galls may occasionally grow on the stems; galls are initially light colored bulges which grow larger and darken; galls may be soft and spongy or hard; if galling is severe and girdles the stem then plants may dry out and die.

Bacterium

Agrobacterium tumefaciens (Smith and Townsend) Conn.

Spread: Disease enters through wounds on plant.

Management

- ✰ Plant disease-free material.
- ✰ Plant rosemary in well-draining soils.
- ✰ Avoid wounding the plants as much as possible.
- ✰ Avoid wounding in the crown and roots during cultivation.
- ✰ Use the biocontrol agent, *Agrobacterium radiobacter* K84.

23. Safed Musli (*Chlorophytum borivilianum* Santapau and R.R.Fern.)

1. Leaf Blight

Symptoms

- ✰ Disease appears as minute pinhead size, circular, reddish brown spots on the leaves.
- ✰ It forms longitudinal streaks along the midribs, veins and margins of infected leaves.

- ☆ Severely affected leaves dry prematurely and such plants fail to produce healthy tubers.
- ☆ The disease is observed after the onset of rainy season.

Fungus

Colletotrichum capsici (Syd. and P. Syd.).

Alternaria alternata (Fr.) Keissl. also causes leaf blight.

Management

Spray carbendazim 0.1 per cent two times at 25 days interval.

2. Tuber Rot

Symptoms

- ☆ Tubers show rotting.
- ☆ Roots are found to rot and disintegrate.
- ☆ Drooping of leaves, wilting and death of the plant occurs.

Fungi

Fusarium solani (Mart.) Sacc. and *Rhizoctonia bataticola* Kuhn.

Mode of survival: As chlamydospores or sclerotia in tubers and roots.

Primary spread: From chlamydospores or sclerotia

Secondary spread: Through irrigation water.

Management

Treat tubers with thiram or captan 4 g/kg of tuber.

3. Collar Rot

Symptoms

- ☆ Under storage conditions the fleshy roots were found to become hollow.
- ☆ Roots are found to rot and disintegrate.

Fungi

Sclerotium rolfsii Sacc.

Mode of survival: As chlamydospores or sclerotia in tubers and roots.

Primary spread: From chlamydospores or sclerotia.

Secondary spread: Through irrigation water.

Management

The roots when treated with thiram and captan at 4g/kg before storage.

24. Saffron (*Crocus sativus* L.)

1. Corm Neck Rot/Corm Decay

Symptoms

- ☆ White spots appear on corms followed by yellowing of leaves and rotting of roots.
- ☆ Pale red or reddish violet wrinkles are seen on the corms.
- ☆ Finally corm rot and the diseased plants die.

Fungus

Rhizoctonia crocorum (Pers.) DC. and *R. violacea* Tul. and C. Tul. (Sometimes)

Mode of survival: As sclerotia in infected corms.

Primary spread: From soil-borne sclerotia.

Secondary spread: Through irrigation water.

Management

- ☆ Follow five year crop rotation.
- ☆ Dip corms in 5 per cent copper sulphate solution for 3 min.
- ☆ Plant disease-free corms.
- ☆ Remove one or two layers of protective sheath before planting.
- ☆ Remove and burn infected plants from the field.
- ☆ Drench the soil in root region with carbendazim 0.1 per cent.

2. Corm Rot

Symptoms

- ☆ Drooping of plants occur during flowering.
- ☆ Yellowing and wilting of leaves which can be easily pulled out.
- ☆ Brown discolouration central portion of corm extends into the flesh is very common.
- ☆ In severe cases, discolouration extends to surface of the corm at the nodes and surface lesions on corms are common.
- ☆ Lesions are dark brown, hard rough sunken.
- ☆ In some cases white or yellowish fungal mass is seen.
- ☆ Roots of infected corms become brown and decay.
- ☆ Flower size is reduced and the quality of stigma is also reduced.
- ☆ Severely infected corms rot before they sprout in the field and become black powdery mass with outer fibrous scales in position.

Fungus

Fusarium moniliforme var. *intermedium* Neish and M.Legg., *F. oxysporum* f.sp. *gladioli* (Massey), *F. solani* (Mart.) Sacc., *F. pallidoroseum* (Cooke) Sacc., *Rhizoctonia solani* Kuhn, *Macrophomina phaseolina* (Tassi) Goid., *Mucor* sp. and *Penicillium* sp.

Mode of survival: As macro- and micro-conidia or chlamydospore or sclerotia in infected rotten corms or in infected plant debris.

Primary spread: From infected planting materials.

Secondary spread: Through conidia or sclerotia or chlamydospores through irrigation water.

Favourable conditions: Excessive soil moisture, poor drainage, monoculturing of saffron and excessive rainfall during crop period.

Management

- ☆ Adopt soil solarisation.
- ☆ Follow deep summer ploughing.
- ☆ Plant only healthy corm (corms free from lesions).
- ☆ Soaking corms in fungicidal solution.
- ☆ Treat the corms with carboxin 300-500 g/100 kg corms before planting or treat the corms with *Trichoderma viride.*
- ☆ Treat corms in mancozeb 0.3 per cent (150 g) + carbendazim 0.1 per cent with FYM for 5-10 min in 50 lit of water before planting. Dry the corms and plant.
- ☆ Collect and destroy rotten corms or infected plants.

3. Bulb Rot

Symptoms

- ☆ The disease occur in patches.
- ☆ Brown to dark brown, sunken, irregular patches are seen below the corm scales.
- ☆ The lesions are 1 mm deep with raised margins.
- ☆ Severely infected corms show drying of foliage from tip.
- ☆ Corms rot at later stages.
- ☆ Presence of mycelium and sclerotia on the corms.

Fungus

Sclerotium rolfsii Sacc. [Sexual stage*: Athelia rolfsii* (Curzi)].

It is a soil living saprophytic pathogen, produces ribbon like hyphae with clamp connections. Basidia are club-shaped, bearing four smooth, ellipsoid basidiospores. Small, brownish sclerotia (hyphal propagules) are also formed.

Mode of survival: As sclerotia and mycelium in infected corms.

Primary spread: From infected corms.

Secondary spread: Through irrigation water, farm implements and rain splashes.

Favourable conditions: Moisture stress, high temperature and monocropping of saffron.

Management

- ✰ Follow deep summer ploughing.
- ✰ Adopt crop rotation with non-host crops.
- ✰ Plant disease free corms.

4. Rust

Symptoms

- ✰ Rust pustules are seen on leaves.
- ✰ Corms are affected by teleutospores in soil.
- ✰ Diseased leaves dry.

Fungus

Uromyces croci Pass. (called as *Fumago* sp.)

Mode of survival: As mycelium in infected corms.

Primary spread: From infected corms.

Management

Destroy diseased leaves and corms.

5. Bacterial Rot

Symptoms

- ✰ Disease appears just before flowering.
- ✰ Sheath wrapping emerging shoots show brown lesions.
- ✰ Leaves and flowers rot.
- ✰ Lesions are circular, brown with reddish brown halos.
- ✰ Reddish brown spots with chlorotic halos appear on times.
- ✰ Affected leaves bend and distal parts wither.
- ✰ Yellowing of leaves and death of plant occur.
- ✰ Reduced flowering.
- ✰ Corms are wrinkled and rot.
- ✰ Loss is heavy when diseased corms are used for planting.

Bacterium

Burkholderia gladioli pv. *gladioli* (Severini 1913) Yabuuchi *et al.*

Mode of survival: As infected plant parts like leaves and corms.

Primary spread: From infected plant parts.

Secondary spread: Through drifting soil and rain splashes.

Favourable conditions: Warm temperature, excessive irrigation or heavy rain.

Management

- ☆ Plant disease free corms in the field.
- ☆ Regulate irrigation water.

Chapter 7

Medicinal Plants

1. Aloe (*Aloe barbadensis* Mill.)

1. Rust

Symptoms

- ☆ Appearance of small and pale yellow spots on the upper surfaces of the succulent leaves.
- ☆ As the spots increase in size, masses of orange coloured spores on the undersurface.
- ☆ The lesions on the undersides of the leaves are orange –yellow to red-orange in colour.
- ☆ The centre of the spot dies and turns brown.
- ☆ It usually appears on the lowermost leaves and slowly progresses upwards.

Fungus

Phakopsora pachyrhizi Syd. and P. Syd.

Mode of survival: As overwinters in other crop hosts.

Primary spread: From infected crop debris.

Secondary spread: Through air-borne spores or by rain splashes.

Favourable conditions: Frequent rain, drizzling, dew with cool temperature and high humidity, poor quality seeds used for sowing and poor field sanitation.

Management

- ☆ Practice crop rotation.

2. Basal Stem Rot

Symptoms

- ☆ The disease causes rotting of the stem near soil level.
- ☆ The base of the affected plant will turn dark red or black and then quickly begin to rot.

Fungus

Fusarium sp.

Mode of survival: As chlamydospore in infected plant debris.

Primary spread: From infected plant debris left in field.

Secondary spread: Through soil borne chlamydospore.

Favourable conditions: Damp conditions, prolonged exposure of the stems near the soil to excessively wet condition.

Management

- ☆ Monitor soil conditions and avoid overwintering to prevent basal stem rot.
- ☆ Discard the affected plants and amend the growing conditions before planting.
- ☆ Treat the soil with fungicides to prevent further spread.
- ☆ Plant aloes in areas receiving better air circulation and at least 10 hrs. of sunlight per day.

3. Bacterial Soft Rot

Symptoms

- ☆ Infection initially occurs as water-soaked lesions at the base of the leaves (leafstalk) which is in contact with the soil.
- ☆ Infected leaves are watery and rot.
- ☆ The rotting progresses very fast and the whole plant dies within two to three days.
- ☆ The affected area becomes soft and turns dark in colour.
- ☆ Infected leaves rot internally resulting in wilting of young leaves followed by tip collapse and dieback.
- ☆ The diseased plant often has a foul colour.

Bacterium

Pectobacterium chrysanthemi (Burkholder *et al.,* 1953)

The bacterium is a motile, Gram-negative, non-sporing, straight rod with rounded ends, and occurs singly or in pairs, peritrichous flagellae.

Primary spread: From rain splashes.

Favourable conditions: Presence of high moisture or rain fall.

Management

- ☆ Provide proper drainage.
- ☆ Control nematodes and other pests that serve as vectors (carriers) of the bacteria to invade the plant tissues.
- ☆ Avoid accumulation of water in the crown region.
- ☆ Avoid overhead watering.
- ☆ Disinfect pruning tools to prevent spread of soft rot.
- ☆ Avoid plant injury during weeding.
- ☆ Remove infected plants and weed hosts immediately.
- ☆ Remove plant residues after harvest.

- ☆ Practice crop rotation with non-host crops like soybean, forage legumes and small grains.
- ☆ Spray streptomycin sulphate and tetracycline 300 g/ha + copper oxychloride 1.25 kg/ha.

4. Sooty Mould

Symptoms

- ☆ Honey dew like excretions secreted by insect pests like aphid or mealy bugs or whiteflies attracts fungal growth on leaf surface.
- ☆ Black ugly patches of growth appear on the plant surface.
- ☆ The sooty growth blocks the sunlight and inhibits photosynthesis. It makes the plant ugly but will not kill it.

Fungi

Species of *Capnodium, Cladosporium, Aureobasidium, Antennariella, etc.*

Mode of survival: As infected plant debris.

Primary spread: From movement of insects carries the inoculum.

Secondary spread: Through air-borne ascospores.

Favourable conditions: dry summer.

Management

- ☆ Control aphid or mealy bug or whiteflies with insecticidal soap and minimize sooty mould growth by using suitable fungicide.
- ☆ Use sticky traps to capture winged adults.

5. *Alternaria* Leaf Blight

Symptoms

- ☆ The disease occurs in India. *e.g.*: Symptoms appear as small, water-soaked lesion on the leaf.
- ☆ The spots turn brown and then black with yellow border.
- ☆ Spots grow rapids and show concentric rings.
- ☆ Many spots coalesce and cover considerable area of the leaf.
- ☆ When spots are formed at leaf tip it will lead to tip drying.
- ☆ Stems also show patches of brown spots with concentric ring.
- ☆ The disease reduces growth, yield and quality of gel.

Fungus

Alternaria alternata (Fr.) Keissl.

Mode of survival: As dormant mycelium in infected plant debris.

Primary spread: From infected plant debris.

Secondary spread: Through air-borne conidia.

Management

- ☆ Collect and destroy diseased leaves.
- ☆ Spray mancozeb 0.25 per cent or copper oxychloride 0.25 per cent or chlorothalonil.

6. *Drechslera* Leaf Blight

Symptoms

- ☆ Large number of circular or irregular reddish brown spots appears on the leaves.
- ☆ Each spot is surrounded by a yellow halo.
- ☆ The spots increase in size, join together and cause blighting of major leaf area.

Fungus

Drechslera halodes (Drechsler) Subramanian *et al.*

Management

- ☆ Field sanitations like removal and destruction of badly affected leaves.
- ☆ Spray mancozeb 0.2 per cent or copper oxychloride 0.25 per cent.

7. Leaf Spot

Symptoms

- ☆ Small, minute, brown spots appear on the leaves.
- ☆ The spots increase in size. They will have light centre and brown margin.
- ☆ In severe infection the spots coalesce and cause blighting and drying of leaves.

Fungus

Fusarium solani (Mart.) Sacc., *Fusarium proliferatum* (Matsushima)

Mode of survival: As chlamydospore in the soil and plant debris.

Favourable conditions: Sandy loam soil, moist weather.

Management

- ☆ Spray mancozeb 0.2 per cent or carbendazim 0.1 per cent or copper oxychloride 0.25 per cent.

2. Ashwagandha/Indian Ginseng [*Withania somnifera* (L.) Dunal]

1. Damping Off and Seedling Mortality

Symptoms

- ☆ Seedling disease.
- ☆ Brown spots appear at collar region and affected tissues soften and the seedling collapse at ground level.
- ☆ Young seedlings of 2.5 -7.5 cm are usually affected.

Fungi

Rhizoctonia solani Kuhn. and *Chromista: Pythium* sp.

Mode of survival: As sclerotia or oospores in the soil.

Primary spread: From infected plant debris.

Secondary spread: Through irrigation water.

Favourable conditions: Dampness in nursery; high seed rate; poor drainage; rainy season.

Management

- ☆ Provide proper drainage and delay sowing till rainy season.
- ☆ Use recommended seed rate.
- ☆ Seed treatment with carbendazim 2 g/kg or *Trichoderma viride* 4 g/kg or *Pseudomonas fluorescens* 10 g/kg.
- ☆ Practice crop rotation.

2. Root Rot

Symptoms

- ☆ Rotting of collar portion of the plants with necrotic water soaked lesions.
- ☆ Complete collapse the seedlings were observed in later stages.

Fungus

Fusarium proliferatum (Matsushima) Nirenberg; *Rhizoctonia solani* Kuhn.

Mode of survival: In infected plant debris.

Spread: Through air-borne conidia.

Favourable conditions: High seed rate; poor drainage; rainy season.

Management

- ☆ Collect and use seeds from disease free plants.

- ☆ Grow resistant varieties like WS90-134.
- ☆ Spray mancozeb 0.2 per cent or Rovral 0.2 per cent.
- ☆ Treat seeds with carbendazim 2 g/kg.
- ☆ Spray mancozeb 0.2 per cent or Rovral 0.2 per cent.
- ☆ Grow resistant varieties like WS 90-134.

3. Leaf Blight

Symptoms

- ☆ Light brown zonate leaf spots on leaves and flowers which coalesce and cause blighting. Necrosis of tender twigs and drying of twigs tip downwards.
- ☆ Dried twigs are greyish brown or straw coloured.

Fungus

Alternaria tenuis Nees

Mode of survival: In infected plant debris.

Spread: Through air-borne conidia.

Favourable conditions: Temperature 22-25°C.

Management

- ☆ Collect and use seeds from disease free plants.
- ☆ Grow resistant varieties like WS90-134.
- ☆ Spray mancozeb 0.2 per cent or Rovral 0.2 per cent.
- ☆ Treat seeds with carbendazim 2 g/kg.
- ☆ Spray mancozeb 0.2 per cent or Rovral 0.2 per cent.
- ☆ Grow resistant varieties like WS 90-134.

4. Die back

Symptoms

- ☆ Branches or top of the plant wither and dry.
- ☆ Necrosis of tender twigs from tip downwards.
- ☆ Dead branches become straw coloured or greyish brown.

Fungus

Colletotrichum sp.

Mode of survival: In diseased twigs.

Spread: Through air-borne conidia.

Management

- ☆ Remove and destroy diseased and dried twigs.
- ☆ Spray carbendazim 0.1 per cent or mancozeb 0.2 per cent or zineb 0.2 per cent or Rovral 0.2 per cent at 10 days interval.

5. Leaf Spot

Symptoms

- ☆ Dirty yellow to brown, water soaked spots on leaves.
- ☆ Each spot has brown to violet margin with a yellow halo.
- ☆ Spots enlarge, coalesce, become papery, brittle and show cracking.
- ☆ Minute black fruit bodies called sporodochia are prominently seen at centre of spots on the lower surface of leaves.

Fungus

Myrothecium roridum Tode; *Alternaria alternata* (Fr.) Keissler.

Mode of survival: In infected leaves.

Spread: Through air-borne conidia.

Favourable conditions: Rainy season.

Management

- ☆ Remove and destroy diseased and dried twigs.
- ☆ Spray carbendazim 0.1 per cent or mancozeb 0.2 per cent or zineb 0.2 per cent or Rovral 0.2 per cent at 10 days interval.

6. Leaf Curl

Symptoms

- ☆ Vein clearing, curling, puckering, wrinkling and rolling of leaves.
- ☆ Affected leaves become malformed and asymmetric.
- ☆ Stunting of plants.
- ☆ Wrinkling is common but leaves are not brittle.
- ☆ Flowering and seed setting reduced.

Causal Agent

Tobacco leaf curl virus

Management

- ☆ Rogue out virus diseased plants.
- ☆ Spray dimethoate 0.05 per cent or acephate 0.05 per cent or dimethoate 0.05 per cent or neem oil to control vector and reduce disease spread.

7. Witches' Broom

Symptoms

- ☆ Proliferation auxiliary buds.
- ☆ Production of large number of small leaves.
- ☆ Excessive branching and shortening of internodes.
- ☆ Stunting of plants and premature drying of twigs.

Causal Agent

Candidatus Phytoplasma strains

'*Withania somnifera*' phytoplasma.

'*Withania somnifera*' witches'-broom phytoplasma.

Spread: Through grafting.

Favourable conditions: Winter season.

Management

- ☆ Rogue out diseased plants.
- ☆ Foliar spray with tetracycline suppresses the disease symptoms after 15-20 days.

3. Asparagus (*Asparagus racemosus* L.)

1. Fusarium Crown, Root and Lower Stem Rot

Symptoms

- ☆ Infection more likely when plants are stressed by drought.
- ☆ Russet-colored lesions on roots, lower stems and or crowns.
- ☆ Yellowing of fern of infected plants.
- ☆ Damping-off of seedlings in crown nurseries.
- ☆ Crown death in nursery and commercial fields.

Fungi

Fusarium oxysporum f. sp. *asparagi* Cohen; *Fusarium proliferatum* (Mats.) Nireb.

Spread: Soil, crop debris, seed or nursery crowns.

Favourable conditions: Light sandy, well-drained soil, pH 6.5-7.5.

Management

- ☆ Avoid root injury during intercultural operations.
- ☆ Affected debris destruction.
- ☆ Summer ploughing.

- ☆ Crop rotation.
- ☆ Soil sterilization.
- ☆ Apply *Pseudomonas fluorescens* 2.5 kg/ha along with 50 kg of well decomposed FYM or neem cake.

2. Leaf spot

Symptoms

- ☆ Infection is characterized by elongated greyish-white to greyish-brown spots originating from leaf tips.

Fungus

Colletotrichum asparagi Sacc.

Acervuli produce numerous setae with pointed tips. Conidia are single celled, oblong, hyaline with obtuse ends, guttulate.

Management

- ☆ Avoid closer planting.
- ☆ Spray carbendazim 0.2 per cent or chlorothanil 0.2 per cent.

3. Cercospora Canker and Leaf-Spot

Symptoms

- ☆ Brown, irregular spots with indistinct margins appear on the stem and leaves.

Fungus

Cercospora asparagi Sacc.

Conidiophores are amphigenous, borne in clusters, dark, simple, geniculate. Conidia are hyaline, multicelled with many transverse septa.

Favourable condition: Mild winter climate.

Management

- ☆ Avoid closer planting.
- ☆ Spray carbendazim 0.2 per cent.

4. Phomopsis Canker

Symptoms

- ☆ Dirty white, circular to irregular spots with distinct, wavy, blackish-brown margins appear on stems and leaves.
- ☆ The infected plants wilt and die.

Fungus

Phomopsis asparagi Sacc.

Conidia are many, hyaline and of two types *viz.*, alpha-conidia and beta-conidia.

Mode of survival: As infected stems buried in the soil.

Spread: Infected plant debris.

Management

- Plant disease free seedlings.
- Spray with mancozeb 0.2 per cent in the nursery.
- Spray difolatan 0.2 per cent or carbendazim 0.1 per cent in main field.

5. Purple Spot

Symptoms

- Sunken, purple, oval-shaped lesions that develop on asparagus spears.
- Tan to brown lesions on the fern, including the needlelike leaves (cladophylls).
- May expand, coalesce and cause defoliation.

Fungus

Stemphylium vesicarium (Wallr); [Sexual stage *Pleospora herbarum* (Pers.)]

Sexual stage of the fungus produces pseudothecia.

Primary spread: Pseudothecia release ascospores via rain splash and wind.

Secondary spread: Infected plant debris.

Management

- Burn crop debris in late fall or winter.
- Till to bury crop debris.
- Spray difolatan 0.2 per cent or carbendazim 0.1 per cent or mancozeb 0.2 per cent or propiconazole 0.1 per cent in main field.

6. Asparagus Virus I and II

Symptoms

- Asparagus viruses I and II produce no distinct symptoms unless both viruses infect the plant.
- When both are present in the same plant, survival and vigor are severely reduced, especially in young plants.

Viruses

Asparagus virus I and *Asparagus virus II.*

Transmission

Aphids (both viruses); *Asparagus virus II* is also transmitted through seed and may be transmitted in pollen from male plants to seed produced by female plants.

Management

- ✰ Plant virus free seed grown from healthy plants or plant transplants grown from tissue culture to eliminate the viruses.

4. Bael or Wood Apple (*Aegle marmelos* Correa)

1. Bacterial Shot Hole or Fruit Canker

Symptoms

- ✰ The spots on the leaves circular, water soaked, surrounded by yellow halo.
- ✰ Later the spots enlarge and form saucer like depression on the centre.
- ✰ The presence of shining beads and scales of bacterial ooze on both sides of leaf spots.
- ✰ The cankerous spots are with raised margin and a sunken centre (crater-like).

Bacterium

Xanthomonas campestris pv. *bilvae* Patel

Management

- ✰ To get rid from this disease prune twigs, branches and burn them.
- ✰ Application of streptomycin sulphate (20g/100 lit of water) + copper oxychloride (0.3 per cent) at the interval of 10-15 days.

2. Shell Soft Rot

Symptoms

- ✰ The affected fruits rotted quickly and not fit for consumption.
- ✰ Lesions developed as water-soaked light brown rot patches with dark brown margins and a gelatinous texture.
- ✰ The rot developed quickly, and on the affected portion, the gelatinous layer was easily removable.

Fungi

Syncephalastrum racemosum Cohn

Management

- ✰ To get rid from this disease prune twigs, branches and burn them.

3. Leaf Blight

Symptoms

- Initially irregular and light brown colour spot on the leaves, later turned into dark brown to gray.
- Several spots coalesce to form large necrotic spots and cover large area.

Bacterium

Alternaria alternata (Fr.) Keissl.

Management

- Spray propiconazole 0.1 per cent or Difenconazole 0.1 per cent or Copper oxychloride 0.2 per cent or Mancozeb at 0.2 per cent.

4. Black Mildew

Symptoms

- Black coloured fungal encrustations appear on the both surface of the leaves.
- Several spots coalesce to form large necrotic spots and cover large area.

Bacterium

Schiffnerula girijae Hosag. and Archana

Management

- Spray wettable sulphur + chlorpyriphos/methyl parathion + gum acacia (0.2 + 0.1 + 0.3 per cent).

5. Belladonna/Deadly Nightshade (*Atropa belladonna* L.)

1. *Pythium* Root Rot

Symptoms

- Sudden wilting and death of plant.
- Basal stem and roots become brown and disintegrate rapidly.
- Roots rot and become black and show mycelium on their surface.

Fungus

Pythium butleri Subraman.

Mode of survival: As oospores.

Primary spread: From oospores in the soil.

Secondary spread: Through air-borne sporangia and zoospores.

Favourable conditions: High soil moisture.

Management

- ☆ Follow crop rotation.
- ☆ Fumigate soil with methyl bromide.
- ☆ Treat seed with captan 2 g/kg.
- ☆ Drench nursery soil with copper oxychloride 0.3 per cent.
- ☆ Avoid excess irrigation.

2. *Fusarium* Root Rot

Symptoms

- ☆ More destructive in grown up crop compared to nursery.
- ☆ Pre-and post-emergence damping off.
- ☆ Wilting of plants just before flowering or after flowering.
- ☆ Drooping, yellowing, wilting and death of plants in the field.
- ☆ Browning of collar region spreading up and down.
- ☆ Extensive disintegration of critical fields and down resulting in wilting of the plants.

Fungus

Fusarium solani (Mart.) Sacc.

Mode of survival: As cchlamydospores in plant debris in soil.

Primary spread: From soil borne chlamydospores.

Secondary spread: Through irrigation water.

Favourable conditions: Sandy loam soil, temperature of 30-32°C and relative humidity of 90-92 per cent.

Management

- ☆ Follow crop rotation.
- ☆ Pull out diseased plants with roots.
- ☆ Drench the soil with carbendazim 0.1 per cent.

3. *Phytophthora* Root Rot

Symptoms

- ☆ Drooping of younger leaves and branches.
- ☆ Yellowing of older leaves in affected plants.
- ☆ Complete or partial wilting and drying of leaves.

- Infection at collar region spreads up the stem and down the roots and dirty brown discolouration of roots.

Fungus

Phytophthora nicotianae var. *nicotianae* Breda de Haan

Mode of survival: As oospores in infected plant debris in soil.

Primary spread: From ooospores in soil.

Secondary spread: Through irrigation water.

Management

- Grow seedlings in raised bed.
- Follow crop rotation.
- Treat seeds with captan 2 g/kg.
- Avoid excessive irrigation.
- Drench Bordeaux mixture 1 per cent.

4. Collar Rot

Symptoms

- Collar rot is a diverse especially in the nursery.
- Elongated water-soaked lesions at collar region causes girdling of stem and toppling of seedling.
- Decay of roots-drooping of foliage and wilting and death of seedling.

Fungus

Rhizoctonia solani Kuhn

Mode of survival: As sclerotia in infected plant debris.

Primary spread: From sclerotia in infected plant debris in the soil.

Secondary spread: Through irrigation water and farm implements.

Favourable conditions: Warm weather and temperature of 28-32°C.

Management

- Deep summer ploughing.
- Avoid disease prone area for nursery area.
- Use optimum seed rate.
- Treat seeds with captan 2g/kg.
- Follow crop rotation.
- Soil drenching with carbendazim 0.1 per cent.

5. Damping Off

Symptoms

- ☆ Pre-emergence and post-emergence damping off.
- ☆ Rotting of seeds and death of seedlings before emergence.
- ☆ Water soaked brown lesions on the hypocotyl region, falling of seedling on the ground and death.

Fungi

Pythium ultimum Trow, *P. debaryanum* R. Hesse

Phytophthora parasitica (Pers.) de Bary and *Rhizoctonia solani* Kuhn.

Mode of survival: As Oospores or sclerotia.

Primary spread: From soil-borne Oospores or sclerotia.

Secondary spread: Through irrigation water.

Favourable conditions: Crowding of seedlings; Waterlogging in the nursery.

Management

- ☆ Sow healthy seeds in the raised beds in the nursery.
- ☆ Adopt optimum seed rate.
- ☆ Treat seeds with captan or thiram or metalaxyl.

6. Leaf Spots

a. *Ascochyta* Leaf Spot (Leaf Necrosis)

- ☆ Greyish white, irregular spots with a slight depression on the upper surface of leaves.
- ☆ Spots increase in size, coalesce and become necrotic.
- ☆ Diseased leaves dry and fall off prematurely.
- ☆ Pin head sized pycnidia are found on the leaf spots.

b. *Cercospora* Leaf Spot

- ☆ Brown spots with chestnut brown margin appear on the leaves.
- ☆ Spots may be circular or angular.
- ☆ Severely infected leaves show yellowing and then drying.

Fungi

Ascochyta atropae Bres. and *Cercospora atropae* Kwashn

Mode of survival: As mmycelium and spores on leaves.

Primary spread: From Pycnidia or conidia.

Secondary spread: Through air-borne pycnidiospores or conidia.

Favourable conditions: Poor nutritional management.

Management

- Spray copper oxychloride 0.25 per cent or carbendazim 0.1 per cent 6 times at 15 days interval.

7. Belladonna Mosaic

Symptoms

- Mosaic symptoms of green and light green patches on the leaves.

Causal Agent

Belladonna mosaic virus

Primary spread: From diseased seedlings.

Secondary spread: Through sap and through nematode.

Management

- Use disease free seedlings.
- Roughening and destruction of diseased plants.
- Avoid wounds to plants during cultivation.

8. Belladonna Mottle

Symptoms

- Severe mosaic symptoms.
- Mottling and vein banding.
- Blistering and deformation of leaves.

Virus

Belladonna mottles virus (BML)

Primary spread: From diseased planting material.

Secondary spread: Through sap and by beetle.

Management

- Use disease free seedlings.
- Roughening and destruction of diseased plants.
- Avoid wounds to plants during cultivation.

6. Black Nightshade/Makoi (*Solanum nigrum* L.)

1. Powdery Mildew

Symptoms

- ☆ The symptom appears on leaves as powdery patches which enlarge and cover the whole leaf resulting in drying up of leaves.

Fungus

Leveillula taurica (Lév.) G. Arnaud.

Mode of survival: As cleistothecia in infected plants.

Primary spread: From infected plants.

Secondary spread: Through air-borne conidia.

Management

- ☆ Spray dinocap (Karathane) 0.05 per cent or wettable sulphur 0.3.

Other Diseases of Black Shadenight

2. **Tomato mosaic virus (ToMV)**
3. **Potato virus Y (PVY)**
4. **Chilli veinal mottle virus**

7. Cinchona (*Cinchona officinalis* L.)

1. Seedling Blight

Symptoms

- ☆ It is a destructive disease in the nursery.
- ☆ Disease starts as discolouration in collar region of seedling.
- ☆ Disease gradually spreads to cotyledons which become flaccid and turn downwards.
- ☆ Leaves become yellow, lose turgidity and fall down.
- ☆ Discoloured area rots, seedling droops and die.
- ☆ Under moist conditions copious mycelial growth is seen.

Chromista

Phytophthora palmivora (E. J. Butler)

Mode of survival: In infected plant debris as oospores.

Primary spread: From ooospores in soil.

Secondary spread: Through zoospores in irrigation water.

Favourable conditions: High soil moisture and temperature of 24°C.

Management

- ✰ Remove and destroy diseased seedling.
- ✰ Spray at weekly intervals with copper oxychloride.
- ✰ Drench the soil with Bordeaux mixture 1 per cent or copper oxychloride 0.3 per cent.
- ✰ Shift the nursery site.

2. Die back/Tip Blight

Symptoms

- ✰ Necrosis and death of tissues the tip portion (tip blight or die back) and leaves of young seedlings.

Chromista

Phytophthora parasitica Dastur

Mode of survival: As dormant mycelium and oospores in soil.

Primary spread: From infected plant debris.

Secondary spread: Through soil borne oospores.

Favourable conditions: High humidity and temperature..

Management

- ✰ Cut and remove diseased and dried tips.
- ✰ Spray Bordeaux mixture 1 per cent.

3. Pink Disease

Symptoms

- ✰ Small twigs and branches are affected.
- ✰ Leaves become yellow.
- ✰ Branches wither and dry and are covered with rose or white crust. Crust is more pronounced in shaded sides.
- ✰ Crackings are seen on the under surface of infected branches.
- ✰ Pink layer becomes ochraceous or white with age.
- ✰ Canker may be covered in pinkish fruiting structures.

Fungus

Corticium salmonicolor Berk. and Broome., [Syn. *Pellicularia salmonicolor* (Berk. and Br.) Dastur], *Erythricium salmonicolor* Berk. and Broom.

Mode of survival: As dormant mycelium around canker or in diseased bark.

Primary spread: From dormant mycelium in plant debris and soil.

Secondary spread: Through fruiting structures and spores spread by air.

Favourable conditions: Humid atmosphere and shady conditions.

Management

- Free air circulation, reduced shade, thin planting and good drainage.
- Cut diseased branches and destroy by fire.
- Prophylactic spray with Bordeaux mixture 1 per cent.
- Timely pruning of trees.

4. Canker

a. Stripe Canker

Symptoms

- Infection starts in the roots and progress upward through the union and into the stem of the tree.
- Vertical, linear and slightly sunken stripes of dead and dying inner barks are spectacular.
- Stripe length and with vary but in advanced stage stripes extend to entire length of the stem.
- Multiple stripes develop on the stem.
- When the disease spreads upwards, droplets of dark sticky exudate is seen on the surface of infected bark.
- Multiple infection on root is common.
- Leaves become yellow.

Chromista

Phytophthora cinnamomi Rands

Mode of survival: As survives in infected plant materials in soil.

Primary spread: From dormant mycelium in soil.

Secondary spread: Through soil borne oospores.

Favourable conditions: Water stagnation in field.

b. Top Blight Girdle Canker

Symptoms

- Infection starts through leaves and petioles of small shoots at the base of scion. The disease spread with bark of main stem causing girdling.
- The pathogen girdles the stem.
- Bark of the scion above the graft union is killed and become sunken in a horizontal band.

- Sunken appearance is accentuated by inrolling of callus tissue above and below dead area.
- Canker spread upto the stem scion.
- Leaves turn yellow twigs and branches show gradual die- leading to death of trees.
- In some cases dead shoots remain attacked to scion at the site of girdled lesion.

Chromista

Phytophthora parasitica Dastur

Mode of survival: As survives in infected plant materials.

Management

- In both stripe and girdle cankers destroy infected trees.
- Spray prophylactically with copper oxychloride.
- Combination of cropping techniques (*e.g.* Cover crops) should be practiced to avoid infestation.
- *Phytophthora cinnamomi* can be avoided by grafting on *C. pubescens* root stock.

5. Damping Off

Symptoms

- Rotting of roots and water soaked lesion just below the soil line on the stem.
- Reducing dropping and defoliation of leaves.
- Infected roots become black and die.
- Root barks slough off.
- Mycelium is often seen on cortex and even in xylem vessels.

Chromista

Pythium vexans de Bary., *Phytophthora* sp.

Fungi

Rhizoctonia solani Kuhn, *Macrophomina phaseolina* (Tassi) Goid.

Fomes lamoensis (Murr.) Sacc. and Trott.

Mode of survival: As sclerotia or oospores in soil.

Primary spread: From infected plant debris in soil.

Secondary spread: Through irrigation water.

Favourable conditions: Poor drainage, excess seed rate.

Management

- ✰ Avoid excess shades and seed rate.
- ✰ Provide prop and drainage.
- ✰ Remove and destroy diseased seedlings.
- ✰ Treat seeds with fungicides like captan or thiram.
- ✰ Spray with Bordeaux mixture, and fumigation with chloropicrin.
- ✰ Soil application of thiram 20 kg/ha.
- ✰ Spray seedlings with Zineb 0.2 per cent or other copper fungicides 0.2 per cent.
- ✰ Soil drenching with copper oxychloride 0.3 per cent.
- ✰ Shift nursery site to a new location.

6. Root and Collar Rot

Symptoms

- ✰ Seedlings in the nursery are affected.
- ✰ Roots and collar regions are affected and shows cottony web of mycelial growth with brown, round sclerotia.
- ✰ Diseased seedlings die rapidly.
- ✰ Collar rot in young trees.

Fungus

Sclerotium rolfsii Sacc. [Sexual stage: *Athelia rolfsii* (Curzi)].

It is a soil living saprophytic pathogen, produces ribbon like hyphae with clamp connections. Basidia are club-shaped, bearing four smooth, ellipsoid basidiospores. Small, brownish sclerotia (hyphal propagules) are also formed.

Mode of survival: As sclerotia in infected plant materials.

Primary spread: From infected plant material in soil.

Secondary spread: Through irrigation water.

Favourable conditions: High soil moisture, more undecomposed organic wastes in the soil.

Management

- ✰ Remove and destroy diseased seedlings.

8. Costus (*Costus speciosus* Koen ex. Retz)

1. Rhizome Rot

Symptoms

- ✰ Initially, the symptoms start from the tip of the injured portion of the rhizomes.

- ✰ Subsequently, the rhizome becomes light brown and gives off an offensive odour.

Fungus

Fusarium solani

Management

- ✰ Properly harvested rhizomes without injuries can be stored for a long time.

2. Pythium Rhizome Rot

Symptoms

- ✰ Infection starts from the injured portion of the rhizome and in later stages the rhizomes turn dirty brown, the leaves become yellowish-brown and finally dry off.

Fungus

Pythium spirosum

Management

- ✰ The prudent selection of rhizomes from the healthy crop and dipping the rhizomes in copper oxychloride or mancozeb.

3. Leaf Blight

Curvularia paradissi

- ✰ The symptoms on the leaf laminae as small, spherical to irregular spots, light brown in colour and in the advanced stages they become dark brown.
- ✰ The upward cutting of young leaves is very common. In the advanced stages all the leaves fall off, leaving bare stems in the fields.

Drechslera maydis

The symptoms as water soaked spots, which gradually enlarge and become necrotic area.

Management

Spray copper oxychloride or mancozeb at fortnight intervals.

9. Dioscorea (*Dioscorea alata* L.)

1. Damping Off

Symptoms

- ✰ Pre-emergence and post-emergence damping off of seedlings.

- ☆ Tubers or crowns fail to germinate.
- ☆ Rotting of tubers is also seen.

Fungus

Pythium spp.

Mode of survival: As oospores in infected plant materials.

Primary spread: From oospores and mycelia in plant debris.

Secondary spread: Through irrigation water, wind-blown sporangia or zoospores.

Favourable conditions: Water logging and wet soil, planting material from weak plants.

Management

- ☆ Use healthy tuber.
- ☆ Follow raised bed nursery.
- ☆ Provide proper drainage especially in planting.
- ☆ Dip planting material in carbendazim solution.
- ☆ Drench the soil with copper oxychloride 0.3 per cent.

2. Tuber Rot

Symptoms

- ☆ Tuber rot is a disease in the field and store house.
- ☆ Purplish brown spots on the tubers cause necrosis of tissues and rotting.
- ☆ Rotten tissues of tuber become soft and wet.
- ☆ Under humid conditions rotting is rapid and develop thick mass of black conidia of the fungus.

Fungus

Aspergillus niger Tiegh

Mode of survival: As infected tubers.

Primary spread: From infected tubers.

Secondary spread: Through contact with diseased tubers and wind-borne conidia.

Favourable conditions: Wounding of tubers, storing healthy tubers with diseased tubers.

Management

- ☆ Prevent wounds on tubers during farm operations and during harvesting.
- ☆ Remove diseased tubers after harvest and store only healthy tubers.

- ✰ Maintain proper hygiene of store house.
- ✰ Dip tubers in carbendazim 0.1 per cent for 5 min.

3. Tuber Rot

Symptoms

Rotting of tubers in the field.

Fungus

Penicillium purpurogenum Stoll

Mode of survival: In infected plant debris.

Primary spread: From diseased plant debris in nursery.

Secondary spread: Through wind-borne spores.

Favourable condition: Wounding, temperature 10-35°C and relative humidity of 90-100 per cent.

Management

- ✰ Avoid wounding to tubers.
- ✰ Remove and destroy of infected plant materials.

4. Leaf Spots

a. *Cercospora* Leaf Spot

Symptoms

- ✰ Dark brown, irregularly angular spots with indefinite margin are scattered on the leaves.
- ✰ Severe infection causes premature defoliation.
- ✰ Size and yield of tubers are reduced.

Fungus

Cercospora dioscoreae-pyrifoliae J.M. Yen

Mode of survival: As infected plant debris.

Primary spread: From infected plant debris.

Secondary spread: Through wind-borne conidia.

Management

Spray carbendazim 0.1 per cent or zineb 0.2 per cent.

b. *Drechslera* Leaf Spot

Symptoms

- ✰ Transparent spots appear on the leaves which later turn brown with dark brown margin on the seedlings.

- ☆ Spots increase in size and coalesce to cause drying of leaves and premature defoliation.

Fungus

Drechslera sorokiniana (Sacc.) Subram. and B.L. Jain

[Syn.: *Cochliobolus sativus* (S. Ito and Kurib.)]

Mode of survival: As infected plant parts.

Primary spread: From infected pant parts.

Secondary spread: Through air-borne conidia.

Management

Spray carbendazim 0.1 per cent or zineb 0.2 per cent.

5. Leaf Blight

Symptoms

- ☆ Light brown, small, circular spots appear on the leaves.
- ☆ Spots increase in size, coalesce and firm reddish or dark brown to black patches and blight the leaves.
- ☆ Diseased leaves dry.
- ☆ Under humid conditions brown to black masses of conidia are seen on the spots.

Fungus

Glomerella cingulata Stoneman

Mode of survival: As infected plant parts.

Primary spread: From infected plant parts.

Secondary spread: Through wind-borne conidia.

Management

Spray mancozeb or zineb 0.2 per cent or copper oxychloride 0.3 per cent.

6. Leaf Fall/Die Back

Symptoms

- ☆ Drying of young leaves and branches from the tip downwards.
- ☆ Heavy defoliation with die back symptom.
- ☆ Severe infection causes death of the plant.
- ☆ Numerous, small, black acervuli are seen on the barks of dried stem.

Fungus

Colletotrichum sp.

Mode of survival: In infected plant debris.

Primary spread: From infected plant debris.

Secondary spread: Through wind-borne conidia.

Management

- ☆ Maintain proper spacing.
- ☆ Cut and remove diseased branches.
- ☆ Spray mancozeb 0.2 per cent or carbendazim 0.1 per cent at 15 days interval.

7. Dioscorea Latent Virus (DLV)

Symptoms

- ☆ DLV with dioscorea green banding virus causes mosaic and green banding in leaves.

Causal Agent

Dioscorea latent virus (DLV)

Transmission

Through sap and not through seed or insect vector.

Management

Remove and destroy diseased plants.

10. Fox-Glove [*Digitalis purpurea* L. (*Digitalis lanata, D. grandiflora* and *D. thapsi*)]

1. Anthracnose

Symptoms

- ☆ Purple brown, circular or angular spots of 1-4 mm dia occur on the leaves. Sometimes the spots are irregular.
- ☆ On veins and petioles lesions are small, sunken fusiform and blackish brown in colour.
- ☆ The leaf spots increase in size and more number of spots may be formed on the leaves.
- ☆ Older spots are light brown and are studded with acervuli.
- ☆ Diseased leaves turn yellow, wither and drop off.
- ☆ Often the diseased plants are stunted and die before harvest.
- ☆ Infected seeds do not emerge from soil.

- ☆ Numerous small, black fruiting bodies (acervuli) are seen at the centre of spots.

Fungus

Colletotrichum fuscum Laubert.

Mode of survival: In seeds and infected plant parts.

Primary spread: From infected plant debris.

Secondary spread: Through air-blown rain splashes.

Favourable conditions: Warm wet weather.

Management

- ☆ Sow disease free seeds.
- ☆ Treat seed in hot water at 55°C for 15 minutes.
- ☆ Avoid denser planting and overhead irrigation or adjust irrigation.
- ☆ Maintain plant vigour.
- ☆ Remove infected plant materials and burn.
- ☆ Spray azoxystrobin or mancozeb 0.2 per cent.
- ☆ Adopt crop rotation.

2. Leaf Spot

Symptoms

- ☆ Attacks seedlings and on young plants spots with purple margin occur on leaves.
- ☆ They become irregular and necrotic with numerous minute pycnidia on the spots.
- ☆ The disease is severe on cotyledonary stage of seedlings on young plants.

Fungus

Septoria digitalis Pass.

Mode of survival: In infected plant materials.

Primary spread: From infected plant debris.

Secondary spread: Through air-borne conidia.

Favourable conditions: Warm wet weather.

Management

- ☆ Spray propiconazole 0.1 per cent.

3. Leaf Spot

Symptoms

- ☆ The disease attacks lower leaves and move upwards.
- ☆ Spots are irregular, snuff, brown with a reddish border and 6 mm in dia.
- ☆ Heavy incidence results in drying of leaves and death of plants.
- ☆ Spots may show white mouldy appearance due to abundant conidia.

Fungus

Ramularia variabilis Fuckel.

Mode of survival: In infected plant materials.

Primary spread: From infected plant debris.

Secondary spread: Through air-borne conidia.

Favourable conditions: Prolonged wetting in the leaf surface.

Management

- ☆ Remove and destroy diseased plants.
- ☆ Practice crop rotation.
- ☆ Avoid dense plantings, and time irrigation to minimize the time leaves are wet.
- ☆ Spray chlorothalonil or thiophanate methyl.

4. Leaf Blight

Symptoms

- ☆ Brown spots appear on the leaves, which enlarge in size.
- ☆ Coalesce and cause blighting and drying of leaves.

Fungus

Alternaria sp.

Spread: Through wind-borne conidia.

Management

Spray copper oxychloride 0.25 per cent or copper hydroxide 0.1 per cent.

5. Downy Mildew

Symptoms

- ☆ Light brown rectangular areas are found vein-limited on the leaves.
- ☆ The spots become chlorotic and the corresponding lower leaf surface will exhibit purple-grey fungal mycelium and sporangia.

- ☆ Fungal growth can at times seen on upper leaf surface.
- ☆ The spots increase, coalesce and cover entire leaf area and turn necrotic.

Chromista

Peronospora digitalidis Gaum

Conidiophores are hyaline, aseptate, branching dichotomously. Conidia are hyaline to gray, ovoid to ellipsoid.

Mode of survival: As infected plant debris.

Primary spread: From oospores in infected plant debris.

Secondary spread: Through sporangia/Zoospores are spread by air or rain splashes.

Favourable conditions: Cool wet weather.

Management

- ☆ Sow certified disease free seed.
- ☆ Avoid overhead irrigation.
- ☆ Provide optimum spacing for good air circulation.
- ☆ Remove and destroy diseased plants and plant debris.
- ☆ Spray mefonoxam or insignia.

6. Ribgrass Mosaic Virus (RMV)

Symptoms

- ☆ Chlorotic mottling and ring- like marking.
- ☆ Chlorotic streaks along the veins and slight twisting of the petioles.

Causal Agent

Ribgrass mosaic virus.

Spread: Mechanical transmission. No vector.

Favourable conditions: High temperature or summer season.

Management

- ☆ Remove and destroy diseased plants.

11. Glory Lily (*Gloriosa superba* L.)

1. Tuber Rot

Symptoms

- ☆ It affects the underground tuber resulting in death of the plant.
- ☆ Infected tubers become soft and the foliage exhibits yellowing.

- In advanced stages, the whole tuber gets infected giving an appearance of discoloured mass and plant dies off.

Fungus

Sclerotium spp.

Mode of survival: As sclerotia in soil and plant debris.

Primary spread: From ascospores.

Secondary spread: Through soil borne sclerotia.

Favourable conditions: Warm weather.

Management

- Provide good drainage facilities.
- Decrease the frequency of irrigation.
- Soil application of *Trichoderma viride at* 5 kg/ha. Tuber treatment with *pseudomonas fluorescens* 0.2 per cent for 20 minutes.
- Soil drenching with carbendazim 0.1 per cent or *Bacillus subtilis* 2 g/l at 30 days after sowing.

2. Leaf Blight

Symptoms

- Small, pale to brown, irregular or round spots appear on the leaves.
- Each spot has a central necrotic lesion with concentric rings.
- In advance stages, several spots coalesce together to form large blighted areas.

Fungus

Alternaria alternata (Fr.) Keissl.

Mode of survival: In infected plants and collateral hosts.

Primary spread: From infected plants and collateral hosts.

Secondary spread: Through air-borne conidia.

Favourable conditions: Moist warm weather.

Management

- Remove the infected plant debris from the field.
- Spray mancozeb 0.2 per cent or propiconazole 0.1 per cent or hexaconazole 0.1 per cent at the time of infection.
- Spray 0.5 per cent ZnSo4 two times from 45 DAP at fortnightly interval.

3. Leaf Blight

Symptoms

- Small narrow elongated spots appear on leaves.
- Later the spots become long elliptical necrotic with straw coloured center with brown margins.
- The lesions can be several centimetres long and wide.
- In advance stages, several spots coalesce; destroy large areas giving the crop a blasted appearance.

Fungus

Curvularia lunata (Wakker) Boedijn

Favourable conditions: Cloudy days couples with high humidity.

Management

- Remove the infected plant debris from the field.
- Spray mancozeb 0.2 per cent or propiconazole or hexaconazole 0.1 per cent.
- Spray ZnSo4 0.5 per cent two times from 45 DAP at fortnightly interval.

4. Mosaic

Symptoms

- The affected leaves become chlorotic with reduced chlorophyll content.
- It also affects the flower and changes the original colour.

Causal Agent

Glorylily mosaic virus

Primary spread: From infected tubers.

Secondary spread: Through aphid vector.

Favourable conditions: Summer season.

Management

- Remove and destroy infected plants.
- Spray insecticides to control vector.

12. Indian Coleus (*Coleus forskohlii* L.)

1. Root Rot/Wilt

Symptoms

- Gradual yellowing marginal necrosis and withering of leaves followed by loss in vigour and premature death.

- ☆ Infected plants show discolouration of roots and complete decaying of tap and lateral root system.
- ☆ The bark of such plants is easily peeled off.
- ☆ Affected plants are finally killed due to severe root and collar rots.
- ☆ The infected tubers showing rotting and emit bad odour.

Fungus

Fusarium chlamydosporum Wollenw. and Reinking, *Fusarium solani* (Mart.) Sacc.

Mode of survival: As chlamydospore in the soil.

Primary spread: From infected plant debris.

Secondary spread: Through soil borne chlamydospore.

Favourable conditions: Humid weather.

Management

- ☆ Select the coleus cuttings from disease free plants.
- ☆ Treat the cuttings in carbendazim 0.1 per cent before planting.
- ☆ Drench soil with carbendazim or propiconazole 0.1 per cent.
- ☆ Apply FYM 12.5 ton/ha + 500 g neem cake/ha + *T.viride* 2.5 kg/ha before planting.

2. Root Rot

Symptoms

- ☆ Yellowing and drooping of the leaves followed by blackening of the stem, rotting of the roots and basal stem.
- ☆ Shredding of stem and root bark.
- ☆ Appearance of black sclerotia on the rotten portion.

Fungus

Macrophomina phaseolina (Tassi) Goid.

Mode of survival: As sclerotia in infected roots in soil.

Primary spread: From soil-borne sclerotia.

Secondary spread: Through irrigation water.

Management

- ☆ Select the coleus cuttings from disease free plants.
- ☆ Treat the cuttings in carbendazim 0.1 per cent before planting.
- ☆ Drench soil with carbendazim or propiconazole 0.1 per cent.
- ☆ Apply FYM 12.5 ton/ha + 500 g neem cake/ha + *T.viride* 2.5 kg/ha before planting.

3. Leaf Spot

Symptoms

- ☆ Elliptical or circular to irregular pale brown or brown coloured spots appear on leaves.
- ☆ These spots are limited by dark brown margin. They coalesce to cause extensive necrosis and yellowing.

Fungus

Corynespora cassiicola (Berk. and M.A. Curtis); *Botryodiplodia theobromae* (Pat.).

4. Leaf Blight

Symptoms

- ☆ Brown water soaked spots appear on leaves that increase rapidly in size become necrotic.
- ☆ Severe infection results in defoliation and death of the plants.

Fungus

Rhizoctonia solani Kuhn.

Mode of survival: In soil and plant debris as sclerotia.

Primary spread: From infested soil.

Secondary spread: Through irrigation water and farm implements.

Favourable conditions: Cool, cloudy weather, high humidity, wet and compact soil and overcrowding.

Management

- ☆ Select the coleus cuttings from disease free plants.
- ☆ Treat the cuttings in carbendazim 0.1 per cent before planting.
- ☆ Drench soil with carbendazim or propiconazole 0.1 per cent.
- ☆ Apply FYM 12.5 ton/ha + 500 g neem cake/ha + *T.viride* 2.5 kg/ha before planting.

5. Bacterial Wilt

Symptoms

The symptoms include initially brown later on becomes black roots due to decaying, oozing and putrefaction of roots.

Bacterium

Ralstonia solanacearum (Yubucchi *et al.*)

[Syn. *Pseudomonas solanacearum* (Smith), *Burkholderia solanacearum* (Smith)].

Mode of survival: As survives for several years in soil.

Primary spread: From soil and seed infestation.

Secondary spread: Through irrigation water.

Favourable conditions: Humid weather and high temperature.

Management

- ✰ Remove and destroy infected plants.
- ✰ Spot drench with streptocycline 300 ppm concentration.

13. Indian Pennywort (*Centella asiatica* L.)

1. Leaf Spot

Symptoms

- ✰ The foliar disease symptom appeared as minute circular to semicircular, light-brown spots which gradually occupied the entire leaf lamina; spots were dark-brown surrounded by chestbrown zone, leaves tuned yellow and defoliated prematurely.

Fungus

Cercospora centellae Manohar., Kunwar and Sharath

Mode of survival: As infected leaves and twigs.

Primary spread: From infected debris.

Secondary spread: Through air-borne conidia.

Management

- ✰ Prune affected twigs and remove severely infected and dried fruits.
- ✰ Spray thiophanate methyl at 0.1 per cent or mancozeb or chlorothalonil at 0.2 per cent.

Other Diseases of Indian Pennywort

2. **Leaf spot – *Alternaria* sp.**

3. **White rot - *Sclerotinia sclerotiorum***

4. **Southern bacterial wilt - *Ralstonia solanacearum***

14. Isabgol/Ispaghul (*Plantago ovate* Forssk.)

1. Downy Mildew

Symptoms

- ✰ Pre-emergence and post-emergence damping off are common.

- Germinating seeds rot completely in pre-emergence damping-off.
- In post-emergence damping-off the pathogen attacks the seedling and causes water-soaked brown spots at the collar regions.
- The tissues in collar region are invaded rapidly by the pathogen causing the seedlings to topple down, rot and die shortly.
- Unusual rain after seedling in the nursery favours the disease.

Chromista

Pythium ultimum Trow

Fungi

Rhizoctonia solani Kuhn and *Fusarium oxysporum* Schltdl

Favourable conditions: Poor drainage.

Management

- Seed treatment with biocontrol agent *Trichoderma harzianum* at 5 g/kg of seed controls damping off.
- Seed treatment with Chlorothalonil at 2 g/kg, Metalaxyl 5 g/kg protects the seedlings.
- Application of chlorothanil 0.2 per cent or copper oxychloride 0.3 per cent in the nursery using rose can.
- Seed treatment with fenaminosulf at 5 g/kg was effective but phytotoxic.

2. Wilt

Symptoms

- Wilt becomes an important disease in places of monoculture.
- Pre-emergence damping off or a typical wilting of plants at seed setting.
- Wilting is observed n circular patches in a field.
- Wilting starts from outer leaves. Green leaves turn into silvery colour.
- Wilting of whole plant is rapid.
- Leaves become prostrate within 30 to 48 hrs. of incipient infection and wilting.
- Black discolouration of tap roots starting from root tip is very common.
- Preserve of hyphae in xylem elements of infected roots and vascular discolouration are very common.

Fungi

Fusarium oxysporum Schlecht ex Fr. and *F. solani* (Mart.) Sacc.

Mode of survival: As chlamydospores in soil and plant debris.

Spread: Through irrigation water.

Favourable conditions: Moist wet weather.

Management

- ✰ Adopt proper crop rotation. Deep summer ploughing.
- ✰ Treat seeds with carbendazim 2 g/kg.
- ✰ Drench the soil with carbendazim 0.1 per cent or copper oxychloride 0.3 per cent.
- ✰ Avoid low lying areas for Isabgul cultivation.
- ✰ Uproot diseased plants along with roots, collect and destroy the diseased plants.

3. Downy Mildew

a. *Peronospora plantaginis* Underw

Symptoms

- ✰ Ashy-white frost like mycelial growth on the lower leaf surface.
- ✰ Corresponding upper leaf surface show chlorotic areas.
- ✰ Leaves become chlorotic and then brown.
- ✰ Leaves curl, shrink and dry.
- ✰ Seed yield highly reduced.

b. *Peronospora alta* Fuckel

- ✰ Leaves develop chlorotic streaks extending along the midrib of leaves.
- ✰ Profuse downy growth is seen on the lower surface.
- ✰ Later the whole leaf turns necrotic and dry.

Chromista

Peronospora alta and *P. plantaginis, Pseudoperonospora plantaginis* (Underw.) also cause downy mildew.

Mode of survival: As oospores in soil plant debris and in it is also seed borne.

Primary spread: From seed and soil-borne inoculum.

Secondary spread: Through irrigation water and air-blown sporangia.

Favourable conditions: Closer spacing.

Management

- ✰ Deep summer ploughing. Use optimum seed rate.
- ✰ Use disease free seeds as it is seed borne.
- ✰ Grow resistant varieties like EC 124345.
- ✰ December sown crops escape from disease.

- ☆ Seed treatment metalaxyl 5 g/kg.
- ☆ Spray metalaxyl-MZ 0.2 per cent thrice sprayed at fortnightly interval.
- ☆ Avoid frequent irrigation.

4. Powdery Mildew

Symptoms

- ☆ Powdery mildew occurs usually more at flowering stage.
- ☆ Small white or greyish powdery talcum like coating in patches seen on the upper leaf surface.
- ☆ They gradually enlarge, coalesce cover entire leaf surface become brown and cause drying of leaves.

Fungus

Erysiphe cichoracearum DC.

Mode of survival: As cleistothecia in infected plant debris.

Primary spread: From infected plant debris.

Secondary spread: Through air-borne conidia.

Favourable conditions: Disease is severe at flowering.

Management

- ☆ Spray dinocap 0.05 per cent or wettable sulphur 0.3 per cent or Rovral 0.25 per cent.

5. Leaf Spot

Symptoms

- ☆ Irregular necrotic spots appear on leaves.
- ☆ Spots increase in size, coalesce and cause drying of leaves.
- ☆ Old and matured leaves are first attacked.
- ☆ Leaves become black due to blighting symptom.
- ☆ Shrivelling and blackening of seed.

Fungus

Alternaria alternata (Fr.) Keissl.

Secondary spread: Through air-borne conidia.

Favourable conditions: Warm humid weather.

Management

- ☆ Treat seeds with carbendazim 2.5 g/kg.
- ☆ Spray mancozeb 0.2 per cent at 15 days interval.

15. Kalmegh [*Andrographis paniculata* (Burm.f.) Wall. ex Nees]

1. Wart

Symptoms

- ☆ The stem holds galls, they are small, transparent protuberance or bump and then gradually enlarged and transformed into a characteristic greenish yellow gall.
- ☆ When yellow sori converted into resting spore, the colour of galls turned into brown to dark brown.
- ☆ The inflorescence became deformed and sterile.
- ☆ The diseased plants were, deformed, stunted growth and greenish yellow to yellow in colour.
- ☆ Finally, the surface of diseased parts turned into roughed, corky, malformed, grey colour and covered with dead tissues.

Fungus

Synchytrium lepidagathidis

Mode of survival: In soil and plant debris.

Primary spread: From infested soil.

Secondary spread: Through irrigation water and farm implements.

Favourable conditions: Cool, cloudy weather, high humidity.

Management

- ☆ Avoid excessive irrigation to reduce humidity around the plants.
- ☆ Discard the diseased seedlings.

2. Leaf Web Blight

Symptoms

- ☆ The disease first appeared as black lesion either on the leaves or any tender parts generally near the ground level.
- ☆ Generally, the whole leaf became blighted and dropped.

Fungus

Rhizoctonia solani Kuhn

Mode of survival: In soil and plant debris as sclerotia.

Primary spread: From sclerotia.

Favourable conditions: Moist and humid condition.

Management

- ☆ Remove and destroy infected plants.
- ☆ Discard the diseased seedlings.

3. Root Rot

Symptoms

- ☆ The symptoms appeared as yellowing, slow wilting and gradual death of the plants. The roots became rotted and turned into black colour.

Fungus

Macrophomina phaseolina

Management

- ☆ Remove and destroy infected plants.
- ☆ Discard the diseased seedlings.

4. Yellow Vein Leaf Curl Disease

Symptoms

- ☆ The symptomatic plants that initially showed yellow veins on younger leaves.
- ☆ Later upward leaf curling, vein clearing, chlorosis, reduced leaf size, poor inflorescence, and stunted growth leading to drastic reduction in herb yield.

Virus

Eclipta yellow vein virus along with a Betasatellite associated.

Transmission

Bemesia tabaci (whitefly).

Management

- ☆ Spray any systemic insecticide.

Other Diseases of Kalmegh

5. **Witches' Broom - *Phytoplasma* (16SrII-D)**

6. **Leaf spot - *Cercospora* sp.**

16. Neem (*Azadirachta indica* A. Juss.)

1. Damping Off

Symptoms

- ☆ Nursery damage.
- ☆ Pre-emergence damping off and post-emergence damping off.

Chromista

Pythium sp., *Phytophthora* sp.

Fungi

Fusarium oxysporum and *Rhizoctonia solani* Kuhn.

Mode of survival: As oospores or chlamydospores or sclerotia in soil.

Primary spread: From soil-borne spores or sclerotia.

Secondary spread: Through irrigation water.

Favourable conditions: High soil moisture, high pH, shade, clay soil and poor drainage.

Management

Follow solar heating of potting mixture.

2. Leaf Web-Blight

Symptoms

- ✰ Greyish brown blotches appear on leaves.
- ✰ Leaves adjoining the infected leaves also get infected and webs can be seen on them.
- ✰ Along with mycelium, dark brown sclerotia are seen.
- ✰ Leaflets fall off prematurely.

Fungus

Rhizoctonia solani Kuhn.

Mode of survival: As sclerotia in infected plant debris in soil.

Primary spread: From sclerotia among the web.

Secondary spread: Through contact of overlapping leaves with infected leaves or water-borne mycelia.

Management

- ✰ Follow field sanitation.
- ✰ Spray carbendazim 0.1 per cent.

3. Root Rot

Symptoms

- ✰ Reddish brown lesions seen in the seedlings at the collar region just above soil-line.
- ✰ Death of plants with shredding of barks containing large number of small sclerotia.

Fungus

Macrophomina phaseolina (Tassi) Goid.

Mode of survival: As soil-borne sclerotia.

Primary spread: From sclerotia in soil.

Secondary spread: Through irrigation water.

Management

- Apply *Trichoderma viride* 2.5 kg + FYM 12.5 tons/ha.
- Drench the soil with copper oxychloride 0.3 per cent or carbendazim 0.1 per cent or propiconazole 0.1 per cent.

4. *Ganoderma* Wilt

Symptoms

- Drooping of branches, wilting and drying of plants.
- Presence of sporophore at the basal trunk portion of the tree after the death.

Fungus

Ganoderma lucidum (Curtis) P. Karst. and *Ganoderma applanatum*

Mode of survival: As soil-borne mycelium and basidiospores.

Primary spread: From soil-borne basidiospores.

Secondary spread: Through root contact and irrigation water.

Management

Drench the soil with copper oxychloride 0.3 per cent.

5. *Colletotrichum* Leaf Blight

Symptoms

- Blighting of leaves and premature defoliation.

Fungus

Colletotrichum gloeosporioides Pat. and Penz.

Mode of survival: In infected plant materials.

Primary spread: From infected plant debris.

Secondary spread: Through air-borne conidia.

Management

Spray copper oxychloride 0.3 per cent.

6. Leaf Spot

Symptoms

- ✰ Brown, sub-circular or irregular lesions appear with dark brown border on the mature leaves.
- ✰ Spots increase in size and coalesce turn.
- ✰ Dark brown necrotic areas and become chlorotic.
- ✰ Greyish areas with conidia are seen on the lower leaf surface during rainy season.

Fungus

Pseudocercospora subsessilis (Syd. and P. Syd.) Deighton

Mode of survival: In dormant lesions in infected plant parts.

Primary spread: From infected plant parts.

Secondary spread: Through wind-borne conidia.

Favourable conditions: Warm weather and young leaves.

Management

Spray mancozeb and Brestan.

7. Powdery Mildew

Symptoms

- ✰ Greyish powdery growth appears in patches on upper leaf surface of young leaves.
- ✰ Patches increase in size, coalesce and cover the entire leaf surface.
- ✰ Such leaves wrinkle, dry and mature leaves.
- ✰ Brown powdery patches also seen on mature leaves, twigs, flowers and fruits.
- ✰ Reduced photosynthesis and flower and fruit production.
- ✰ Shedding of young fruits.

Fungus

Oidium azadiractae A. Juss.

Mode of survival: As mycelium and cleistothecia.

Primary spread: From ascospores from cleistothecium on the infected plant parts.

Secondary spread: Through wind-borne conidia.

Favourable condition: Warm weather.

Management

- ✰ Adopt proper spacing.
- ✰ Provide adequate nutrients.
- ✰ Spray wettable sulphur 0.3 per cent or carbendazim 0.1 per cent at 10-15 days interval.

8. *Alternaria* Leaf Blight

Symptoms

- ✰ Brown, concentric spots appear on older leaves.
- ✰ Spots increase in size covering larger area of leaves causing blighting and falling of leaves.

Fungus

Alternaria alternata (Fr.) Keissl.

Primary spread: From infected plant debris.

Secondary spread: Through air-borne conidia.

Management

Spray copper oxychloride 0.3 per cent at 15 days interval.

9. Angular Leaf Spot/Bacterial Blight

Symptoms

- ✰ Water-soaked, angular spots appear on the leaves causing the leaves to become pale yellow with oily appearance.
- ✰ Shot hole and vein blight symptoms are also seen.
- ✰ Severely infected leaves dry and fall off.

Bacteria

Pseudomonas azadirachtae; *Xanthomonas azadirachtii* Moniz and Raj.

Primary spread: From infected plant debris.

Secondary spread: Through rain splashes.

Management

- ✰ Plant only disease-free saplings.
- ✰ Cut and burn severely infected plant parts.
- ✰ Apply N and P judiciously and increased K application.
- ✰ Avoid sprinkler irrigation.
- ✰ Spray copper oxichloride 0.3 per cent.

10. Die Back

Symptoms

- ☆ Young growing shoots are completely withered and dried.
- ☆ Disease results in inflorescence blight and fruit rot resulting in almost 100 per cent yield loss.

Fungus

Phomopsis azadirachtae Sateesh, Bhat and Devaki.

Mycelium is immersed, branched, septate, profuse and colourless. Conidiomata is pycnidial, solitary or aggregate. The pathogen produces two types of spores: Alpha (α) - conidia and Beta (β)- conidia. α - conidia are fertile.

Primary spread: From conidia spread thorugh rain droplets and insects.

Secondary spread: Through seed-borne conidia.

Favourable conditions: On-set of rainy season and early winter season.

Management

Spray carbendazim at 0.25 ppm and thiophanate methyl at 0.75 ppm.

Other Diseases in Neem

11. Pink disease - *Corticium salmonicolor* Berk. and Broome.

12. Twig canker and shot-hole - *Phoma jolyana* Piroz. and Morgan-Jones

13. Blight and stem rot - *Sclerotium rolfsii* Sacc.

14. Damping-off - *Fusarium oxysporum* Schltdl

15. Collar-rot - *Fusarium semitectum* Berk. and

16. Stem-rot - *Sclerotinia sclerotiorum* (Lib.) de Bary.

17. Periwinkle [*Catharanthus roseus* (L.) G.Don]

1. Leaf Blight/Web Blight

Symptoms

- ☆ Water-soaked lesions appear on the sides of midrib at the leaf petiole junction.
- ☆ Spots spread to tip of the leaf.
- ☆ Spots develop concentric light and dark brown rings and coalesce.
- ☆ Infected leaves are quickly blighted and killed.
- ☆ Mats of dried tan-coloured dead leaves cling to the green shorts.

- In severe infection, all but youngest leaves are blighted and the diseased plant dies.
- Numerous brown to black sclerotia are seen.

Fungus

Rhizoctonia solani Kuhn [*Thanatephorus cucumeris* (A.B. Frank) Donk]

Mode of survival: As sclerotia and mycelium infected plant debris.

Primary spread: From sclerotia in soil.

Secondary spread: Through irrigation water.

Favourable condition: Warm and humid conditions.

Management

- Rotation with non-hosts crops.
- Follow deep summer ploughing.
- Grow resistant varieties.
- Plant disease-free plants.
- Adopt proper spacing for good air circulation.
- Apply balanced plant nutrition.
- Spray thiophanate methyl.

2. *Phytophthora* Aerial Blight

Symptoms

- Sudden wilting of a branch or entire plant.
- Water-soaked, grey lesions appear at the base of the wicked stem.
- Reddish brown, sunken cankers causes girdling of main stem or lateral branches.
- Infected plants die within 1-2 weeks of symptoms development.

Chromista

Phytophthora parasitica Dastur

Mode of survival: As oospores and mycelium in infected plant material in soil.

Primary spread: From soil-borne oospores.

Secondary spread: Through irrigation water.

Favourable conditions: Heavy fertilization, hot, wet weather condition and overhead irrigation.

Management

- Avoid overhead irrigation.
- Spray azoxystrobin or fosetyl –Al 2-3 times at monthly interval.

3. *Phytophthora* Root Rot

Symptoms

- ☆ Small, irregular, water-soaked spots appear on the leaves which enlarge and cover entire leaf.
- ☆ Leaves become olive green to brown and rot.
- ☆ Rotting spreads to branches and main stem.
- ☆ Diseased branch or entire plant wilts.
- ☆ Leaves are shed to leave bare branches.
- ☆ Finer roots and larger roots rot.
- ☆ Plants are stunted and weak.

Chromista

Phytophthora nicatianae Brede de Haan.

Mode of survival: As chlamydospores in plant debris in soil.

Primary spread: From soil-borne chlamydospores.

Favourable conditions: Warm temperature of 25-28°C.

Management

- ☆ Sow disease free clean seeds and transplant disease free seedlings.
- ☆ Raise seedlings in raised bed nursery.
- ☆ Improve drainage.
- ☆ Uproot and burn diseased plants.
- ☆ Avoid heavy fertilization and overhead irrigation.
- ☆ Apply etridiazole or fosetyl-Al.

4. *Pythium* Root Rot

Symptoms

- ☆ Tan to brown, water-soaked or slimy spots appear on the stem.
- ☆ Outer root cortex disintegrate and slough-off or stripped from the thread-like inner root when pulled out.
- ☆ Leaves become chlorotic, gradually become yellow, and wilt.

Chromista

Pythium aphanidermatum (Edson) Fitzp. and *P. ultimum* Trow.

Mode of survival: As oospores in infected stubbles.

Primary spread: From oospores in soil.

Secondary spread: Through irrigation water or movement of infested soil.

Favourable condition: Ill drained conditions.

Management

- ☆ Provide good drainage.
- ☆ Drench the soil with copper oxychloride 0.3 per cent.
- ☆ Regulate irrigation to have optimum moisture.

5. Damping Off

Symptoms

- ☆ Rotting of seeds occur prior to germination.
- ☆ Seedlings rot just below soil.
- ☆ In grown up seedlings, roots rot and the plants die.

Chromista

Pythium sp, *Phytophthora nicotianae* var. *parasitica* (Dastur) Waterhouse and fungi.

Rhizoctonia solani Kuhn.

Spread: Low soil temperature before germination and high and overcrowding of seedlings.

Management

- ☆ Raise seedlings in raised beds.
- ☆ Follow recommended seed rate.
- ☆ Treat seeds with metalaxyl.
- ☆ Avoid excessive overhead irrigation.
- ☆ Follow strict sanitation.
- ☆ Drench soil with copper oxychloride 0.3 per cent or fosetyl-Al 0.1 per cent.

6. Foot Rot

Symptoms

- ☆ Yellowing of leaves, drying of leaf tips.
- ☆ Wilting of plants and rotting of bark at collar region.
- ☆ White or brown, mustard seed-like sclerotia with white mycelial mat seen at the base of affected plants, on stem and soil.

Fungus

Sclerotium rolfsii Sacc. [Sexual stage*: Athelia rolfsii* (Curzi)].

It is a soil living saprophytic pathogen, produces ribbon like hyphae with clamp connections. Basidia are club-shaped, bearing four smooth, ellipsoid basidiospores. Small, brownish sclerotia (hyphal propagules) are also formed.

Mode of survival: As sclerotia in infected plant debris.

Primary spread: From soil-borne sclerotia.

Secondary spread: Through irrigation water.

Favourable condition: Excessive moisture.

Management

- ☆ Regulate irrigation.
- ☆ Remove diseased plants with roots.
- ☆ Drench the soil in root region with copper oxychloride 0.3 per cent.

7. Die back

Symptoms

- ☆ Withering of terminal buds of young branches.
- ☆ Wilting and drying of young leaves.
- ☆ Infection spreads to entire branch which leads to wilting and dry of plants.
- ☆ Minute, black, fruiting bodies (acervuli) are seen on dried buds and branches.

Fungus

Colletotrichum dematium (Pers.) Grove.

Mode of survival: As mycelia and acervuli in infected plant parts.

Primary spread: From diseased plant parts.

Secondary spread: Through air-borne conidia.

Management

- ☆ Mulch with citronella or palmarosa straw.
- ☆ Grow resistant variety like WH-40.
- ☆ Spray carbendazim 0.1 per cent or captafol or mancozeb 0.2 per cent at 15 days interval.

8. Rust

Symptoms

- ☆ Affected leaves turn chlorotic.
- ☆ Lower surface of leaf exhibit brown uredinia and dark brown telia.
- ☆ Infected shoots remain sterile, stand erect and leaves become abnormally small.
- ☆ It is autoecious rust.

Fungus

Puccinia vincae Berk. (*Uredo vincae* DC)

Mode of survival: As infected plant debris.

Spread: Air-borne uredospores.

Management

- ☆ Remove and destroy infected plant parts.
- ☆ Avoid overhead irrigation.
- ☆ Spray triflumizole.

9. Powdery Mildew

Symptoms

- ☆ White powdery spots appear on leaves which later spread and occupy the entire leaf.
- ☆ Infected leaves become brown and dry.
- ☆ Mixed infection of rust and powdery mildew disease often noticed.

Fungus

Golovinomyces orontii (Castagne) V.P. Heluta

Mode of survival: As infected plant debris.

Spread: Through wind-borne conidia.

Management

Spray wettable sulphur 0.3 per cent.

10. Stem Rot

Symptoms

- ☆ Green, water soaked lesions appear below flower head or at collar region in the stem.
- ☆ Slimy wet rot of cortical tissues occurs.
- ☆ Wet rot region becomes pale, then yellowish brown and dry.
- ☆ Severe infection leads to death of the plants.

Fungus

Rhizopus stolonifer Vuillemin

Spread: Through rain splashes and overhead irrigation.

Favourable conditions: High relative humidity.

Management

- Provide good drainage.
- Adopt proper spacing for free air circulation.
- Avoid overhead irrigation.
- Spray copper oxychloride 0.3 per cent

11. *Botrytis* Blight

Symptoms

- Affects weakened or injured plant parts or senescent plant tissues like old flowers and leaves or stems.
- Water-soaked spots are seen on petals.
- Tan spots become greyish brown and circular blotches with a concentric ring pattern on leaves when infected flower petals lie against the leaf.
- Blighted tissues are often covered with fuzzy, greyish brown hyphae and clusters of the fungal conidia.

Fungus

Botrytis cinerea Pers. Ex Fr.

The mycelium is septate, grey colored. Sclerotia were dark and irregular. Conidia were smooth, ash colored, ovoid in shape.

Mode of survival: As survives on and within infected plant debris (leaves and flowers).

Primary spread: From infected plant debris.

Secondary spread: Through wind-borne conidia.

Favourable conditions: Mild to warm temperatures coupled with high relative humidity, extended periods of cloudy, humid and wet weather.

Management

- Adopt strict sanitary measures to remove infected dead or dying flowers or leaves from the soil surface and plant and soil debris.
- Follow recommended spacing for enough air circulation.
- Spray iprodione or captan 0.2 per cent.

12. Leaf Spots

Symptoms

- Brown to black spots (pinhead to 3mm) appear on leaves, stems and petioles.
- Disease appears first on lower leaves and then progress upward through the plant.

- ✰ Spots increase in size and are with alternating light and dark rings of dead tissues giving a target spot or bull's eye appearance.
- ✰ More spots on leaves cause yellowing and falling of leaves.
- ✰ In severe cases, all leaves except top few younger leaves fall prematurely.

Fungi

Alternaria alternata (Fr.) Keissl., and *Ulocladium* sp.

Favourable condition: Warm temperature, wet conditions and low soil fertility.

Secondary spread: Through overhead irrigation.

Management

- ✰ Remove severely infected plant parts.
- ✰ Minimize leaf wetness by avoiding overhead sprinkler irrigation.
- ✰ Adjust soil fertility and pH to improve plant growth.
- ✰ Spray chlorothalonil, azoxystrolin, ipodione or fludioxonil at every 7-10 days interval.

13. Black Root Rot

Symptoms

- ✰ Poor growth, yellowing of leaves, wilting and death of the plant.
- ✰ Initially black or brown spots or bands occur on the secondary feeder roots.
- ✰ Later disease spreads and affect the entire root system.
- ✰ Root system become soft and mushy black due to production of black resting spores and show salt and pepper appearance.

Fungus

Thielaviopsis basicola (Berk. and Br.) [*Chalara elegans* Nag Raj and W.B. Kendr.]

Favourable conditions: High temperature and pH abovc 6.0.

Mode of survival: As survive as spores in infected plant materials in soil.

Primary spread: From infected plant materials.

Secondary spread: Through irrigation water.

Management

- ✰ Adjust the planting date and pH.
- ✰ Avoid overhead watering.
- ✰ Do not over fertilizer plants and avoid fertilizers with high ammonium as a nitrogen source.
- ✰ Discard infected plants.

14. Stem Blight/Stem Canker

Symptoms

- Shoot tips of infected plants become dark brown, wilt and black to the soil.
- Dark brown to black grinding lesions on stem which extent to entire length in few days.
- Small, black, dot-like pycnidia are produced in infected tissues.
- Affected stem wilts, turn dark brown to black and die.
- After several weeks, entire clump may wither and die.
- Stem lesions spread into the leaf petiole and leaves results in withering and premature defoliation.

Fungus

Phoma exigua var. *exigua* Desmazières

Mode of survival: As dormant mycelia and pycnidia.

Primary spread: From infected plant parts.

Secondary spread: Through irrigation water.

Favourable conditions: Denser planting, Temperature of 10-18°C, prolonged periods of wet weather, plants grown under moist propagation.

Management

- Use disease free vigorous seedlings.
- Remove fallen leaves infected plants including old runners which are hidden among other plants.
- Use optimum fertilizer dose.
- Avoid over-watering.
- Water the soil rather than plants.
- Clean out dead leaves and prune the overhanging branches.
- Work with plants when they are dry to avoid further spread.
- Spray azoxstrobin at 7-10 days interval.

15. Tomato Spotted Wilt

Symptoms

- Infected plant becomes yellow and stunted.
- Leaves are distorted with small black concentric rings or spots or line patters.
- Purple to black discolouration in "V" shaped pattern spreading from petiole to the main leaf vein.

- ✰ Black lesions turn into tan to brown cankers on infected stems.
- ✰ Infected plants collapse due to stem lesions or cankers and are severely stunted.
- ✰ Flowers are discoloured and malformed.

Causal Agent

Tomato spotted wilt virus (TSWV)

Primary spread: From planting infected seedlings.

Secondary spread: Through western flower thrips and tobacco thrips.

Favourable conditions: Alternate host crops and weeds.

Management

- ✰ Use disease free seedlings.
- ✰ Remove and destroy diseased plants and weed hosts.
- ✰ Do not grow near vegetable crops as many of them are hosts of TSWV.
- ✰ Provide fine-mesh screening.
- ✰ Monitor and control thrips by spraying suitable insecticides.

16. Phyllody

Symptoms

- ✰ Severe reduction in leaf size.
- ✰ Shortening of internodes.
- ✰ Extensive stunting and branching with virescent flowers.
- ✰ Infected plants buds like a bush with leaves of narrow and pointed tips.
- ✰ They do not produce normal flowers or seeds.
- ✰ Reduced alkaloid content.

Causal Agent

Candidatus Phytoplasma trifolii; *Candidatus* Phytoplasma malaysianum.

Transmission

Craft and leafhopper.

Management

- ✰ Remove and destroy diseased plants.
- ✰ Spray oxytetracycline to suppress the symptoms.
- ✰ Spray methyl demeton 0.025 per cent to control insect vectors.

18. Sarpagantha/Indian Snakeroot (*Rauvolfia serpentina* L.)

1. Wilt

Symptoms

- ☆ Wilting of branches followed by wilting of entire plant.
- ☆ In advance stages, collar region near ground level turn dark coloured.
- ☆ Cortex of root disintegrate in seedlings and young plants.
- ☆ Leaves dry and drop off followed by upward curling.
- ☆ At the end infected plant die.

Fungus

Fusarium oxysporum f.sp. *rauvolfii* Janardh., Ganguly and A. Husain

Mode of survival: In the corm/cormels and soil.

Primary spread: From infected corm and soil.

Secondary spread: Through irrigation water.

Favourable conditions: Poor soil aeration, flooding or hot weather.

Management

- ☆ Grow resistant variety.
- ☆ Avoid low-lying areas for growing the crop.
- ☆ Treat seeds with thiram 1.5 g + Carbendazim 1.5 g/kg of seed.

2. Powdery Mildew

Symptoms

- ☆ Lower surface of leaves show whitish growth in irregular patches.
- ☆ The fungal growth increases and coalesce to cover the entire leaf.
- ☆ Affected leaves roll.

Fungus

Leveillula taurica (Lév.) G. Arnaud.

Mode of survival: As cleistothecia in infected plants.

Primary spread: From infected plants.

Secondary spread: Through air-borne conidia.

Management

- ☆ Spray dinocap (Karathane) 0.05 per cent or wettable sulphur 0.3 per cent at the disease initiation.

3. Leaf Blight and Bud Rot

Symptoms

- ☆ The disease appear as brown circular spots on one year old plants.
- ☆ Spots with yellowish margin on the lower surface of tender leaves.
- ☆ Spots enlarge and form dark brown lesions, coalesce and cover the entire leaf leading to blighting of leaf.
- ☆ Affected lead dry and drop.
- ☆ Tips of branches are also infected resulting in death of auxiliary buds.
- ☆ Stem just below affected leaf bud turn purplish black.
- ☆ Young flower buds and fruits are also attacked.

Fungus

Alternaria alternata (Fr.) Keissl.

Mode of survival: As dormant mycelium in infected plant debris.

Primary spread: From infected plant debris.

Secondary spread: Through air-borne conidia.

Favourable conditions: Dry humid weather.

Management

- ☆ Spray mancozeb 0.3 per cent or copper oxychloride or Rovral 0.2 per cent.
- ☆ Treat seed with thiram 1.5g/kg + carbendazim 1.5 g/kg of seed.

4. Leaf Spot

Symptoms

- ☆ Disease appears and minute, yellow spots which gradually increase in size and become dark brown patches.
- ☆ Purple blotches appear on lower surface of leaves which become dark brown in a week.
- ☆ Affected leaves dry and fall off.

Fungus

Cercospora rauvolfiae Chupp and Muller and *Mycosphaerella rauvolfiae* Ram. and Ram. other fungus.

Cercospora serpentine.

Management

- ☆ Treat seeds with carbendazim 1.5 g/kg.
- ☆ Spray carbendazim 0.1 per cent or mancozeb 0.2 per cent at 10-15 days interval.

5. Target Spot

Symptoms

- ☆ Small brown spots appear on both the surfaces of leaves.
- ☆ The spots increase to circular spots with concentric zones.
- ☆ Spots on the leaves are dark brown on the upper surface and yellowish brown on lower surface and are usually surrounded by pale yellow margin.
- ☆ The disease appears to upper leaves also.
- ☆ Infected leaves fall off leaving only few tender leaves at the top of infected plant.

Fungus

Corynespora cassiicola (Berk. and M.A. Curtis)

Management

- ☆ Six sprays of captan 0.25 per cent at monthly intervals.

6. Anthracnose

Symptoms

- ☆ Numerous circular spots appear scattered on the upper surface of leaves.
- ☆ Spots enlarge to circular patches studded with black, dot-like acervuli.

Fungus

Colletotrichum gloeosporioides Pat and Penz.

Conidia are single celled, subhyaline, cylindrical, straight with obtuse ends with one or two oil globules and in mass looks pink.

Mode of survival: As infected plant parts *viz.*, inflorescence and buttons.

Primary spread: From infected plant parts.

Secondary spread: Through air-borne conidia and rain splashes.

Favourable conditions: Dry periods

Management

- ☆ Remove heavily diseased leaves.
- ☆ Avoid excess irrigation.
- ☆ Spray zineb 0.2 per cent.

7. Die back

Symptoms

- ☆ Spots appear on leaves and stems.
- ☆ The spots enlarge, coalesce and form circular necrotic patches.
- ☆ Severe infection results in drying defoliation and decapitation.

Fungus

Colletotrichum dematium (Pers.) Grove

Mode of survival: As mycelium and acervuli in diseased plant debris.

Primary spread: From diseased plant debris.

Secondary spread: Through wind-borne conidia.

Management

- ☆ Spray zineb 0.2 per cent.
- ☆ Remove infected leaves and prune infected branches.

8 Leaf Spot

Symptoms

- ☆ Spots over the leaf surface.
- ☆ In severe infection, it causes premature defoliation.

Fungus

Curvularia lunata (Wakker) Boedijn

Favourable conditions: Cloudy days couples with high humidity.

Management

- ☆ Spray Bordeaux mixture 1 per cent at weekly intervals for 4-6 weeks.

9. Sooty Mould

Symptoms

- ☆ Black coloured fungal growth is seen on the upper leaf surface.
- ☆ Black fungal growth on stems, flowers and fruits results in reduced photosynthesis and vigour.
- ☆ Diseased fruits fetch poor market price.

Fungus

Capnodium sp.

Management

Spray dimethoate 0.05 per cent to control insects like aphids which secrete sweet honeydew harvesting sooty mould fungus.

10. Mosaic

Symptoms

- ☆ Alternate patches of dark green and light green patches appear as mosaic symptom on the leaves.
- ☆ Alkaloid content is reduced.

Causal Agent

Tobacco mosaic virus

Spread: Mosaic is sap transmissible, no vector.

Favourable conditions: Heavy wind and rainfall.

Management

- ✰ Rogue out diseased plant as and when noticed.
- ✰ Remove collateral weed host periodically.
- ✰ Adopt proper sanitary measures to avoid wounding of plant parts.

11. Phyllody

Symptoms

- ✰ Stunting of plants with shortened internodes.
- ✰ Production of small, pale leaves in a cluster.
- ✰ Floral parts converted into leaf-like structures leading to complete sterility.
- ✰ Alkaloid content is reduced.

Causal Agent

Candidatus Phytoplasma

Management

- ✰ Rogue out diseased plants as and when noticed.
- ✰ Remove collateral weed hosts periodically.

19. Senna (*Senna alexandrina* Mill.)

1. Damping Off and Wilt

Symptoms

- ✰ Water soaked black lesions appear on the collar region of the seedling.
- ✰ Blackening extends upwards and downwards.
- ✰ Softening of affected tissues due to rotting.
- ✰ Roots decay and become dark brown or black.
- ✰ Collapse of affected seedlings.
- ✰ In grown up plants blackening at collar region near soil line leading to wilting and death of plants.

Fungus

Macrophomina phaseolina (Tassi) Goid.

Mode of survival: As infected plant materials in soil.

Primary spread: From sclerotia in infected debris in soil.

Secondary spread: Through mycelium and sclerotia in soil.

Favourable conditions: Warm weather, soil temperature 28-32°C with optimum moisture and neutral pH.

Management

- ☆ Collection and destruction of diseased seedlings.
- ☆ Deep ploughing during summer.
- ☆ Providing proper drainage in the nursery.
- ☆ Follow 2 year crop rotation with non-host crops including green manures.
- ☆ Seed treatment with thiram or captan 2.5 g/kg protects the seedling from the disease.
- ☆ Removal of diseased plants in the field and soil drenching with carbendazim (0.1 per cent).

2. Leaf Blight

Symptoms

a. *Alternaria alternata* (Fr.) Keissl.

- ☆ Minute, dark brown, circular spots especially at the tips and margins of leaves.
- ☆ Spots enlarge become black and irregular.
- ☆ Lower leaves are more susceptible.
- ☆ Diseased leaves turn chlorotic dry and fall off.
- ☆ Diseased leaves contain reduced level of sennoside.
- ☆ *Cassia acutifolia* is less susceptible than *C. angustifolia.*

b. *Alternaria cassiae* Jurair and Khan

- ☆ Small, pale to dark brown spots appear on tips and margins of leaflets.
- ☆ They increase in size, coalesce and caused complete.
- ☆ Dark irregular spots appear on stems causing coalesce of stem and death plants.
- ☆ Infected pods show irregular brown in black discolouration.

Fungi

Alternaria alternata (Fr.) Keissl.and *A. cassia* Jurair and Khan.

Primary spread: From dormant mycelia and ascospores in perithecia in diseased plant debris.

Secondary spread: Through air-borne conidia.

Favourable conditions: Nutritionally poor soil, rainy season, temperature of 28-35°C and relative humidity of 85-90 per cent.

Management

- Collect and destroy the infected plant parts.
- Use disease free seed.
- Amend the nutritional status of the soil.
- Maintain optimum plant population.
- Spray mancozeb (0.2 per cent) or zineb or Rovral 0.2 per cent at 15 days interval for a period of 5-6 weeks.

3. Leaf Spot

Symptoms

- Small, brown spots appear on lower leaves.
- More number of spots cause yellowing and premature leaf fall.

Fungus

Cercospora sp. and *Phyllostica* sp.

Spread: Air borne conidia.

Favourable conditions: Cloudy weather, poor management of the crop (Neglected crop), poor nutrition.

Management

- Collect diseased leaves and burn.
- Adopt optimum spacing.
- Apply optimum dose of fertilizers.
- Spray mancozeb 0.2 per cent.

4. Wilt

Symptoms

- Plant shows yellowing and marginal necrosis of leaves.
- The root system is reduced with the discoloration of vascular tissues and dark brown colored banda on the stem starting fro root zone.

Fungus

Fusarium semitectum Berk.

Spread: Soil borne macroconidia and chlamydospores.

Favourable conditions: Poor management of the crop (Neglected crop), poor nutrition.

Management

- ✰ Collect diseased leaves and burn.
- ✰ Adopt optimum spacing.
- ✰ Apply optimum dose of fertilizers.
- ✰ Drench with FYM + *Pseudomonas fluorescens* and *Bacillus subtilis.*

20. Sweet Basil (*Ocimum basilicum* L.)

1. Downy Mildew

Symptoms

- ✰ Symptoms initially appear as yellowing of leaves and are typically concentrated around the middle vein.
- ✰ The discolored area may cover most of the leaf surface.
- ✰ On the underside of leaves, a gray fuzzy growth of the pathogen may be apparent by visual inspection.
- ✰ The chlorotic areas turn to dark brown quickly.

Chromista

Peronospora belbahrii Thine.

Spread: Sporangia are wind-blown and rain splashed.

Favourable conditions: During wet or humid weather.

Management

- ✰ Remove and destroy the infected plant immediately.
- ✰ Spray azoxystrobin or potassium bicarbonate or mefenoxam or metalaxyl-M + copper hydroxide.

2. Anthracnose/Black Spot

Symptoms

- ✰ On leaves, lesions appear as either circular or irregular necrotic spots that often expand and coalesce.
- ✰ Older lesions appear desiccated, with shredded tissue at the center.

Fungi

Colletotrichum gloeosporioides (Penz.) Penz. and Sacc.,

Colletotrichum capsici (Syd. and P. Syd.).

Mode of survival: As dead twigs and branches.

Primary spread: From infected twigs and branches.

Secondary spread: Through air-borne conidia from acervulus and rain splashes.

Favourable conditions: Disease is favoured by warm weather at temperature of 30-32°C, cloudy weather and presence of susceptible host.

Management

- Remove and destroy the infected plants.
- Spray carbendazim 0.1 per cent or captafol 0.2 per cent or mancozeb 0.3 per cent.

3. Leaf Spot

Symptoms

- Dark spots form on leaves and the dead tissue within leaf spots may drop out, causing a shot-hole symptom.
- The disease can cause defoliation, tip dieback, stem lesions, and sometimes loss of entire plants.

Fungi

Cercospora canescens Ellis and G. Martin, *Cercospora omicola, Glomerella cingulata* (Stoneman), *Phyllosticta ocimicola, Corynespora cassicola* Berk.

Spread: Spores are water-splashed from diseased tissue.

Management

- Increase plant spacing to increase air movement.
- Avoid overhead irrigation.
- Remove diseased plants to reduce inoculum levels.
- Spray carbendazim 0.1 per cent or chlorothalanil 0.2 per cent.

4. Fusarium Wilt

Symptoms

- Initial disease symptoms are yellowing of the shoots, distorted young leaves, and internal vascular discoloration of stems.
- As the disease advances, plants wilt and die.

Fungus

Fusarium oxysporum f. sp. basilicum Dzidzariya

Spread: Soil and seed-borne.

Survival: In the soil for many years.

Management

- Remove and destroy infected plants.
- Seed may be disinfested with carbendazim 0.1 per cent.

5. Bacterial Leaf Spot

Symptoms

- ✰ Spots on leaves are water-soaked and dark.
- ✰ They may be both angular and delineated by small veinsin the leaves or irregular in shape.

Bacterium

Pseudomonas cichorii (Swingle) Stapp

Spread: Splashing water.

Survival: As seed-borne bacterium.

Favourable conditions: Wet and humid conditions.

Management

- ✰ Remove diseased leaves and plants.
- ✰ Avoid overhead irrigation.
- ✰ Use clean, sterile equipment.

21. Sweet Flag (*Acorus calamus* L.)

1. Basal Rot

Symptoms

- ✰ The leaves resulting gradual dried of total leaves.
- ✰ The rhizomes of the infected plants were rotted.
- ✰ In severe infection, white mat of mycelia and occasionally sclerotia of the pathogen were found to cover the base of leaves and on the adjacent soil.

Fungus

Sclerotium rolfsii Sacc. [Sexual stage: *Athelia rolfsii* (Curzi)].

Mode of survival: In plant debris as sclerotia.

Primary spread: From soil-borne sclerotia.

Secondary spread: Through irrigation water.

Favourable conditions: Warm climate, high moisture and high temperature.

Management

- ✰ Follow crop rotation with cereals.
- ✰ Removal of weed hosts.
- ✰ Deep summer ploughing.
- ✰ Avoid use of infected infected FYM, compost, soil, *etc.*

- ☆ Use disease free seeds/tolerant varieties.
- ☆ Apply green manure followed by *Trichoderma viride* or *Pseudomonas fluorescens* 2.5 kg/ha in soil.

2. Leaf Spot

Symptoms

- ☆ The leaf spot disease was characterized by dark brown to black necrotic lesions surrounded by yellow hallow with lighter grey centre.
- ☆ Gradually, the lesions became enlarged and coalesced to form blighted areas. Severely blighted leaves turned withered and dry.
- ☆ The conspicuous spots with whitish mycelia were also present even on the dried leaf spot.

Fungus

Passalora acori (J.M. Yen) U. Braun and Crous [*Cercospora acori* J.M. Yen]

Mode of survival: In plant debris as mycelia.

Favourable conditions: Moderate to low temperatures.

Management

- ☆ Use of disease-free seeds.
- ☆ Adopt crop rotation.
- ☆ Spray carbendazim 0.2 per cent or copper oxychloride 0.25 per cent.

3. Leaf Blight or Anthracnose

Symptoms

- ☆ In leaf blight disease, the symptoms first appear at the edge and/or the tip of leaves as dark blighted lesion with yellow margin and gradually spread to the middle.
- ☆ The affected leaves are gradually withered.
- ☆ Heavily infested field looks burning symptoms.
- ☆ Older leaves are much more affected than the younger leaves.

Fungus

Colletotrichum gloeosporioides Penz and Sacc.

Mode of survival: In plant debris as acervuli.

Favourable conditions: Moderate to low temperatures.

Management

- ☆ Use of disease-free seeds.

- ☆ Adopt crop rotation.
- ☆ Spray carbendazim 0.2 per cent or chlorothanil 0.1 per cent.

4. Rust

Symptoms

- ☆ The rust appears as dark orange to brown colour pustules on both surface of leaves with ruptured epidermis and crop of uredospores.
- ☆ In severe case, the whole plant is covered with reddish brown colour by the uredospores and in advance stage the plants become bronzy in appearance.

Fungi

Uromyces acori [T.S. Ramakrishnan and Rangaswami]

Uredia were observed during study while telia, sperogonia and aecia were not observed till now.

Favourable conditions: Moderate to low temperatures.

Management

- ☆ Use of disease-free seeds.
- ☆ Adopt crop rotation.
- ☆ Spray oxycarboxin 0.2 per cent.

22. Sweet Leaf/Stevia/Charkarai Tulsi (*Stevia rebandiana* Bertoni)

1. Leaf Spot

Symptoms

- ☆ Small, light brown spots appear on leaves.
- ☆ Later, these spots become dark brown to grey with concentric zones with 2-18 mm in dia.
- ☆ Several spots coalesce to form large necrotic areas.
- ☆ On older leaves concentric spots are more common at the tips.

Fungus

Alternaria alternata (Fr.) Keissl.

Primary spread: From dormant mycelia in diseased plant debris.

Secondary spread: Through air-borne conidia.

Favourable conditions: Nutritionally poor soil, rainy season, temperature of 28-35°C and relative humidity of 85-90 per cent.

Management

- ☆ Remove the affected plant parts.
- ☆ Spray mancozeb 0.2 per cent.

2. Leaf Spot

Symptoms

- ☆ Angular, shiny olive-grey depressed lesions bounded by veins appear on the foliage that rapidly coalesce and are surrounded by a chlorotic halo.
- ☆ Affected leaves become necrotic and drop prematurely.
- ☆ They are initially formed circular leaves and gradually move upward.
- ☆ Pycnidia as minute black starches are seen at the spots.

Fungus

Septoria steviae Ishiba

Mode of survival: As infected plant debris.

Spread: Air-blown conidia and conidia through rain splashes.

Management

- ☆ Adopt strict sanitation.
- ☆ Spray mancozeb 0.2 per cent.

3. Charcoal Rot

Symptoms

- ☆ Occurs mostly during early stage or at flowering.
- ☆ Necrosis and wilting of plants.
- ☆ Rotting of roots with minute sclerotia on the root barks.

Fungus

Macrophomina phaseolina (Tassi) Goid.

Mode of survival: As sclerotia in infected plant material and soil.

Primary spread: From sclerotia in soil.

Secondary spread: Through irrigation water.

Favourable conditions: Hot dry weather and temperature between 28 and 35°C.

Management

- ☆ Pull out with roots and burn by fire.
- ☆ Deep summer ploughing.
- ☆ Adopt rotation with non-hosts.

4. *Pythium* Root Rot

Symptoms

- ☆ Primarily a seedling disease.
- ☆ Diseased seedlings are stunted and leaves show curling.
- ☆ Root lesions and necrotic dead roots.

Fungi

Pythium myriotylum Drechsler, *P.irregulare* Buisman and

P. aphanidermatum (Edson) Fitzp.

Mode of survival: As ascospores in infected plant material.

Primary spread: Through irrigation water/overhead irrigation.

Favourable conditions: High soil moisture.

Management

- ☆ Adopt strict sanitary measures.

5. Stem and Root Rot

Symptoms

- ☆ Moderate wilting or flagging of stems similar to symptoms of drought stress.
- ☆ Wilting is followed by rapid necrosis and death of stems.
- ☆ At soil line, abundant white to brown sclerotia are seen at the base of stems along with white cord-like mycelial growth.
- ☆ Plants die in a period of 7 days.

Fungus

Sclerotium rolfsii Sacc. [Sexual stage*: Athelia rolfsii* (Curzi)].

It is a soil living saprophytic pathogen, produces ribbon like hyphae with clamp connections. Basidia are club-shaped, bearing four smooth, ellipsoid basidiospores. Small, brownish sclerotia (hyphal propagules) are also formed.

Mode of survival: As sclerotia in infected plant material and in soil.

Primary spread: Through irrigation water and implements.

Favourable conditions: High soil moisture.

Management

- ☆ Rotation with corn or small grains.
- ☆ Removal of plants with roots and mycelia and sclerotia with soil at root region.

6. Stem Rot

Symptoms

- ☆ Wilting of plants with bleaching at the stem base.
- ☆ Later stems and leaves become necrotic
- ☆ Presence of tufts of white mycelium on the stem.
- ☆ At the base of the stem irregularly shaped, 2-8 mm sized black sclerotia are seen.

Fungus

Sclerotinia sclerotiorum (Lib.) de bary.

Mode of survival: As sclerotia in infected plant material and soil.

Primary spread: From ascospores in asci in apothecia on the stem.

Secondary spread: Through irrigation water.

Favourable conditions: Frost

Management

- ☆ Adopt strict sanitation

Minor Diseases

7. **Root rot - *Rhizoctonia solani***
8. **Powdery mildew - *Erysiphe cichoracearum***
9. **Stem rot - *Sclerotium rolfsii, Sclerotium dephinii***

23. Thorn Apple/Devil's Trumpets (*Datura stramonium* L.)

1. *Alternaria* Leaf Spot

Symptoms

- ☆ Dark brown, round to oval or slightly irregular spots appear on leaves.
- ☆ Spots are with concentric rings.
- ☆ Spots increase and covers the entire leaves causing blighting of leaves.

Fungus

Alternaria tenuissima (Nees) Wiltshire

Mode of survival: As dormant mycelium in infected leaves.

Spread: Air borne conidia.

Favourable conditions: Nutritionally poor soil.

Management

- ☆ Collect and destroy affected leaves.
- ☆ Increase nutritional status of the soil by judicious fertilizer application.
- ☆ Adopt proper spacing.
- ☆ Spray mancozeb 0.2 per cent.

2. Leaf Blight

Symptoms

- ☆ Water soaked, circular spots appear on leaves.
- ☆ Later they coalesce, become dull yellow and turn light brown due to necrosis.
- ☆ Older spots exhibit typical spots with concentric rings.
- ☆ Single spot when abnormally increase in size, such leaf shows curling.
- ☆ When two or three leaves show blighting wilting of the branch occurs.
- ☆ Severely affected leaves show curling, blighting and premature defoliation.

Fungus

Alternaria alternata (Fr.) Keissl.

Secondary spread: Through air-borne conidia.

Favourable conditions: Older leaves are highly susceptible.

Management

- ☆ Spray mancozeb 0.2 per cent or zineb 0.2 per cent.

3. Leaf Blight and Fruit Rot

Symptoms

- ☆ Small, necrotic spots appear on leaves.
- ☆ These spots increase in size, coalesce and cause blighting of leaves.
- ☆ Infection spreads to fruits. Diseased fruits blight and rot.
- ☆ Reduced hyascyanine and scopolamine content in fruits.

Fungus

Alternaria crassa (Saac.) Rands

Primary spread: From seed-borne inoculum.

Secondary spread: Through air-borne conidia.

Favourable conditions: High humidity.

Management

- ✰ Treat the seeds with thiram suspension.
- ✰ Spray mancozeb 0.2 per cent.

4. Wilt/Root Rot

Symptoms

- ✰ Yellowing and gradual wilting of plants.
- ✰ Roots rot and diseased plant dies.
- ✰ Cottony white mycelium with brown sclerotia are found in the root region of the plant after death.

Fungus

Sclerotium rolfsii Sacc. [Sexual stage*: Athelia rolfsii* (Curzi)].

It is a soil living saprophytic pathogen, produces ribbon like hyphae with clamp connections. Basidia are club-shaped, bearing four smooth, ellipsoid basidiospores. Small, brownish sclerotia (hyphal propagules) are also formed.

Mode of survival: As sclerotia in infected plant debris in soil.

Primary spread: From mmycelium and sclerotia in soil.

Secondary spread: Through irrigation water and agricultural operations.

Favourable conditions: Irrigation moisture.

Management

- ✰ Deep summer ploughing.
- ✰ Destroy infected plant materials by burning.
- ✰ Adopt crop rotation for 3-4 years.
- ✰ Drench soil in root region with copper oxychloride 0.3 per cent or carbendazim 0.1 per cent.

5. Root and Foot Rot

Symptoms

- ✰ Disease appears in all stages but more destructive on younger plants.
- ✰ Seedlings show damping off symptoms and are killed within a week.
- ✰ In older plants, matured leaves show brownish black, discolouration and fall off prematurely.
- ✰ Root tissues disintegrate followed by death of the plant.

Fungus

Corticium solani (Prill. and Delacr.)

Mode of survival: As sclerotia in infected plant material.

Primary spread: From sclerotia.

Secondary spread: Through mmycelial strands and basidiospores.

Favourable conditions: Cool weather, temperature of 18-22°C and relative humidity of 75-90 per cent.

Management

- ☆ Deep summer ploughing.
- ☆ Adopt crop rotation.
- ☆ Soil drench with carbendazim 0.1 per cent.

6. Root Rot

Symptoms

- ☆ Reddish brown lesion appear on the stem above soil level.
- ☆ Leaves and branches of affected plant suddenly droop, wilt and die.
- ☆ Roots rot and barks shred.
- ☆ Large number of minute, black or dark brown sclerotia are seen on the bark and stem tissues.

Fungus

Macrophomina phaseolina (Tassi) Goid.

Mode of survival: As sclerotia in infected plant material in soil.

Primary spread: From sclerotia in soil.

Secondary spread: Through irrigation water.

Management

- ☆ Drench the soil with carbendazim 0.1 per cent or propiconazole 0.1 per cent.
- ☆ Soil application of *Pseudomonas fluorescens* or *Trichoderma viride* 2.5 kg/ha + FYM 12.5 t/ha.

7. Distortion Mosaic

Symptoms

- ☆ Mosaic symptoms with puckering and leaf distortion.
- ☆ Yellowing of veins followed by inward rolling and distortion of leaves.

Causal Agent

Cucumber mosaic virus (CMV)

Management

- ☆ Rogue out and destroy diseased plants.

8. Little Leaf

Symptoms

- ☆ Diseased plants are stunted due to reduced intermodal length.
- ☆ Leaves become small narrow, soft and pale.
- ☆ Auxiliary buds induced to grow. Affected plants give bushy appearance.
- ☆ Infected plants rarely flower or produce fruits.
- ☆ If flowers are produced floral parts are converted into leaf-like structures.
- ☆ Plants become sterile.

Causal Agent

Ca. Phytoplasma

Mode of survival: In infected *Datura* spp. and weed hosts.

Primary spread: From infected host including weed plants.

Secondary spread: Through lleafhopper vectors.

Management

- ☆ Rogue out and destroy diseased plants.
- ☆ Apply recommended NPK dose.
- ☆ Deep summer ploughing.
- ☆ Spray dimethoate 0.05 per cent or Imidacloprid 0.05 per cent to control vector and reduce disease incidence.

24. Tropical Soda Apple (*Solanum viarum* Dunal)

1. Wilt

Symptoms

- ☆ The crop is attacked at all stages in the field.
- ☆ Wilt occurs usually after the rain in a season.
- ☆ Yellowing, drooping and death of the plant occur.
- ☆ The disease spreads killing many plants in a field within 10-15 days.

Fungus

Fusarium oxysporum f. sp. *lacimiti*

Mode of survival: As soil-borne.

Management

- ☆ Follow direct sowing instead of transplanting of seedlings.
- ☆ Follow early planting (in October-November) to escape from the disease.

- ✰ Apply *Trichoderma viride* at 2.5 kg/ha along with 50 kg FYM along plant rows.
- ✰ Remove and destroy diseased plants.
- ✰ Drench the soil on the root region with carbendazim 0.1 per cent.

2. Powdery Mildew

Symptoms

- ✰ White powdery patches are seen on the upper side of the leaves.
- ✰ The fungus spreads with age and covers the entire surface of the leaf with powdery coating consisting of conidia, conidiophore and mycelia.
- ✰ Later the white coating becomes brown and the affected leaves dry in due course.

Fungus

Oidium sp.

Spread: Through air-borne conidia.

Management

Spray wettable sulphur 0.3 per cent covering the entire foliage in the plant.

3. Bacterial Blight

Symptoms

- ✰ The disease appears as yellow streaks alternated with green stripes.

Bacterium

Xanthomonas campestris pv. *campestris* (Pammel) Dowson.

Management

- ✰ Spray streptocycline and copper oxychloride solution (Dissolve 150 g streptocycline and 750 g copper oxychloride each in 2.5 lit and mix and add 750 lit water and spray).

References

Agrios, G.N. (2005). Plant Pathology, Fifth Edition, Elsevier Academic Press, p. 922.

Amin, K.S., B.A. Ullasa and H.S. Sohi (1982). Quantitative Determination of Resistance in Powdery Mildew and Downy Mildew in Muskmelon, Indian J. Agric Sci., 52: 601–604.

Arya, A., (2009). Diseases of Fruit Trees, International Book Distributing Co., Lucknow.

Barkai-Golan, R. (2001). Postharvest diseases of fruits and vegetables: development and control. Elsevier.

Barnett, H.L. and B.B. Hunter. (1998). Illustrated genera of imperfect fungi. APS Press. 3340 Pilot Knob Rd., St. Paul, MN 55121.

Beckman, C.H. (1964). Host Responses to Vascular Infection, Ann. Rev. Phytopath., 2: 231-252.

Beckman, C.H. (1969). The Mechanics of Gel Formation by Swelling of Simulated Plant Cell Wall Membranes and Perforation Plates of Banana Root Vessels, Phytopath., 59: 837-843.

Beckman, C.H. (1987). The Nature of Wilt Diseases of Plants, APS Press, St.Paul, Minnesota.

Beckman, C.H. and S. Halmos (1962). Relation of Vascular Occluding Reaction in Banana Roots to Pathogenicity of Root-invading Fungi, Phytopath., 52: 893-897.

Beckman, C.H., S.Halmos and M.E. Mace. (1962). The Interaction of Host, Pathogen and Soil Temperature in Relation to Susceptibility to Fusarium Wilt of Bananas, Phytopath., 52: 134-140.

Bélanger, R. R., W. R. Bushnell, A. J. Dik, and T. L. Carver. (2002). The powdery mildews: a comprehensive treatise. American Phytopathological Society (APS Press).

Bent, K.G. (1969). Fungicides in Perspective, Endeavour, 28: 129 – 134.

Bhardwaj, L.N., I.M. Sharma and V.K. Gupta. (1997). Field Evaluation of Grape Cultivars to Powdery Mildew under Temperate Conditions of Himachal Pradesh, J. Mycol. Pl. Pathol., 27:341.

Blancard, D. (2019). Tomato diseases: identification, biology and control: a colour handbook. CRC Press.

Blumer, S. (1967). chte Mehltaupilze (Erysiphaceae). Ein Bestimmungsbuch Für dic in Europa vorkommenden Arten, Jena, Fisher, p. 436.

Bock, K.R. (1962). Control of coffee leaf rust in Kenya colony, Trans. Brit. Mycol. Soc, 45:301-313.

Borkar, S.G. and Yumlembam, R.A. (2016). Bacterial diseases of crop plants. CRC Press.

Braun U. (1987). A Monograph of the Erysiphales (Powdery Mildews), Beihefte zur nova, Hedwigia, 89: 1–700.

Brian, P.W. (1967). Obligate parasitism in fungi, Proc. Roy. Soc., London, 168(B): 101-118.

Brundza, K. (1933). Beitrage zur kenntnis der Erysiphacea Litanens, Zemes Ukio. Akad. Matrascio., 2: 107–197.

Bushnell, W.R. and A.P. Roelfs (Eds.) (1984). The Cereal Rusts Vol. 1, Academic Press, New York.

Butani, D. K., (1978). Pests and diseases of jackfruit in India and their control (a review). Fruits, 33(5). 351–357.

Butler, E.J. (1918). Fungi and Disease in Plants, Thacker Spink and Co., Calcutta, p. 547.

Chakrabarti, D. K., S. Kumar & Chand, G., (2010). A Guide Book for Diseases of Horticultural Crops (Diseases of Fruits, Ornamentals, Plantation, Spices, Medicinal, Forest and Vegetable Crops) (pp. 1–152). Narendra Publishing House, Maliwara, Delhi–11 IIHR, (1987). Diseases of Fruit and Vegetable Crops and their Control (p. 37). IIHR, Bangalore.

Chambers, H.L. and M.E. Corden. (1963). Semeiography of Fusarium wilt of tomato, Phytopathology, 53:1006-1010.

Chand, G. and S. Kumar, eds., (2016). Crop Diseases and Their Management: Integrated Approaches. CRC Press.

Chandrasekhara Rao, K. (1989). Comparative Efficacy of Fungicides for the Control of Grape Powdery Mildew (*Uncinula necator*), Indian J. Mycol Plant Pathol., 19: 200–201.

Chavan, S.B., S.V. Khandge, M.C. Varshney and J.D. Patil, (1995). Influence of Weather Parameters on Conidia Formation in Powdery Mildew of Grapes, Indian Phytopath., 19: 200–201.

Chen, T. M., & S.F. Chen, (1982). Diseases of tea and their control in the People's Republic of China. Plant Disease, 961–965.

Cherewick, W.J. (1944). Studies on the Biology of *Erysiphe graminis* DC, Can. J. Res. Sect. C. 22: 52–86.

Chupp, C. and A. F. Sherf, (1960). Vegetable Diseases and Their Control, Ronald Press, New York, pp. 436–437.

Cortesi, P., D.M. Gadowy, R.C. Seem and R.C. Pearson, (1995). Distribution and Retention of Cleistothecia of *Uncinula necator* on the Bark of Grapevine, Plant Dis., 9: 15–19.

Crowe, T.J. (1963). Possible insect vectors of the uredospores of *Hemileia vastatrix* in Kenya, Trans. Brit. Mycol. Soc., 46: 24-26.

Daayf, F., A. Schmith and R. R. Belanger (1995). The Effect of Plant Extracts of *Reynautria sachalinensis* on Powdery Mildew Development and Leaf Physiology of Long English Cucumber, Plant Dis., 79: 577.

Davis, R. M., (2004). Carrot Diseases and their Management. In: Naqvi, S. A. M. H., (ed.). Diseases of Fruits and Vegetables (Vol. 1, pp. 397–439). Kluwer Academic Publishers, The Netherlands.

Dik, A. J., M.A.Verhaar and R.R. Belanger, (1998). Comparison of Three Biological Control Agents against Cucumber Powdery Mildew in Semi-Commercial-Scale Glasshouse Trials, Eur. J. Plant Pathol., 104: 413.

Dimond, A.E. (1955). Pathogenesis in the Wilt Disease, Ann. Rev. Pl. Physiol., 6: 329-350.

Dimond, A.E. (1967). Physiology of Wilt Disease, in: The Dynamic Role of Molecular Constituents in Plant Parasite Interaction, C.J. Mirocha and I. Uritani (Eds.). Am. Phytopathol. Soc., St. Paul, Minnesota,

Dutt, B. L., (1979). Bacterial and Fungal Diseases of Potato (pp. 1–17). ICAR, New Delhi.

Ehrlich. H. G. and M. A. Ehrlich, (1963). Electron Microscopy of the Host–Parasite Relationships in Stem Rust of Wheat, Ann. J. Bot., 50: 123–130.

English, L. G., A. P. Norton, D. M. Gadoury, R. C. Seem and W. F. Wilcox, (1999). Control of Powdery Mildew in Wild and Cultivated Grapes by a Tydeid Mite, Biological Control, 14: 97.

Eriksson, J. (1894). Ueber die specialisierung des Parasitismus bei den Getreiderostpilzen, Berlin Deut. Botan. Ges., 12: 292-331.

Falk, S. P., D. M. Gadowap, R. C. Pearson and R. C. Seem (1995). Partial Control of Grape Powdery Mildew by the Mycoparasite *Ampelomyces quisqualis*, Plant Dis., 79: 483 – 490.

Fincham, J.R.S. and P.R. Day (1971). Fungal Genetics, Third edition, Blackwell, Oxford Davis, Philadelphia, p. 402.

Forsberg, J.L. (1979). Diseases of ornamental plants. Special Publication No. 3. University of Illinois at Urbana-Champaign. 222 pp.

Gaumann, E.A. and F.L. Wynd (1952). The Fungi (Translated by F.L. Wynd). Hafner Publ. Co., New York, p. 420.

Gleason, M.L., Daughtrey, M.L., Chase, A.R., Moorman, G.W., and Mueller, D.S. (2009). Diseases of herbaceous perennials. APS Press, St. Paul, MN. 281 pp (over 700 color photographs).

Gottlieb, D. (1943). The Presence of a Toxin in Tomato Wilt, Phytopathology, 43: 535.

Gottlieb, D. (1944). The Mechanism of Wilting Caused by *Fusarium bulbigenum* var. *lycopersici*, Phytopath., 34: 41-59.

Gupta, G. K., (1996). Fungal disease of temperate fruit in India, In: Singh, S. J., (ed.).

Gupta, S. K. and G. K. Gupta (1991). Eradicative Activity of Fungicides against Apple Powdery Mildew, Indian J. Mycol. Plant Pathol., 21:70–72.

Gupta, V. K., and S. K., Sharma (1999). Diseases of Fruit Crops (p. 3–27). Kalyani Publisher, Ludhiana.

Gupta, S.K. and T.S. Thind, (2018). Disease problems in vegetable production. Scientific Publishers.

Hadidi, A., Barba, M., Candresse, T., & Jelkmann, W., (2011). Virus and Virus-Like Diseases of Pome and Stone Fruits. St. Paul, MN: APS Press.

Hammarlund, C. (1925). Zur Genetick, Biologie and Physiologie einiger Erysiphaceen, Hereditas, 6: 1–126.

Heffer, V., M. L. Powelson, K. B. Johnson and N. Shishkoff (2006). Identification of Powdery Mildew Fungi, in: Laboratory Exercises in Plant Pathology. The American Phytopathological Society.

Helyer, N., N.D. Cattlin, and K.C. Brown, (2014). Biological control in plant protection: a colour handbook. CRC Press.

Hirata, K (1986). Host Range and Geographical Distribution of the Powdery Mildew Fungi, 2nd ed. Japan Scientific Society Press, Tokyo, Japan.

Hiroe, I., and S. Nishimura (1956). Pathological Studies on Watermelon Wilt-II: On Some Physiological and Chemical Properties of Phytonivein Wilt Toxin Produced by the Causal Fungus and its Metabolic Physiology, J. Agr. Chem. Soc., 30: 528-531.

Homma, Y. (1937). Erysiphaceae of Japan, J. Fac. Agr., Hokkaido Imp. Univ. 38: 183–461.

Horsfall J.G. and A.E. Dimond (Eds.). Academic Press, London and New York, 1: 349-390.

Horst, R.K. (2013). Westcott's plant disease handbook. Springer Science & Business Media.

Husain, A. and A. Kelman (1958). Relation of Slime Production to Mechanism of Wilting and Pathogenicity of *Pseudomonas solanacearum*, Phytopath., 48: 155-165.

Jaczewki (1927). Karmannii Opredelitel gribov, Leningrad, p. 626.

Jarvis, W.R. 1992. Managing diseases in greenhouse crops. APS Press, St. Paul, MN. 288 pp.

Jhooty, J. S. and G.D. Munshi (1990). Mode of Perpetuation of Powdery Mildew of Pea in north-western India, Indian Phytopath., 43: 455–456.

Jhooty, J.S. (1967). Identity of Powdery Mildew of Cucurbits in India, Pl. Dis. Reptr., 51: 1077–1080.

Jhooty, J.S. and D.S. Behar (1972). Control of Powdery Mildew of Pea and Pumpkin by Seed-Treatment with Benomyl, Set 16, pp. 152–159.

Kabitarani, K (1993). Sexuality and Pathological Specialization in *Sphacelotheca fuliginea*, Indian Phytopath., 46: 50–53.

Kalyan, K.M and J.P. Verma (2002). Biological control of cotton diseases, In: Biological Control of Crop Diseases, Eds. S.S. Gnanamanickam, CRC Press, U.S.A.

Kapoor, J. N. (1967). *Sphacelotheca fuliginea, E. cichoracearum*: C.M.I. Descriptions of Pathogenic Fungi and Bacteria, Set 16, pp. 152–159.

Khan, A. M., M. W. Khan, M. Akram and A. M. Khan (1972). Powdery Mildew Flora of Aligarh, Proc. Indian Sci. Congress (1972). p. 569.

Khan, A. M., M. W. Khan, and M. Akram (1971). Status of Cucurbit Powdery Mildew in India, Second Int. Symp. Pl. Path., IARI, New Delhi, pp. 144–145.

Khan, M. W. and A. M. Khan (1970). Studies on the Powdery Mildew I. Perithecial Production in Cucurbit Powdery Mildew in North India, Indian Phytopath., 23: 497–500.

Khan, M. W. and G. K. Sharma (1993). Races of *Sphaerotheca fuligi*nea Infecting Cucurbits in India and Host Differentials for Their Identification, Indian Phytopath., 46: 236–242.

Khurana, S. M. P., & Singh, M. N., (1986). Viral and mycoplasmal diseases of potato. Rev. Trop. Plant Pathol., 33, 123–184.

Kirk, P.M., P.F. Cannon, D.W. Minter and J.A. Stalpers (2008). Ainsworth and Bisby's Dictionary of the Fungi, 10th edition, CAB International, Wallingford, Oxon, U.K., p. 771.

Ko, W. H., S. Y. Wang, T. F. Hseih and P. J. Ann (2003). Effect of Sunflower Oil on Tomato Powdery Mildew caused by *Oidium neolycopersici*, J. Phytopathol., 151: 144–148.

Koike, S.T., P. Gladders, and A.O. Paulus (2007). Vegetable diseases: a color handbook. Gulf Professional Publishing.

Kolte, S.J. (1994). Diseases of Annual Edible Oilseed Crops, Vol. 1 Peanut Diseases, CRC Press Inc., Boca Raton, Florida, p. 142.

Kranz, J. S., & Werner, K., (1978). Diseases, Pests and Weeds in Tropical Crops (pp. 194–195). Wiley Publications.

Kumar, S., M. Sahai and A. B. Ray (1985). Chemical Constituents of the Leaves of *Flueggea microcarpa*, Plant Med., 5: 46.

Lakra, B.S. (1992). Correlation of Infection Intensities of Black Scurf with Yield Components of Potato, Indian J. Mycol. Pl. Pathol., 202: 203-204.

Lakshaman, P. and S. Mohan (1989). A Novel Method to Control Panama Disease of Banana Incited by *Fusarium oxysporum* f. sp *cubense*, Indian J. Mycol. Plant Pathol., 19: 93-97.

Laudon, G.F. (1973). Uredinales, in: The Fungi-An Advanced Treatise, IVB, G.C. Ainsworth, F.K. Sparrow and A.S. Sussman (Eds.). Academic Press, London and New York, pp. 247-279.

Lebeda, A., G.J. Holmes, B. Mauch-Mani, and M.J. Jeger (2011). The Downy Mildews-Biology, Mechanisms of Resistance, Population Ecology. European journal of plant pathology, 129(2).

Littlefield, L.J. (1981). The Biology of Plant Rusts, Iowa State University Press, Ames.

Ludwig, R.A. (1952). Studies on the physiology of hydromycotic wilting in the tomato plants, Macdonald Coll. Tech. Bull., 20: 1-40.

Ludwig, R.A. (1960). Toxin, in: Pl. path., J.G. Horsfall and A.E. Dimond (Eds). Academic Press, New York, 2: 315-357.

Management (p. 379). Scientific Publishers, Jodhpur, India.

Marchal, E. (1902–1903). De la specialisation de la parasitism chez.—I: Erysiphe graminis, Compt. Rend. Acad: des. Sci., Paris, 135: 210–212, 1067–1068; 136; 1280–1281.

Mathur, K., K. S.Shekhawat and L. K.Bhatia (1991). Performance of Pea Varieties against Powdery Mildew, Indian Phytopath., 44: 137.

Mehrotra, R. S., & Aggrawal, A., (2008). Plant Pathology Tata McGraw-Hill Publishing Co. Pvt. Ltd (p. 846). New Delhi – 110008. 0006 (India).

Mishra, A. P. and A. Krishna (1990). Effect of Calixin and Karathane on Development of Powdery Mildew Epidemics on Pea, Indian Phytopath., 43: 85–89.

Misra, A.P. and M. Mahmood (1960b). Effect of Nitrogen and Carbon Nutrition on Growth and Sporulation of *Colletotrichum capsici*, J. Indian Bot. Soc., 37: 314-321.

Misra, A.P. and M. Mahmood (1961). Effect of Vitamins and Hormones in Growth and Sporulation *Colletotrichum capsici*, Indian Phytopath., 14: 20-22.

Mundkur, B. B. (1949). Fungi and Plant Diseases, Macmillan, London, p. 246.

Munjal, R. L., V. V. Chenulu and T. S. Hora (1963). Assessments of Losses due to Powdery Mildew *Erysiphe polygoni* on Pea, Indian Phytopath., 16: 260–267.

Munshi, G. D. and T. Singh (1994). Effect of Temperature and Relative Humidity on the Development of *Uncinula necator* on Grapes, Indian J. Mycol. Plant Pathol., 24(3): 179–182.

Muthappa, B.N. and N. Kumari (1976). Comparative efficacy of four fungicides for control of coffee rust in South India, Plant Dis. Rept., 60: 879-883.

Muthappa, B.N. and N. Kumari (1981). Efficacy of Vitavax 75 WP for control of coffee rust, J. Coffee Res., 11(2): 47-48.

Naqvi, S.A.M.H., (2004). Diseases of fruits and vegetables. Springer.

Ogawa, J. M., & English, H., (1991). Diseases of Temperate Zone Tree Fruit and But Crops (p. 461). Pub. 3345. Univ. Calif. Div. Agric. Natural Resources, Oakland.

Panagopoulos, C. G., (2000). Diseases of vegetable crops. In: Vegetable Disease (pp. 15–189).

Parmeter, J.R.Jr. (1970). *Rhizoctonia solani*- Biology and Pathology, University of California Press, Berkeley, Los Angeles and London, p. 255.

Parthasarathy, S., G. Thiribhuvanamala and K. Prabakar (2020). Diseases of Field Crops and their Management, CRC Press, London.

Parthasarathy, S., G. Thiribhuvanamala, P. Muthulakshmi and K. Angappan (2020). Diseases of Forest Trees and their Management, CRC Press, London.

Pathak, K. D. and L. M. Joshi (1972). Chemical Control of Powdery Mildew of Wheat, Indian Phytopath., 25: 139–140.

Pathak, V. N., (1986). Diseases of Fruit Crops (p. 309). Oxford & IBHCo. New Delhi.

Pearson, R. C., D. G. Riegel and D. M. Gadowry (1994). Control of Powdery Mildew in Vineyards using Single Application, Vapour Action Treatments of Triazole Fungicides, Plant Dis., 78: 164–168.

Pegg, G.F. (1985). Life in a Black Hole – The Micro-environment of the Vascular Pathogen, Trans. Brit. Mycol. Soc., 85: 1-20.

Peralta, G. J. (1949). Report by the Plant Pathologist for the Year 1946–1947,. Dep. Agric. Malta, pp. 124–129.

Persley, D., Cooke, T., & House, S., (2010). Diseases of Vegetable Crops in Australia. CSIRO Publishing: Collingwood, Victoria.

Person, C. (1959). Gene for gene relationships in host parasite systems, Can. J. Bot., 37: 1101-1130.

Peterson, R.A. (1984). Mango Diseases, Proc. First Australian Mango Res. Workshop, p. 233.

Phadtare, S.G., P.H. Singh and G.S. Shekhawat (1990). Wart Disease of Potato in Darjeeling Hills, C. P. R. I., Technical Bulletin No. 19., p. 47,.

Pirone, P.P. 1978. Diseases and pests of ornamental plants. John Wiley & Sons, NY. 566pp.

Prabhavathy, K. G., & Reddy, S. R., (1995). Post-harvest fungal disease of chili (C. annuum) from Andhra Pradesh. Indian Phytopathology, 48(4). 492.

Prakash, O., (1996). Principal diseases of mango causes and control. In: Advances in Diseases of Crops in India (pp. 191–256). Kalyani Publisher, Ludhiana.

Prakash, Om and A.K. Misra (2001). Diseases of Mango, in: Diseases of Fruits and Vegetables and Their Management, T.S. Thind (Ed.). Kalyani Publishers, Ludhiana, pp. 47-72.

Rajan, P. P., Gupta, S. R., Sarma, Y. R., & Jackson, G. V. H., (2002). Diseases of ginger and their control with *Trichoderma harzianum*. Indian Phytopathology, 55(2). 173–177.

Ram, S. and L.D. Bist (1984). Occurrence of Malformin-like Substances in Malformed Panicles and Control of Floral Malformation in Mango, Scientia Hort., 24: 331-336.

Rama, T. S., Krishnan, & Sowmini, C. K., (1954). Rhizome and root rot of turmeric caused by *Pythium graminicolum*. Indian Phytopathology, 7, 152–159.

Rawal, R. D. (1998). Major Fungal Diseases of Tropical Fruits, in: Pathological Problem of Economic Crop Plants and Their Management, S. M. Paul Khurana (Ed.). Scientific Publications, Jodhpur (India). pp. 379–414.

Reddy, M. S. and A. Apparao (1971). A Systemic Fungicide Highly Effective against Powdery Mildew, Indian Phytopath., 24: 196–197.

Reddy, P. P., (2008). Diseases of Horticultural Crops: Nematode Problem and Their

Reddy, P. P., (2010). Bacterial and viral diseases and their management. In: Plant Protection in Horticulture (Vol. 3, p. 288). Scientific Publishers, Jodhpur, India.

Reddy, P. P., (2010). Plant Protection in Horticulture: Fungal Diseases and Their Management (Vol. 2, p. 359). Scientific Publishers, Jodhpur, India.

Reuveni, M. and R. Reuveni (1998). Foliar Applications of Mono-Potassium Phosphate Fertilizers Inhibit Powdery Mildew Development in Nectarines, Can. J. Plant Path., 20(3): 253.

Reuveni, M., H. Cohen, T. Zahavi and A. Venezian (2000). Polar-A Potent Polyoxin B Compound for Controlling Powdery Mildew in Apple and Nectarine Trees and Grapevines, Crop Protection, 19(6): 393.

Roy, A. K. (1973). Occurrence and Control of Powdery Mildews on a Few Plants in Assam, Indian Phytopath., 26: 578–581.

Salmon, E. S. (1900). A Monograph of the Erysiphaceae, Mem. Torrey Bot. Club., 91: 1–292.

Saharan, G.S., N.K. Mehta, and P.D. Meena (2019). Powdery Mildew Disease of Crucifers: Biology, Ecology and Disease Management. Springer Singapore.

Saharan, G.S., N.K. Mehta, and P.D. Meena (2017). Downy mildew disease of crucifers: biology, ecology and disease management. Singapore: Springer.

Sawada, K. (1914). On the Classification of Erysiphaceae by the Conidial Stage, (mostly in Japanese). Formosa Agr. Exp. Sta. Spec. Rep., 9: 1–102. (Abstract in Rev. Appl. Mycol., 7: 273[1928]).

Scott, K.J. and A.K. Chakravorty (Eds.) (1982). The Rust Fungi, Academic Press, New York.

Scott, K.J. and D.J. Maclean (1969). Culturing of rust fungi, Ann. Rev. Phytopath., 7: 123-146.

Sharma, K. K. and V. K Gupta (1994). Influence of Fungicides on Spore Germination, Sporulation and Control of Apple Powdery Mildew, Indian J. Mycol. Plant Pathol., 24: 24–28.

Sharma, S. K. and V. K. Gupta (1980). Efficiency of Fungicides in Relation to the of Application against Powdery Mildew of Apple, Indian J. Mycol. Plant Pathol., 10: 173–174.

Sharma, S. K., S. K. Gupta and D. S. Kaith, (1992). Efficacy of fungicides in relation to Number of Sprays against Powdery Mildew of Apple, Indian J. Mycol. Plant Pathol., 22: 267–269.

Shaw, M. (1963). The physiology and host-parasite relations of the rusts, A. Rev. Phytopath., 1: 259-294.

Sherf, A. F., & MacNab, A. A., (1986). Vegetable Diseases and Their Control (2nd edn., p. 728). John Wiley & Sons, NY.

Shrivastava, S. and S.B. Saksena (1968). Studies on the Rhizosphere and Rhizoplane Microflora of Potato with Special Reference to Black Scurf and Wilt Disease, Indian Phytopath. Soc. Bull., 4: 107-119.

Singh, D. V. and A. N. Mishra (1975). Control of Powdery Mildew Pea with Fungicides, Indian Phytopath., 28: 414–416.

Singh, R. S. (2007). Plant Diseases, Oxford and IBH Publ. Co. Pvt. Ltd, New Delhi, p. 720.

Singh, R. S., (1999). Diseases of Vegetable Crops (pp. 207–236). Oxford & IBH Publishing Co. Pvt. Ltd. New Delhi – 110001, Calcutta.

Singh, R.S., H.S. Chaube and N.Singh (1972). Studies on the Control of Potato Black Scurf Disease, Indian Phytopath., 25: 343-349.

Singh, D. and R.R. Sharma (2018). Postharvest diseases of fruits and vegetables and their management. In Postharvest disinfection of fruits and vegetables (pp. 1-52). Academic Press.

Sohi, H. S. and S. K. Nayar (1969). Some New Records of Fungi from India-I, Indian Phytopath., 22: 410–412.

Sohi, H. S. and T. S. Sridhir (1978). Evaluation of Some Fungicides for the Control of Powdery Mildew of Citrus, Res Rept. Fruit Res. Workshop, Hyderabad, pp. 63.

Sohi, H. S. and T. S. Sudhir (1973). Powdery Mildew of Mango and its Control, Indian Phytopath., 26: 361–362.

Spencer, D. M et al. (1978). The Powdery Mildews, Academic Press, New York.

Stakman, E.C. and J.G. Harrar (1957). Principles of Plant Pathology, Ronald Press, New York, p. 581.

Stamoulis, Athens. Persley, D., Cooke, T., & House, S., (2010). Diseases of Vegetable Crops in Australia. CSIRO Publishing: Collingwood, Victoria.

Staples, R.C. and W.K. Wynn (1965). The physiology of urediospores of the rust fungi, Bot. Rev., 31: 537-564.

Starker, C. (1954). Dusting, Spraying Conference, Agricultural Chemicals, 9: 39–41.

Suhag, L. S. and R. S. Rana (1984). Efficacy and Cost Benefit Ratio of Different Doses and Number of Sprays in Controlling Pea Powdery Mildew, Indian Phytopath., 37: 193–195.

Thind, T. S., C. Mohan, R. Raj and J. K. Arora (2002). Activity of Strobilurin Fungicides against Some Pathogenic Fungi, J. Mycol. Pl. Pathol., 32(3): 425.

Toyoda, H., H. Hashimoto, R. Utsumi, H. Kobayashi, and S. Ouchi (1988). Detoxification of Fusaric Acid by a Fusaric Acid-resistant Mutant of *Pseudomonas solanacearum* and its Application to Biological Control of Fusarium Wilt of Tomato, Phytopathol., 78: 1307-1311.

Trigiano, R.N., Windham, M.T., and Windham, A.S. 2003. Plant Pathology concepts and laboratory exercises. 2nd Ed. CRC Press, Boca Raton. 558 pp.

Uritani, I. And T. Akazawa (1959). Alteration of the respiratory pattern of infected plants, in: Plant Pathology,

Walker, J. C., (1952). Disease of crucifers. In: Diseases of Vegetable Crops (pp. 150–152). McGraw Hill Book Co., New York.

Waraitch, K. S., G. D. Munshi and J. S. Jhooty (1975). Fungicidal Control of Powdery Mildew of Bottle Gourd, Indian Phytopath., 28: 556–557.

Yarwood, C. E.(1957). Powdery Mildews, Bot. Rev., 23: 235–300.

Yildirim, J., E. Onogur and M. Irshad (2002). Investigations on the Efficacy of Some Natural Chemicals against Powdery Mildew (*Uncinula necator*) of Grapes, J. Phytopath., 150(11–12): 697.

Zaracovitis, C. (1965). Attempts to Identify Powdery Mildew Fungi by Conidial Characters, Trans. Br. Mycol. Soc., 48(4): 553–558.

Index

A

Abelmoschus esculentus 81
Abnormal leaf fall 339
Acidovorax anthurii 160
Acorus calamus 473
Aegle marmelos 421
Agrobacterium tumefaciens 7, 192
Albugo candida 97
Algal spot 31, 43, 53
Allium cepa 285
Allium sativum 276
Allodus podophylli 382
Aloe barbadensis 411
Alternaria alternata 35, 184, 385, 461
Alternaria blight 100, 115, 132, 371
Alternaria brassicae 96
Alternaria burnsii 266
Alternaria dianthi 167, 214
Alternaria flower blight 167
Alternaria hibiscinum 82
Alternaria leaf blight 203, 242, 413, 453
Alternaria leaf spot 35, 82, 85, 95, 167, 174, 183, 188, 255, 478
Alternaria raphani 115
Alternaria rot 152
Alternaria solani 123, 142
Alternaria tenuis 152, 416
Alternaria zinniae 242
American ginseng 367
Amomum subulatum 330
Anabe Roga 301
Anacardium occidentale 306
Ananas comosus 65
Andrographis paniculata 448
Angular leaf spot 76, 117, 453
Annona squamosa 34
Anthracnose 10, 19, 24, 33, 34, 38, 41, 44, 47, 55, 58, 63, 73, 112, 150, 158, 193, 204, 209, 221, 224, 295, 335, 366, 374, 398, 436, 466, 474
Anthurium 157
Anthurium andraeanum 157
Aonla 44

Apium graveolens 105
Apple 3
Areca catechu 301
Arecanut 301
Armillaria mellea 356
Armillaria root rot 356
Aromatic plants 359
Artemisia pallens 365
Artocarpus heterophyllus 45
Ascochyta and septoria leaf spots 174
Ascochyta blight 122
Ascochyta chrysanthemi 183
Ascochyta leaf spot 425
Ascochyta pinodes 122
Ashwagandha 415
Asian ginseng 367
Asparagus 418
Asparagus racemosus 418
Asparagus virus I 420
Aspergillus niger 277, 290
Aspergillus rot 55, 277
Aspermy 185
Aster 161
Aster amellus 161
Aster yellows 164, 176, 193, 215
Atropa belladonna 422
Azadirachta indica 449
Azhukal disease 347

B

Bacterial blight 68, 110, 158, 184, 215, 252, 453, 483
Bacterial brown rot 225
Bacterial canker 147
Bacterial gall 216
Bacterial leaf spot 51, 146, 160, 206, 244, 252, 257, 392, 473
Bacterial leaf stripe 305
Bacterial rot 407
Bacterial scab 200
Bacterial shot hole 421
Bacterial slow wilt 170
Bacterial soft rot 102, 141, 153, 218, 223, 412
Bacterial stem rot 191
Bacterial wilt 14, 89, 116, 128, 145, 159, 170, 282, 443
Balansia sclerotica 380
Banana 9
Banana bract mosaic 18, 350
Banana bunchy top virus 16
Banana mosaic 17
Banana streak mosaic 18
Bangle disease 128
Basal rot 166, 207, 286
Basal stalk rot 107
Basal stem rot 316, 411
Basal wilt 336
Bean 73
Bean common mosaic 78
Beet curly top virus 80
Beetroot 79
Belladonna 422
Belladonna mosaic 426
Belladonna mottle 426
Ber (Jujube) 20
Berry blotch 328
Beta vulgaris sub sp. *vulgaris* 79
Betelvine 249
Bhendi 81
Bipolaris incurvata 322
Bird's eye spot 38, 341
Bitter rot 8

Black canker 36
Black cardamom 330
Black leaf spot 95
Black leg 93
Black mildew 422
Black mold 290
Black mould blemish 277
Black nightshade 427
Black nose disease 158
Black pod 311
Black root rot 356, 461
Black rot 19, 39, 65, 97, 135, 157, 182, 217, 224, 295, 329, 353
Black scurf 125
Black spot 3, 12, 51, 225, 231, 392, 471
Black tip 19, 54
Blight 266
Blister blight 351
Blossom blight 183, 195, 238
Blossom brown necrotic streak 219
Blotch 177, 363
Blue mould 8, 32
Botryis cinerea 8
Botryodiplodia fruit rot 63
Botryodiplodia rot 19
Botryodiplodia theobrumae 35, 55, 63, 69, 232, 312
Botryosphaeria canker 363
Botrytis blight 164, 182, 190, 196, 208, 213, 234, 238, 373, 460
Botrytis cinerea 19, 39, 191, 234, 460
Botrytis flower blight 169
Botrytis soft rot 198
Brassica oleracea 90
Bremia lactucae 195
Brinjal 83
Bronzy wilt 149
Broom rape 394
Brown eye spot 328
Brown leaf spot 103, 307
Brown root disease 344
Brown root rot 356
Brown rot 33, 128, 224
Brown rust 179
Brown scale 209
Bud rot 303, 317, 465
Bulb rot 207, 276, 286, 406
Burkholderia caryophylli 170
Burkholderia gladioli 201
Burkholderia solanacearum 14, 89, 128, 160
Bushy dwarf virus 333
Button shedding 303
Butt rot 65

C

Cabbage 90
Cabbage yellows 94
Cadang-Cadang 324
Callistephus chinensis 171
Camellia sinensis 351
Camphor 361
Canker 28, 51, 429
Canna yellow mottle virus 252
Capnodium citri 26
Capnodium sp. 45, 70, 467
Capnodium theae 355
Capsicum annuum 253
Capsule rot 347, 392
Cardamom mosaic virus 349
Cardamom necrosis disease 350
Cardamom vein clearing 350
Carica papaya 56

Carnation 165
Carrot 99
Cashew 306
Cassava 103
Cassava mosaic 105
Catharanthus roseus 454
Cattleya Leaf-necrosis 219
Cattleya spp. 217
Cauliflower 90
Cauliflower mosaic virus 98
Celery 105
Celery mosaic 110
Centella asiatica 444
Cephaleuros mycoides 337
Cephaleuros virescens 31, 43, 53
Ceratobasidium theobromae 314
Ceratocystis paradoxa 20, 65
Ceratostomella paradoxa 319
Cercospora abelmoschi 82
Cercospora apii 108
Cercospora beticola 79
Cercospora brunkii 367
Cercospora canker 419
Cercospora carotae 100
Cercospora centellae 444
Cercospora foeniculi 269
Cercospora golaghati 152
Cercospora leaf blight 99
Cercospora leaf spot 67, 79, 81, 86, 115, 134, 204, 214, 221, 243, 255, 274, 425, 434
Cercospora punicae 68
Cercospora rauvolfiae 465
Cercospora traversiana 275
Cercosporidium henningsii 103
Certatocystis fimbriata 135
Charcoal pod rot 312
Charcoal rot 59, 344, 476
Charcoal stump rot 356
Charkarai Tulsi 475
Chebulic Myrobalan 375
Chenthal disease 348
Cherelle rot 312
Chilli 253
China aster 171
Chirkey disease 332
Chlorophytum borivilianum 403
Chlorotic streak disease 350
Chrysanthemum 177
Chrysanthemum aspermy virus 185
Chrysanthemum cinerariifolium 228
Chrysanthemum stunt viroid 185
Chrysopogon zizanioides 377
Cigar end rot 13
Cinchona 427
Cinchona officinalis 427
Cinnamomum camphora 361
Cinnamomum zeylanicum 259
Cinnamon 259
Citronella 363
Citrus 21
Citrus greening 29
Citrus tristeza virus 30
Clavibacter michiganensis sub sp. *michiganensis* 147
Clove 261
Club root 92
Clump rot 346
Cocoa 309
Cocoa swollen shoot disease 315
Coconut 316
Cocos nucifera 316

Cocoyam 138
Coffea arabica 326
Coleosporium solidaginis 164
Coleus forskohlii 441
Collar and root rot 400
Collar rot 102, 172, 203, 271, 273, 326, 344, 364, 380, 388, 404, 424
Colletotrichum anilla 295
Colletotrichum asparagi 419
Colletotrichum capsici 254, 293
Colletotrichum circinans 288
Colletotrichum crossandrae 188
Colletotrichum dematium 113, 458
Colletotrichum gloeosporioides 25, 34, 46, 48, 55, 58, 67, 193, 259, 261, 283, 303, 307, 313, 335, 340, 466
Colletotrichum gloeosporioides f. sp. *alatae* 150
Colletotrichum graminicola 378
Colletotrichum leaf blight 451
Colletotrichum leaf disease 340
Colletotrichum leaf spot 279
Colletotrichum lindemuthianum 73
Colletotrichum sp. 63, 416
Colletotrichum zingiberis 280
Colocasia esculenta 138
Common bacterial blight 77
Common scab 130
Condiments 247
Coriander 263
Coriandrum sativum 263
Corm decay 405
Corm neck rot 405
Corm rot 138, 199, 405
Corm soft rot 139
Corticium koleroga 330
Corticium salmonicolor 342
Corynespora cassiicola 343, 443
Costus speciosus 431
Cottony rot 173
Cottony soft rot 402
Crinipellis perniciosa 314
Crocus sativus 405
Crossandra 186
Crown choking 325
Crown gall 6, 192, 235, 403
Crown rot 20, 229, 303
Crucifers 90
Cucumber mosaic 117
Cucumber mosaic virus 111
Cucumber mosaic virus 201, 210, 252
Cucurbita L. 111
Cumin 265
Cuminum cyminum 265
Curcuma longa 292
Curcurbits 111
Curly top 80
Curry leaf 267
Curvularia lunata 384
Cuscuta subinclusa 32
Custard apple 34
Cylindrocladium colhounii 34
Cylindrocladium leaf 34
Cylindrocladium scoparium 306
Cylindrosporium chrysanthemi 178
Cymbidium mosaic disease 296
Cymbidium mosaic virus 223, 297
Cymbidium spp. 220
Cymbopogon flexuosus 377
Cymbopogon martini 395
Cymbopogon winterianus 363

D

Dahlia 189

Damping off 56, 83, 90, 105, 118, 141, 171, 210, 239, 253, 267, 269, 326, 346, 365, 368, 387, 415, 425, 430, 432, 457, 468

Darchini 259

Dasheen 138

Datura stramonium 478

Daucus carota sub sp. *sativus* 99

Davana 365

Deadly nightshade 422

Defoliation disease 144

Dendrobium spp. 220

Dendrophthoe falcata 43, 52

Devil's trumpets 478

Dianthus caryophyllus 165

Diaporthe vexans 87

Didymellina dianthi 168

Die back 24, 231, 254, 259, 308, 400, 416, 454, 458

Digitalis lanata 436

Digitalis purpurea 436

Dioscorea alata 150, 432

Dioscorea bulbifera 150

Dioscorea esculenta 150

Dioscorea latent virus 436

Dioscorea spp. 150

Diplocarpon rosae 231

Diplodia gummosis 23

Diplodia natalensis 51, 104

Diplodia rosarum 232

Diplodia rot 35

Diplodia stem-end rot 50

Disappearing root rot 369

Distortion mosaic 481

Dodder 32

Downy mildew 37, 95, 113, 195, 233, 273, 287, 389, 403, 438, 444, 446

Drechslera halodes 414

Drechslera heveae 341

Drechslera leaf blight 414

Drechslera leaf spot 434

Drechslera sacchari 378

Drumstick 118

Dry root rot 24, 59, 75

Dry rot 344

E

Early blight 108, 123, 142

East Indian rosebay 366

Elettaria cardamomum 345

Elsinoe ampelina 38

Elsinoe fawcettii 27

Erwinia amylovora 5

Erwinia carotovora 15, 153

Erwinia carotovora pv. *carotovora* 103

Erwinia carotovora sub sp. *carotovora* 129, 192, 223

Erwinia chrysanthemi pv. *chrysanthemi* 184

Erwinia papaveris 393

Erwinia tracheiphila 116

Erysiphe cichoracearum 114, 383

Erysiphe cinnamomi 361

Erysiphe polygoni 75, 101, 121, 265, 291, 391

Erythricium salmonicolor 28

Eulophia spp. 224

European anthracnose 227

Exobasidium vexans 351

F

Fennel 268

Fenugreek 272

Finger tip rot 19
Fire blight 5, 64
Flat limb 69
Floral malformation 54
Floury leaf spot 76
Flower blight 174, 190
Flower bud rot 214, 239
Flower bud shedding 261
Flower crops 155
Flower spot 238
Foeniculum vulgare 268
Foliar blight 367
Fomes lamoenis 356
Foorkey disease 333
Foot rot 56, 122, 161, 209, 249, 301, 457, 480
Fox-glove 436
Freckle disease 12
Fruit canker 42, 421
Fruit crops 1
Fruit rot 10, 41, 87, 284, 301, 347, 479
Fruit spot 34, 67
Fusarium crown 418
Fusarium equiseti 238
Fusarium moniliforme 364
Fusarium oxysporum f. sp. *apii* 107
Fusarium oxysporum f. sp. *batatas* 136
Fusarium oxysporum f. sp. *callistephi* 161, 172
Fusarium oxysporum f. sp. *cattleyae* 217
Fusarium oxysporum f. sp. *chrysanthemi* 178
Fusarium oxysporum f. sp. *cumini* 265
Fusarium oxysporum f. sp. *gladioli* 197
Fusarium oxysporum f. sp. *lacimiti* 482
Fusarium oxysporum f. sp. *lycopersici* 143
Fusarium oxysporum f. sp. *raphani* 112
Fusarium pallidoroseum 119
Fusarium proliferatum 415
Fusarium root rot 423
Fusarium solani 186, 423
Fusarium wilt 9, 40, 84, 94, 111, 120, 136, 143, 161, 165, 202, 228, 256, 388, 397, 472
Fusarium yellows 107, 197
Fusicladium dendriticum 3

G

Ganoderma lucidum 302
Ganoderma root rot 23
Ganoderma wilt 316, 451
Garden Geranium 366
Garlic 276
Gerbera 193
Giant mistletoe 43, 52
Ginger 278
Ginseng 367
Gladiolus communis 197
Gloeosporium musarum 11, 19
Gloeosporium psidii 41
Glomerella cingulata 8, 34, 44, 193, 280, 307
Glomerella leaf spot 243
Glomerella lindemuthianum 73
Gloriosa superba 439
Glory Lily 439
Glycyrrhiza glabra 380
Golovinomyces chrysanthemi 181
Golovinomyces cichoracearum 163, 194
Grapevine 36
Grassy shoot 380, 395
Grassy top 202
Green mould 32

Green scurf 31
Grey blight 42, 49, 320, 352, 379
Grey mould 8, 175, 182
Grey mould rot 39
Guava 40
Guignardia bidwellii 39
Guignardia dioscoreae 152
Guignardia musae 13
Gummosis 21
Gymnema 374

H

Haritaki 375
Head 15
Heart mosaic 111
Heart rot 65, 227
Helminthosporium papaveins 391
Hemileia americana 218
Henbane 375
Henbane mosaic 376
Hevea brasiliensis 338
Horse hair blight 284
Huanglongbing 29
Hyaloperonospora brassicae 95
Hyoscyamus niger 375

I

Indian Coleus 441
Indian ginseng 415
Indian gooseberry 44
Indian ipecac 376
Indian pennywort 444
Indian snakeroot 464
Infectious chlorosis 17
Inflorescence blight 222, 309
Inflorescence die back 303
Internal bulb rot 277
Ipomoea batatas 134
Isabgol 444

J

Jackfruit 45
Jasmine 202
Jatropha curcas 398
Justicia adhatoda 382

K

Kalmegh 448
Katte disease 349
Kerala wilt 322
Khus Khus 377
Kokkan disease 18
Kokke Kandu 350

L

Lasiodiplodia leaf blight 320
Lasiodiplodia theobromae 69, 312, 321
Late blight 109, 124, 144
Leaf blight 157, 188, 221, 225, 268, 348, 362, 363, 381, 385, 391, 396, 397, 403, 416, 422, 432, 435, 438, 440, 441, 454, 465, 469, 474, 479
Leaf blotch 280, 293, 396
Leaf curl 60, 61, 66, 148, 245, 258, 417
Leaf fall 435
Leaf roll 131
Leaf rot 249, 262, 321
Leaf rust 326
Leaf sheath 228
Leaf smut 190
Leaf spot 46, 67, 151, 162, 168, 174, 177, 178, 183, 191, 194, 199, 205, 214, 221, 224, 226, 227, 228, 238, 243, 250, 259, 261, 269, 281, 293, 331, 342, 362, 367, 378, 381, 382, 384, 401, 414, 417, 419, 425, 434,

437, 438, 443, 447, 452, 460, 465, 467, 472, 474, 475, 476
Leaf web blight 376, 448, 450
Lemongrass 377
Leptosphaeria coniothyrium 232
Leptosphaeria maculans 93
Lethal yellowing 323, 365
Leveillula taurica 464
Lilium 207
Lily mottle virus 210
Liquorice 380
Little leaf 90, 262, 380, 395, 482
Long smut 379
Lower stem rot 418

M

Macrophomina phaseolina 24, 59, 75, 187, 277, 442, 449, 451
Macrophomina root rot 388
Macrophomina rot 277
Mahali disease 301
Makoi 427
Malabar nut 382
Malformation 54
Malus × domestica 3
Mangifera indica 47
Mango 47
Manihot esculenta 103
Manilkara zapota 69
Marasmius scandens 310
Marble mosaic 349
Marigold 210
Mayapple 382
Medicinal plants 409
Melioa mangiferae 50
Meloidogyne incognita 338
Mentha arvensis 383
Microsphaera alphitoides 361
Mint 383
Moko disease 14
Moniliophthora perniciosa 313
Moringa 118
Mosaic 7, 59, 98, 147, 176, 192, 215, 236, 245, 349, 393, 441
Mosaic diseases 257
Mosaic flower-break 219
Murraya koenigii 267
Musa sp. 9
Mycosphaerella musicola 12
Mycovellosiella phaseoli 77
Myristica fragrans 283
Myrothecium roridum 417

N

Neck rot 276
Nematode and fungal complex 338
Nerium oleander 216
Nilgiri necrosis disease 350
Nilgiri twig disease 328
Nursery leaf spots 345
Nut neg 283

O

Ocimum basilicum 471
Odontoglossum ring spot 219
Oidium caricae-papayae 57
Oidium chrysanthemi 181
Oidium erysiphoides 194
Oidium heveae 339
Oidium mangiferae 47
Oidium piperis 251
Oidium tingitaninum 25
Oleander 216
Onion 285
Opium poppy 387

Opium poppy mosaic 394
Orchids 217
Oriental leaf disease 326

P

Palmarosa 395
Panama disease 9
Papaver somniferum 387
Papaya 56
Papaya mosaic virus 59
Paphiopedilum spp. 225
Patchouli 397
Pea 120
Peach 61
Pear 64
Pectobacterium carotovorum 129
Pectobacterium carotovorum sub sp. *carotovorum* 103, 218
Pelargonium capitatum 401
Pelargonium graveolens 401
Pelargonium × hortorum 366
Pellicularia salmonicolor 45
Pencil point disorder 325
Penicillium corm rot 198
Penicillium digitatum 32
Penicillium expansum 8
Penicillium gladioli 199
Penicillium italicum 32
Pepper ringspot virus 252
Periwinkle 454
Peronospora arborescens 390
Peronospora destructor 287
Peronospora parasitica 95
Peronospora plantaginis 446
Peronospora sparsa 233
Peronospora trigonellae 273
Pestalotiopsis mangiferae 49, 379
Pestalotiopsis palmarum 320
Pestalotiopsis psidii 42
Pestalotiopsis theae 352
Pestalotiopsis versicolor 69
Petal blight 222, 226
Phaeoisariopsis griseola 76
Phakopsora zizyphi-vulgaris 21
Phalaenopsis spp. 224
Phanerogamic parasite 32, 43, 52
Phaseolus vulgaris 73
Phavophleospora indica 69
Pholiota imbricata 226
Phoma lingam 93
Phoma rot 153
Phomopsis annonacearum 36
Phomopsis blight 87
Phomopsis canker 419
Phomopsis citri 33
Phomopsis leaf spot 221
Phomopsis vexans 87
Phragmidium mucronatum 233
Phyllanthus emblica 44
Phyllody 206, 337, 463, 468
Phyllosticta colocasiophila 140
Phyllosticta gerbericola 194
Phyllosticta glycyrrhizae 381
Phyllosticta leaf spot 243, 267
Phyllosticta musarum 13
Phyllosticta zingiberi 282
Physicnut 398
Physiological disorder 54
Phytophhora meadii 295
Phytophthora aerial blight 455
Phytophthora arecae 301, 304
Phytophthora blight 165
Phytophthora cactorum 8, 209

Phytophthora capsici 334
Phytophthora cinnamomi 66, 138, 260
Phytophthora cryptogea 162, 172
Phytophthora dieback 260
Phytophthora foot rot 333
Phytophthora infestans 124, 145, 153
Phytophthora meadii 347
Phytophthora nicotianae var. *parasitica* 141, 157
Phytophthora palmivora 22, 195, 227, 310, 311, 318, 340
Phytophthora parasitica var. *piperina* 249
Phytophthora root rot 367, 423, 456
Phytophthora rot 8, 153
Phytophthora spp. 33
Pineapple 65
Pineapple disease 20
Pink disease 27, 45, 308, 341, 353, 428
Pink root 278, 289
Pink rot 106
Piper betle 249
Piper nigrum 333
Piper yellow mottle virus 252
Pisum sativum 120
Plantago ovate 444
Plantations crops 299
Plasmodiophora brassicae 92
Plasmopara halstedii 195
Plasmopara viticola 37
Plum 66
Podophyllum peltatum 382
Podosphaera leucotricha 4
Podosphaera pannosa 230
Pod rot 311
Pogostemon cablin 397
Polianthes tuberosa 237
Pollu disease 335
Pomegranate 67
Poria hypolateritia 356
Postharvest diseases 8, 19, 32, 39, 55, 63, 152
Potato 123
Powdery mildew 4, 25, 36, 47, 57, 62, 74, 81, 101, 114, 120, 163, 181, 189, 194, 213, 230, 241, 251, 256, 265, 266, 270, 272, 291, 308, 338, 361, 373, 374, 383, 390, 399, 427, 447, 452, 459, 464, 483
Powdery scab 127
Prunus domestica 66
Prunus persica 61
Pseudocercospora contraria 152
Pseudocercospora dendrobii 222
Pseudocercospora musae 12
Pseudomonas caryophylli 170
Pseudomonas cichorii 110
Pseudomonas solanacearum 14, 89, 128, 146, 160, 282
Pseudomonas sp. 362
Pseudomonas syringae pv. *lachrymans* 117
Pseudomonas syringae sub sp. *savastanoi* 216
Pseudoperonospora cubensis 113
Psidium guajava 40
Puccinia chrysanthemi 179
Puccinia horiana 180
Puccinia nakanishikii 379
Puccinia pruni-spinosae 62
Puccinia substriata var. *penicillariae* 87
Punica granatum 67
Purple blotch 285

Purple spot 420
Pyrenochaeta terrestris 289
Pyrethrum 228
Pyrus communis 64
Pythium aphanidermatum 56, 57, 83, 118, 139, 141, 253, 267, 279, 292
Pythium butleri 141
Pythium debaryanum 91, 106
Pythium rhizome rot 432
Pythium root rot 422, 456, 477
Pythium splendens 217
Pythium vexans 346

Q

Quick decline 30
Quick wilt 333

R

Radish 132
Radish mosaic virus 133
Radish phyllody 134
Radopholus similis 338
Ralstonia solanacearum 14, 89, 128, 146, 160, 282, 443
Ramularia foeniculi 268
Raphanus raphanistrum subsp. *sativus* 132
Rauvolfia serpentina 464
Ravenelia emblicae 44
Ray blight 229
Ray (flower) blight 182
Red leaf spot 378, 395
Red root rot 356
Red rust 31, 43, 53, 337, 354
Rhizobium radiobacter 7, 192, 235
Rhizoctonia bataticola 59, 75, 380, 385
Rhizoctonia solani 91, 106, 108, 118, 126, 162, 187, 366, 448
Rhizoctonia stem rot 181
Rhizome rot 278, 292, 346, 431
Rhizopus atrocarpi 46
Rhizopus fruit rot 46
Rhizopus rot 55, 63, 137
Rhizopus stolonifer 63, 459
Rhynchostylis retusa 226
Ribgrass mosaic virus 439
Ring disease 128
Ring spot 60, 167, 168
Ripe fruit rot 254
Root and collar rot 431
Root and crown rot 188
Root and stem rot 162, 294, 296
Root diseases 356
Root gall 402
Root rot 133, 139, 157, 166, 172, 186, 196, 212, 239, 271, 273, 366, 368, 380, 387, 388, 397, 442, 449, 450, 480, 481
Root (Wilt) disease 322
Rose 230
Rosellinia arcuata 356
Rosemary 402
Rose scented geranium 401
Rosette 236
Rosmarinus officinalis 402
Rubber 338
Rust 21, 44, 62, 74, 86, 121, 164, 168, 175, 199, 205, 218, 232, 296, 348, 378, 383, 399, 407, 411, 458, 475
Rusty root rot 371

S

Safed Musli 403
Saffron 405
Sapota 69

Sarpagantha 464
Scab 3, 26, 42, 64
Scab/black spot 3
Scald disease 285
Sclerotial rot 397
Sclerotinia blight 389
Sclerotinia collar rot 229
Sclerotinia rot 173
Sclerotinia sclerotiorum 101, 106, 271
Sclerotinia white rot 370
Sclerotium collar rot 88
Sclerotium foot rot 250
Sclerotium rolfsii 80, 88, 102, 166, 237
Sclerotium root rot 80
Sclerotium wilt 203
Secondary leaf fall 340
Seed decay 368
Seedling blight 260, 306, 375, 387, 427
Seedling die back 309
Seedling mortality 415
Seedling rot 330
Seedling wilt 262, 285
Senna alexandrina 468
Septoria apiicola 109
Septoria chrysanthemella 177, 229
Septoria leaf spot 144, 169, 243
Sett rot 104
Sheath rot 364
Shell soft rot 421
Shot hole 283
Sigatoka leaf spot 11
Slow decline 338
Small cardamom 345
Smudge 288
Smut 396
Sobralia-blight 225
Soft rot 8, 129, 137, 278, 393
Solanum lycopersicum 141
Solanum melongena 83
Solanum nigrum 427
Solanum tuberosum 123
Solanum viarum 482
Sooty mould 26, 42, 45, 49, 70, 329, 355, 413, 467
Southern blight 203, 240
Sphaceloma fawcettii 27
Sphaerotheca fuligena 114
Sphaerotheca pannosa 63
Spices 247
Spilocaea pomi 3
Spindle tuber 131
Stalk end rot 55
Stalk rot 195
Stem bleeding 304, 319
Stem blight 462
Stem canker 125, 174, 227, 252, 354, 400
Stem end rot 33, 55, 65
Stem gall 263
Stemphylium botryosum 392
Stemphylium leaf spot 174
Stem pitting disease 30
Stem rot 56, 104, 136, 166, 180, 187, 211, 237, 459
Stevia 475
Stevia rebandiana 475
Stolon rot 385
Streptomyces scabies 130
Stripe canker 260, 429
Stump rot 344
Stunt 170, 185
Sudden death 261

Sugar apple 34
Surface rot 136
Sweet basil 471
Sweet flag 473
Sweet leaf 475
Sweet potato 134
Syncephalastrum racemosum 421
Synchytrium endobioticum 126
Syzygium aromaticum 261

T

Tabernaemontana coronaria 366
Tagetes erecta 210
Tamarind 291
Tamarindus indicus 291
Tanjore wilt 316
Taphrina deformans 61, 66
Taphrina maculans 280, 293
Tapioca 103
Target spot 466
Taro 138
Taro leaf blight 139
Tea 351
Terminalia chebula 375
Theobroma cacao 309
Thielaviopsis paradoxa 65, 304
Thorn apple 478
Thread blight 284, 329
Tip blight 428
Tip over 15
Tobacco ring spot virus 201
Tomato 141
Tomato bushy stunt virus 336
Tomato spotted wilt 149, 462
Top blight girdle canker 429
Top rot 15
Trigonella foenum-graecum 272
Tristeza 30
Tropical soda apple 482
Tuberose 237
Tuber rot 404, 433, 434, 439
Tulip breaking virus 210
Tumour 263
Turmeric 292
Twig blight 261
Twig canker 119
Twig die back 241, 354
Tylophora indica 376

U

Uncinula necator 37
Uredo elettariae 348
Uromyces joffrini 296
Uromyces phaseoli-typica 74
Uromyces transversalis 199
Uromyces trigonellae 275
Uromyces viciae-fabae 121
Ustulina zonata 345, 356

V

Vanda caerulea 227
Vanilla 294
Vanilla planifolia 294
Vasaka 382
Vascular streak die back 314
Vegetable crops 71
Vegetative malformation 54
Vein clearing 82
Venturia inaequalis 3
Venturia pirina 64
Verticillium dahlia 212
Verticillium theobromae 13

Verticillium wilt 85, 212, 361, 370
Verucosis 26
Viral diseases 152, 201, 210, 219, 223
Vitis vinifera 36

W

Wart 126, 448
Watery soft rot 101
Web blight 454
Whiptail 99
White mould 106, 389
White root disease 343
White rot 101, 289
White rust 96, 133, 179
White thread blight 284, 310
Wilt 171, 178, 186, 189, 211, 217, 220, 237, 249, 264, 265, 270, 274, 386, 445, 464, 482
Wilt and corm rot 197
Wire stem 91
Witches' broom 313, 418
Withania somnifera 415
Wither tip 24
Wood apple 421

X

Xanthomonas axonopodis 28, 51, 68
Xanthomonas axonopodis pv. *dieffenbachiae* 159
Xanthomonas campestris 98, 392
Xanthomonas campestris pv. *phaseoli* 77
Xanthomonas campestris pv. *vesicatoria* 146
Xanthomonas campestris pv. *zinnniae* 244
Xanthomonas vesicatoria 146

Y

Yam 150
Yellow disease 281
Yellow leaf disease 305
Yellow mosaic 394, 401
Yellow vein leaf curl disease 449
Yellow vein mosaic 82

Z

Zingiber officinale 278
Zinnia elegans 239
Zinnia leaf curl virus 245
Ziziphus mauritiana 20

Color Plates

1. Mango malformation

2. Mango powdery mildew

3. Mango anthracnose

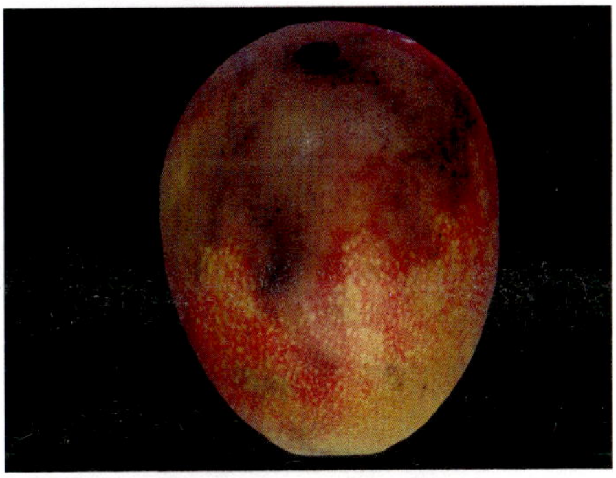

4. Mango stem-end rot

5. Citrus sootymould

6. Citrus green and blue mould

7. Citrus canker

8. Grapevine powdery mildew

9. Guava algal spot

10. Jack fruit rot

11. Banana bunchy top disease

12. Banana sigatoka leaf spot

13. Banana moko wilt

14. Banana fusarium wilt

15. Papaya ring spot disease

16. Papaya mosaic disease

17. Guava fruit rot

18. Peach leaf curl

19. Sapota leaf spot

20. Pomegranate bacterial spot

21. Brinjal little leaf

22. Brinjal phomopsis leaf spot

23. Bhendi yellow vein mosaic virus

24. Bhendi powdery mildew

25. Tomato early blight

26. Tomato septoria leaf spot

27. Tomato spotted wilt

28. Bottle gourd leaf spot

29. Cucumber mosaic virus

30. Chayote yellow mosaic virus

31. Carrot powdery mildew

32. Chilli anthracnose

33. Chilli leaf curl

34. Chilli tospovirus

35. Onion wilt

36. Amaranthus white blister

37. Cassava leaf spot

38. Cassava mosaic virus

39. Beetroot leaf spot

40. Radish black rot

41. Yam leaf spot

42. Chrysanthemum leaf spot

43. Rose botrytis blight

44. Rose powdery mildew

45. Jasmine rust

46. Jasmine leaf spot

47. Curry leaf spot

48. Turmeric rhizome rot

49. Pepper wilt

50. Pepper pollu disease

51. Betelvine stem blight

52. Ginger yellows

53. Coffee rust

54. Tea blister blight

55. Aloe leaf blight

56. Aloe soft rot

57. Periwinkle phyllody

58. Glory lily root rot

59. Coleus root rot

60. Cuscuta parasite